AGRONOMY RESEARCH AND DEVELOPMENTS

PLANTATION FORESTRY IN GHANA

THEORY AND APPLICATIONS

AGRONOMY RESEARCH AND DEVELOPMENTS

Additional books in this series can be found on Nova's website under the Series tab.

Additional E-books in this series can be found on Nova's website under the E-book tab.

AGRONOMY RESEARCH AND DEVELOPMENTS

PLANTATION FORESTRY IN GHANA

THEORY AND APPLICATIONS

DAVID MATEIYENU NANANG

Nova Science Publishers, Inc.
New York

For permission to use material from this book please contact us:
Telephone 631-231-7269; Fax 631-231-8175
Web Site: http://www.novapublishers.com

Additional color graphics may be available in the e-book version of this book.

Library of Congress Cataloging-in-Publication Data

Nanang, David M. (David Mateiyenu)
Plantation forestry in Ghana : theory and applications / author, David Mateiyenu Nanang.
p. cm.
Includes bibliographical references and index.
ISBN 978-1-61324-315-2 (hardcover : alk. paper) 1. Forests and forestry--Ghana. 2. Tree farms--Ghana. I. Title.
SD242.G5.N36 2011
634.9667--dc22

2011010155

Published by Nova Science Publishers, Inc. † New York

CONTENTS

PREFACE

In the process of editing the book "*Natural resources in Ghana: management, policy and economics*," I learnt a lot about renewable natural resources management challenges in Ghana. As well, I envisioned some opportunities that could arise from those challenges. In one of the chapters I authored in the book, I made a strong case for a three-pronged approach to sustainable forest management in Ghana. The three policy options, which are best implemented simultaneously, included arresting the current high rates of deforestation, expanding forest plantations, and developing and implementing a national strategy that would ensure sustainable forest management on present and future forests. I further argued that, of the three prongs, forest plantation development offered one of the best hopes of averting a serious crisis in the forestry sector in Ghana.

But this got me wondering about whether there was sufficient scientific information and tools available to those who either want to develop plantations, or the students who are studying to become foresters to advise clients to successfully manage plantations. Even more basic than that, there seems to be little understanding of the key issues affecting plantation forestry such as the importance of forest plantations, the barriers and opportunities for their development, the impacts of international climate change policy on plantation forestry, and the role of plantations in the national development context.

Plantation forestry in Ghana: theory and applications is a direct response to fill this information vacuum. It achieves this by consolidating and making available existing information and results of recent research on plantation forestry in Ghana. The book provides a wide range of information on plantation forestry in Ghana and will contribute to the understanding of plantation forestry issues and concepts. The topics were carefully chosen to avoid repeating information on plantation silviculture available in existing texts. The scientific information, policy and legal issues, as well as recommendations provided in this book would serve as a useful guide for government and private organisations, communities, and individuals that are interested in developing plantation forestry. The book is particularly appropriate for undergraduate courses in plantation forestry and rural development, though graduate students and other forestry practitioners and general interest readers would find the material as a very useful reference.

Throughout the book, several case studies are presented, using growth and yield information on teak and neem plantations in the guinea savannah and semi-deciduous vegetation zones of Ghana. These two species are used because of the availability of detailed data on them through previous studies. While these case studies are meant to be illustrative, they are useful in demonstrating how the principles described in the theoretical parts of the book are applied. In this regard, I would particularly like to thank Dr. Thompson Nunifu, who has generously agreed to let me use some of his data on teak plantations in some of the case studies presented in the book.

This book has benefited from the ideas and inspiration of several people. I am grateful to Prof. (Emeritus) Robert Day, who supervised my master's degree thesis. He first introduced me to novel ways of thinking about natural forest and plantation silviculture in Ghana. Our time together, riding on motorcycles in the savannah of northern Ghana, carrying out forest inventory and drinking beer at the Picorna Hotel each day after returning from the woodlands have no doubt influenced my own thinking and approach to forestry issues in Ghana. My forestry colleagues from KNUST, Lakehead University, and The University of Alberta have all contributed to my training and continuous interest in Ghana forestry. Special thanks to Dr. Thompson Nunifu, as well as the editors, publisher and other anonymous reviewers who reviewed the manuscript.

I would be forever grateful to my family – for my parents (Nanang Nanyun and Yarpaak [Amina] Konlan) who gave me an opportunity for a formal education while they themselves had none and my brothers and sisters (Rockson, Akua, Jakper, Liyobe, Nanyun, Namka, Bikinteeb and Yennukuan) who supported me in many ways throughout my life. My wife, Bernice Nanang and 2 daughters (Baaloon and Yennumi) have always been supportive of my academic and professional pursuits and to them I say a big thank you. My friends, Martin & Glana Agelinchaab and Lucio & Fidelia Beyere have continued to be sources of inspiration and encouragement. To all my family and friends and everyone who holds dear the conviction to promote sustainable management of natural resources in Ghana and around the world, I dedicate this book.

David Mateiyenu Nanang

SUMMARY

This book contains three main parts: plantation forestry in context; plantation silviculture and management; and plantation forestry economics. The introductory chapters provide the context for plantation development, by highlighting the importance of Ghana's forest resources and their current state, why plantations are a major part of the solution to present and future wood scarcity and the barriers of, and opportunities and incentives for, plantation development in Ghana.

The second part of the book begins with descriptions of the critical silvics of six tree species that have been grown and /or promoted for forest plantation development in Ghana. This is followed by a chapter on plantation silviculture, including descriptions of seed collection and nursery practices, weeding, pruning and thinning. The final chapter in Part II focuses on forest management tools related to forest productivity and growth and yield modelling, and provides example applications of models for teak and neem plantations in Ghana. An important consideration in plantation establishment is the economics of forest plantations vis-à-vis other land uses.

Part III concentrates on the economics of plantations, and describes the tools and data requirements for carrying out economic analyses of plantation forestry investments, the alternative models for determining the optimal rotation age of plantations, and a final chapter on the economics of managing plantations for carbon credits and wood products. The concluding chapter provides recommendations on the conditions that have to be created to accelerate plantation development in Ghana.

Part I. Plantation Forestry in Context

Chapter 1

OVERVIEW OF GHANA'S FORESTRY SECTOR

This first Chapter provides a context for plantation forestry development within Ghana's forestry sector. It provides background information on the importance of the forestry sector, a description of the vegetation zones and ecology of the forests, forest sector governance, rights and benefits to forest resources, forest products trade, the challenges facing the forestry sector, and a forecast of the future demand and supply of wood products in Ghana. The final section of the chapter focuses on plantation forestry in Ghana, by describing a brief history of plantation development, the governance and legislative framework underpinning plantation forestry, and the extent of plantation forests in Ghana.

1.1. IMPORTANCE OF THE FORESTRY SECTOR

The forestry sector continues to play a significant role in Ghana's socioeconomic development agenda. Historically, this sector has accounted for about 5 – 6% of Ghana's gross domestic product (GDP) and approximately 11% of total commodity export earnings (Owusu, 1998). In 2006, forest products exports contributed about 9% to total export earnings and 3% of GDP (Ghana Forestry Commission, 2007; World Bank, 2007). In 2010, Ghana earned US$189 million from forest products exports. This constitutes about 3.2% of total export earnings for that year. Although the contribution of forest products exports to export earnings and GDP has been increasing over the years in absolute terms, its percent contribution has declined in the last few years mainly due to increases in non-traditional exports (fruits, cocoa products, canned fish, etc.) and rising prices of cocoa and minerals, especially gold (Nanang, 2010). For example, the percentage contribution of non-traditional exports to total exports stood at 27.8% in 2007 representing an increase of over 150% since 2001 (Ghana Export Promotion Council, 2008). It is estimated that the formal wood sector employs about 120,000 people across the country and approximately two million people depend on the formal and informal forestry sectors.

Forests and woodlands serve as the primary source of fuelwood energy for at least 75% of the total population of Ghana, and about 90% of the rural population (Benhin and Barbier, 2001). In addition, forests and woodlands provide food, medicines, spiritual, cultural and environmental services to over two million Ghanaians. The forests also play environmental roles such as soil conservation, watershed functions, carbon sequestration and minimising extreme weather damage to human and animal lives and property. Furthermore, Ghana's forests and woodlands are home to several animal and plant species and serve to maintain biodiversity. Non-timber forest products such as bushmeat, wild fruits and tubers, honey, oils and construction materials are derived from Ghana's forests and woodlands. The bushmeat sector is estimated to include about 300,000 hunters at the local community level who produce about 385,000 tonnes of bushmeat annually, mainly from forests, valued at about US$350 million (Adu-Nsiah, 2009), making the trade an important contributor to household and national incomes. Of the total quantity of bushmeat harvested, Ghanaians are estimated to consume 225,287 tonnes, worth US$205 million annually (Adu-Nsiah, 2009).

Ghana's forests and woodlands also support several plant and animal species with enormous recreational and tourism potential. Some of these areas have been developed into national parks and other tourism areas, which generate significant revenues from fees, accommodations, meals, and souvenir shops located in and around them. In addition, they generate employment in surrounding communities and are major contributors to the regional and national economies. In 2008, tourism was the country's fourth highest foreign exchange earner after gold, cocoa and remittances from Ghanaians resident abroad, generating about US$1.3 billion that year (Mensah, 2009).

1.2. Natural Vegetation and Ecology

Ghana occupies an area of about 23.9 million ha and lies north of the equator (between 4°45' and 11°11' North latitude and between 1°14' East and 3°07' West longitude) and wholly within the tropics. The high forest zone (HFZ) is found in the south-western third of the country (Figure 1). This zone covers an area of 8.1 million ha, with four broad ecological types - wet evergreen, moist evergreen, moist semi-deciduous, and the dry semi-deciduous forest types. These zones have been identified to be floristically synonymous with the *Cynometra-Lophira-Tarrientia*, *Lophira-Triplochiton,* *Celtis-Triplochiton*, and the *Antiaris-Chlorophora* associations respectively, recognised by Taylor (1952). According to Taylor (1952), there is no distinct line of demarcation between these associations as one imperceptibly merges into another. The main timber-producing areas are the deciduous and evergreen forests in the southwest.

Within The HFZ, the moist evergreen forest contains about 27% of the commercial/ economic species, whilst the moist semi-deciduous forest has up to 17% of such species; the wet evergreen forest is relatively poor in economic species (only 9%) (FAO, 2002a). The HFZ has a two peak rainfall during April to July and September to November. The rainfall varies between 1,200 - 2,200 mm, with a comparatively short dry season during January and February and a high relative humidity that is seldom below 85% (FAO, 2002a). The soils in the HFZ are highly leached and acidic (pH 4.0 -5.5) due to the high rainfall. They have low cation exchange capacity, available phosphorus, nitrogen and organic matter (FAO, 2002a). In the wettest zones, the soils are very infertile, strongly acidic and often have high aluminium content. FAO (2002a) notes that the relatively short dry seasons coupled with the high humidity of the HFZ reduce the risk of fire in forest plantations.

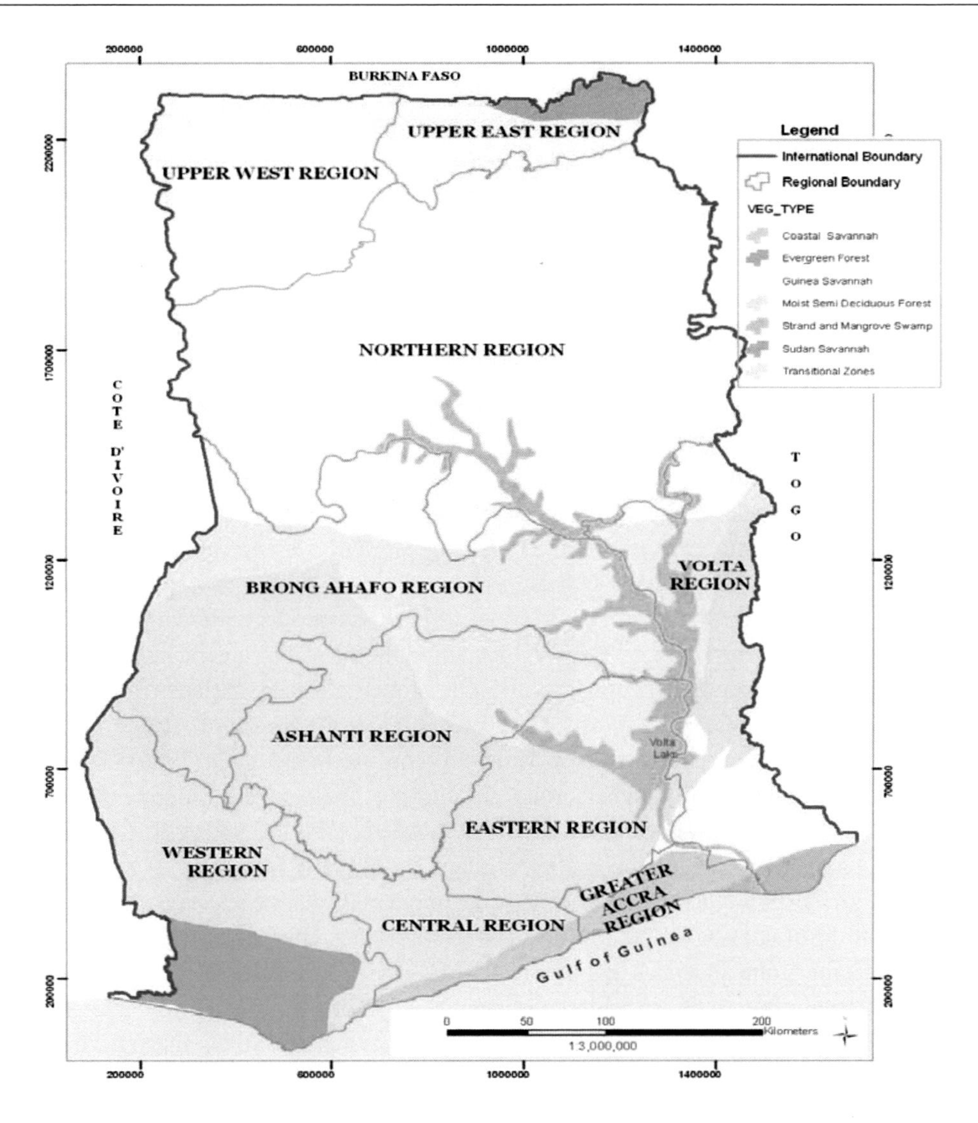

Figure 1. Natural vegetation zones of Ghana.

The savannah zone is classified into the southern (coastal) and the northern savannahs, based on their location. The largest part of the savannah zone is found mainly in the northern part of Ghana and occupies an area of about 14.7 million ha and a forest-savannah transition zone (in the middle belt) of about 1.1 million ha (ITTO, 2005a).The northern savannah is further divided into the Guinea and Sudan savannah zones. The Sudan savannah is restricted to a small area in the north-eastern corner of the Upper East Region.

The characteristic vegetation of the Guinea Savannah zone consists of short deciduous, widely spaced, fire-resistant trees. These do not form a close canopy and overtop an abundant ground flora of grasses and shrubs of varying heights (Taylor, 1952). The most frequent and characteristic tree species are *Isoberlina doka*, *Monites kerstingii*, *Burkea africana*, *Danielia oliveri* and *Terminalia avecinoides*. Two indigenous species, *Vitellaria paradoxa* (shea tree) and *Parkia biglobosa* (dawadawa), are conserved by farmers because of their economic value and are therefore common on farmlands. The ground vegetation, which includes *Panicum maxima*, *Andropogon gayanus* var. *gayanus* and *Cassia mimosoides*, desiccates during the dry season and predisposes the savannah to annual fires which leave the soil surface bare.

The savannah zone is characterised by distinct wet (rainy) and dry seasons of about equal duration. Two air masses of very contrasting characteristics determine the climate in this zone. These are the harmattan winds generally called the North East Trade Winds that usher in the dry season and the South Atlantic Maritime Air Mass referred to as the south west monsoon winds which transport moisture into the area during the rainy season. There is a moderate mean annual rainfall of 960-1200 mm falling in one season from March/April to October and showing a very irregular distribution within a rainy season and great differences from year to year (Fisher, 1984). Maximum rainfall during the year is achieved in July-August. Mean annual temperature is 28.3°C which does not vary significantly during the seasons.

The soils of savannah zone are varied because of the varied nature of the underlying geology. In general, however, two broad groups of soils are recognised: the savannah ochrosols and the groundwater laterites. The savannah ochrosols are found on the Voltaian sandstones (Boateng, 1966). They consist of well-drained, friable, porous loams and are mostly red or reddish-brown in colour. Most of the area covered by these soils has a gently undulating topography. Soils in the depressions are quite thick, but upland soils usually have a zone of ironstone concretions from 10 cm to one metre below the surface (Boateng, 1966). The soils tend to be eroded and form surface crusts under the impact of strong rainfall, but they have only a small capacity to keep water. According to Boateng (1966), despite their deficiency in nutrients, notably phosphorous and nitrogen, these soils are among the best soils in the northern savannah zone and are extensively farmed. A typical soil profile shows a dark greyish humus loam on the surface and subsequent layers show from grey to brownish loam with quartz gravel through light brown clay into moderately compact clays at about 70 cm below around level (Lawson et al., 1968). The groundwater laterites are very extensive and are formed on the Voltaian shales and granites. They consist of a pale-coloured, sandy or silty loam with a depth of up to 65 cm underlain by an iron pan or a mottled clayey layer so rich in iron that it hardens to form an iron pan on exposure (Boateng, 1966). Drainage on these soils is poor; they tend to get waterlogged during the rains and to dry out during the long dry season. These soils, especially those developed on the Voltaian shales are considered to be among the poorest soils in Ghana and little cultivation takes place on them (Boateng, 1966).

1.3. Forest Sector Governance

1.3.1. Forest Sector Institutions

Institutions are the formal and informal rules that govern human behaviour. However, we can also think about institutions as organisations, such as government bureaucracies or non-governmental organisations (NGOs). Within the forestry sector in Ghana, the main institutions that influence forest management and governance include the policy and legislative framework, the Forestry Commission, the Ministry of Lands and Natural Resources, the Ghana Timber Association, NGOs, and other government-wide institutions such as other ministries and the judiciary.

Historically, forest management was the responsibility of the Forestry Department, which was created in 1909 with the aim of encouraging the reservation of 20-25% of the country's land area (Smith, 1999). Since 1999, management of forestry and other natural resources in Ghana became the responsibility of the Ministry of Lands and Natural Resources[1]. Forest management is now led by the Forest Commission (FC), which was created under the *Forestry Commission Act*, 1999 (Act 571). The FC is responsible for regulating the utilisation of forest and wildlife resources, the conservation and management of those resources and the co-ordination of policies related to them. The FC is also tasked to manage the nation's forest reserves and protected areas and assist the private sector and the other bodies with the implementation of forest and wildlife policies. According to the Act, the FC shall undertake the development of forest plantations for the restoration of degraded forests areas, the expansion of the country's forest cover and the increase in the production of industrial timber.

The FC consists of 12 members, including a Chairperson, the Chief Executive Officer of the FC, and representation from the forest and wildlife industries, national house of chiefs, professional foresters' association, NGOs, etc. The FC embodies the various public bodies and agencies that were individually implementing the functions of protection, management and regulation of forest and wildlife resources. These agencies currently form the divisions of the Commission: Forest Services Division, Wildlife Division, Timber Industry Development Division, Wood Industries Training Centre, and Resource Management Support Centre. The FC has responsibilities for law enforcement, monitoring, policy development, forest management and revenue collection, all at the same time.

These functions of the FC in forest sector governance are potentially conflicting in nature and may result in inefficiencies in the implementation of forest sector plans. Planning and managing forests pose a particular challenge because of the multiple benefits produced by forests and also because of the fact that forestry practices have impacts on other land uses. Forest governance in Ghana is centralised within the Forestry Commission, and this centralisation is a consequence of the fact that forests are publicly owned. When the planning agency is the same as the implementation and regulatory agency, then there are no external controls and goals that are set out in the plans can be changed internally at the implementation stage. A case in point was when the FC reduced the annual planting target under the National Forestry Plantation Development Programme from 20,000 ha to 16,250 ha in 2004 and further to 10,000 in 2005. This was possible because the FC had the mandate to plan how much to plant, and the mandate to carry out the planting as well. If a separate agency is responsible for implementation, it would find efficient ways to deliver on them and become accountable to the planning agency. With the same agency responsible for both planning and management, there is no accountability, and this results in inefficiency. As a result, there is a need to separate the functions of planning from those of implementing forest management plans to eliminate such inefficiencies.

The separation of functions may neither be easy to achieve nor cheap to implement. Separation may come at an additional cost in terms of physical infrastructure and personnel. As a result, any opportunities to exploit economies of scale in designing planning and management institutional structures should be utilised. In cases where this separation is not possible, the organisation needs to ensure that the management (implementation) branch has

[1] The name of the ministry responsible for forestry has often changed with changes in governments. As at 2010, it was called the Ministry of Lands and Natural Resources. Between the years 2000 - 2008, it was called the Ministry of Lands and Forestry.

an arms-length relationship with the planning branch, and that the management branch is given clear goals and direction to operate with.

Forestry research is led by the Forestry Research Institute of Ghana (FORIG) and other forestry faculties at the universities. The Faculty of Renewable Natural Resources and the Faculty of Forest Resources Technology, Sunyani, both of which are part of the College of Agriculture and Natural Resources of the Kwame Nkrumah University of Science and Technology (KNUST), offer training in all aspects of forestry at the diploma, undergraduate and graduate levels. The newly created Faculty of Renewable Natural Resources at the University for Development Studies in Tamale also teaches forestry at the undergraduate and graduate levels as well as research. The Faculty of Forest Resources Technology is the only faculty that offers specialisation in forest plantation development and owns 250ha of teak plantation and a field station at Brosanko in the Brong Ahafo Region.

Community participation in forest management is facilitated by the Collaborative Forest Management Unit of the FC. In addition, there are NGOs such as Conservation International, World Vision, Forest Watch Ghana and the Adventist Development and Relief Agency (ADRA), which are active in forestry activities. The Ghana Timber Association represents timber loggers and millers in Ghana and constitutes an active lobby group for the timber industry interests. The Timber and Wood Workers Union of the Trade Union Congress of Ghana is also an important stakeholder in forestry issues in Ghana (ITTO, 2005a).

1.3.2. Policy and Legislative Framework

The forestry sector is governed by the 1992 Constitution of the Republic of Ghana (and subsequent amendments) and a suite of laws, regulations and policies. The main ones include: *Administration of Stool Lands Act*, 1962 (Act 123), *Forest Protection Decree*, 1974 (NRCD 243), *Trees and Timber Management (Amendment) Act*, 1994 (Act 493); *Local Government Act*, 1993 (Act 462), *Forestry Commission Act*, 1999, (Act 571), *Timber Resources Management Act*, 1997 (Act 547), *Timber Resources Management Regulations*, 1998, (LI 1649 and LI 1721 of 2003), *Forest Plantation Development Fund Act,* 2000 (Act 583) and subsequent amendments in 2002 (Act 623), *Forest Protection (Amendment) Act*, 2002 (Act 624), the *Timber Resources Management Act*, 2002 (Act 617) and the Forest and Wildlife Policy (1994).

This long list of legislations and policies may suggest a robust system for managing forest resources. But it also suggests something else – the over-centralisation of forest sector governance within the central government. These policies and laws are largely ineffective in ensuring sustainable forest management due to lack of enforcement. There are significant bottlenecks in implementing many aspects of the policy initiatives. For example, a closer look at these policies and laws shows that they tend to be piecemeal in their approaches, and hence lead to confusion. For consistency and ease of application, it is preferable to clarify and consolidate these laws and regulations. Implementation at the various levels remains a nightmare in most cases, and penalties for breaches of forestry-related laws are ridiculously low.

1.3.3. Recent Policy and Legislative Changes

Forest management has been practised in Ghana since the granting of concessions to companies for timber exploitation began in 1900 but it was not until 1927 that the legal power to enforce reservation was secured through the *Forests Ordinances* (cap 157) (FAO, 1997). Over several decades, successive governments have pursued many legislative and policy initiatives to ensure management and protection of forest resources and optimise the benefits that Ghanaians derive from their forest resource endowments. These changes have not significantly reduced the rate of deforestation nor improved forest management practices.

One of the most comprehensive re-structuring in the industry's history was the phased ban on exports of unprocessed timber logs, which begun in 1979. In fact, the genesis of restricting exports of unprocessed logs can be traced to provisions in the *Trees and Timber Act*, 1974 (NRCD 273) which imposed levies on selected species exported in unprocessed form. Ghanaian governments have sought to control external demand for Ghana's timber using log export bans in an effort to encourage value-added economic activities. In 1979, 14 timber species including the traditional redwood species were banned from export in log form (Richards, 1995). The list of banned species was increased to 18 in 1989. In a follow-up to this policy, the government extended the number of restricted timber species, imposed higher duties on other species, and completely phased out log exports by 1995 (Richards, 1995). The government hoped that with these measures, increased sales of wood products would replace earnings from logs.

In 1994, a new Forest and Wildlife Policy was introduced to replace the previous forest policy that was adopted in 1948. The objectives of the new policy, among other things, were to manage and enhance Ghana's permanent estate of forest and wildlife resources for preservation of vital soil and water resources; promote the development of viable and efficient forest-based industries, particularly in secondary and tertiary processing; and promote public awareness and involvement of rural people in forestry and wildlife conservation so as to maintain life-sustaining systems. The promulgation of the 1994 policy resulted in the development of the Forest Sector Development Master Plan (1996-2020) and a medium-term plan. This Master Plan was prepared by the Ministry of Lands and Forestry to achieve sustainable utilisation and development of forest and wildlife resources, modernisation of the timber industry, and the conservation of the environment thereby ensuring the realisation of the objectives of the 1994 Forest and Wildlife Policy (Ghana Forestry Commission, 1998). Under the medium-term plan, the Ministry implemented a comprehensive sector investment ten-year (1999 to 2009) programme called the Natural Resources Management Programme (NRMP). The NRMP consisted of five components: high forest resource management, savannah resource management, wildlife resource management, biodiversity conservation, and environmental management co-ordination.

The 1994 policy was developed through broad consultations with stakeholders and took five years to complete (1989-1994). This policy is arguably the most comprehensive forest policy Ghana has had. A major shortcoming of the current policy is that while it clearly articulates the objectives of the policy, neither strategies nor specific government agencies were identified to achieve these objectives. Secondly, after 17 years of implementation, the rate of deforestation has actually increased, plantation forestry development is still low, the wood harvesting and processing sectors are still inefficient, and community participation in forest management is minimal. This suggests that Ghana needs to re-examine not only the forest policy, but the entire legislative and institutional framework that underpin the forestry sector. Among the weaknesses of the current forest policy is the fact that it does not provide any incentives to conserve trees on off-reserve lands or farmlands by farmers. It is time to evaluate the policy.

In 1996, an air-dry levy on nine selected species ranging from 10 – 30% was introduced. These nine species accounted for 80% of Ghana's exported forest products (Donkor, 2003). The levy was intended as a disincentive for exporting air-dried sawnwood. Moisture accounts for at least 75% of wood manufacturing problems; therefore, reduction of wood-related problems correlates with a reduction in moisture content (Wengert, 2001), and consequently results in increased value of the wood. The government also embarked on promotion of the use of lesser-used species (Donkor, 2003). These policy initiatives resulted in increased value-added processing, a reduction in export of air-dried sawnwood and an increase in the export of plywood and veneer (Donkor et al., 2006; Nanang, 2010).

Whilst export trade in timber and wood products has been encouraged over the years, little attention has been paid to the supply of wood products to the local market, and this has encouraged illegal logging by chainsaw operators. As a result, in 2005, the government directed sawmills to sell at least 20% of their timber production on the domestic market. The Timber Industry Development Division (TIDD) of the FC was asked to ensure that permits for timber exports are only approved after sawmillers show evidence that they have supplied 20% of their production to the domestic market. However, this 20% is inadequate to meet the demand for wood products in the local market. The need to fill the gap between demand and supply of sawnwood in the local market is one main cause of illegal logging activities (Gayfer et al., 2002; Forest Research Programme, 2006). A few studies have examined the link between illegal logging and export of forest products from, and deforestation in, Ghana (Sarfo-Mensah, 2005; Owusu, 1998). These studies conclude that illegal logging supplies the local timber markets and is not a major source of wood for the export market.

In recognition of the important role of forest plantations in providing wood products to the population, a Forest Plantation Development Fund (FPDF) was established through the *Forest Plantation Development Fund Act*, 2000 (Act 583) to provide financial assistance for the development of private commercial forest plantations and for research and technical advice to persons involved in commercial plantation forestry on specified conditions. In May 2002, this Act was amended through the *Forest Plantation Development Fund (Amendment) Act* (Act 623) to enable plantation growers, both in the *public* and private sectors, to participate in forest plantation investments. Money to support the Fund is derived from the proceeds of the timber export levy imposed under the *Trees and Timber Decree 1974* (NRCD 273) as amended by the *Trees and Timber Management (Amendment) Act*, 1994 (Act 493); grants and loans for encouraging investment in plantation forestry; grants provided by international environmental and other institutions to support forest plantation development projects for social and environmental benefits; and moneys provided by Parliament for private forest plantation purposes.

The most recent policy decision that has implications for forest governance in Ghana is the European Union initiated-process to curb illegal logging through Forest Law Enforcement, Governance and Trade (FLEGT) Action Plan introduced in 2003. The cornerstone of this policy is the FLEGT Voluntary Partnership Agreement (VPA). In November 2009, Ghana was the first country in the world to formally sign the FLEGT-VPA, which is a bilateral agreement between the European Union (EU) and wood exporting countries, and aims to improve forest governance and ensure that the wood imported into the EU has complied with the legal requirements of the partner country. Under this agreement Ghana will develop systems to verify the legality of their timber exports to the EU (European Union Commission/VPA, 2009). Ghana expects that the VPA will help further its governance reforms of the forestry sector, contribute to sustainable forest management, provide conditions that encourage investment in forest restoration and thus improve the resource base, realise the full economic value of forests and ensure that the forest sector contributes to poverty alleviation (European Union Commission/VPA, 2009). Ghana decided to enter into a VPA to demonstrate its commitment to good forest governance, and as a means to maintain access to valued markets and open up new ones. With the VPA Ghana will also promote investment in the sector to ensure the future viability of its industry (European Union Commission/VPA, 2009). The FLEGT licensing system was scheduled to be operational in December 2010 when the first FLEGT licences were to be issued. However, due to the delays in the development of the licensing system, implementation has been delayed until December, 2011.

1.4. Rights and Benefits to Forest Resources

The 1992 Constitution provides under section 267 that all forest lands shall be vested in the appropriate stool on behalf of and in trust for the subjects of the stool in accordance with customary law and usage. Until 1994, rights to exploiting forest resources were transferred to timber firms in the form of conventional forest concessions. This system followed basic forest-management requirements such as harvest planning, standards for road-building and tree marking, pre-harvest operations, environmental conservation and enrichment planting (ITTO, 2005a).

The Forest and Wildlife Policy of 1994 abolished the concession system and replaced it with a new timber utilisation contract (TUC) system (ITTO, 2005a). The right to harvest trees is granted by the Forestry Commission in the form of permits that detail the area, volume and species to be harvested. There are three types of harvesting permits: Timber Utilisation Contracts (TUCs); Timber Utilisation Permits (TUPs); and Salvage Permits (SPs). TUCs are issued by the government for the commercial exploitation of timber. Contract holders must conclude a Social Responsibility Agreement with land-owning communities. TUPs are harvesting rights given to communities for the exploitation of timber for non-commercial and development purposes.

The annual allowable cut (AAC) is based on a polycyclic selection felling system using a cutting cycle of 40 years in natural forests. In 2005, concerns about the sustainability of the forest resource motivated the Ministry of Lands and Forestry to direct that Ghana's AAC be reduced from 1.2 million to 1.0 million m^3 (for both on and off reserve). The Forest Services Division (FSD) now has the mandate to set the AAC each year. The Forestry Master Plan (1996 – 2020) directs that only 0.50 million m^3 and 0.30 to 0.50 million m^3 of the resource can be harvested annually from the reserves and off-reserves, respectively. Unofficial estimates however, show that more than twice the AAC is harvested every year.

Due to the inefficiencies of the concession system, a competitive bidding system was introduced in 2002 to allocate timber-resource-use rights. This move was intended to promote efficiency, transparency and accountability in timber resource management and revenue collection. This reform was based on the *Timber Resource Management Act* (Amendment) 2002 (Act 617) and its subsidiary legislation the *Timber Resource Management (Amendment) Regulations,* 2003 (LI 1721). Under the new system, allocation of timber utilisation contracts (TUCs) is based on public bidding for rights to harvest timber in each area on the basis of an annual Timber Rights Fee (Ghana Forestry Commission, 2008). Act 617 also excluded from its application, land with private forest plantation, provided for the maximum duration of timber rights, and for incentives and benefits applicable to investors in forestry and wildlife sectors.

According to the 1992 Constitution of the Republic of Ghana, revenues accruing to forest resources in forest reserves would be distributed as 40% to the stools and 60% to the State (through the Forestry Commission) and 60% to stools and 40% to the State, respectively on off-reserve forest lands. With 60% of the revenue from on-reserve forest resources going to the government, communities and landowners had always considered the distribution to be inequitable and this had implications for the government's ability to protect forest reserves from encroachment and illegal logging. This concern was mitigated by the introduction of social-responsibility agreements between TUC holders and the communities where they operate to provide them with negotiated social amenities. The cost of such amenities should not exceed 5% of the annual royalty accruing from the operations under the TUC according to the *Timber Resources Management Regulations*, 1998 (LI 1649, section 13(1) b). Secondly, in 2006, a recommendation was made to change the 60/40 stumpage sharing ratio between the FC and other stakeholders to a 50/50 ratio. This recommendation was adopted and implemented by the then Minister for Lands and Forestry. On December 20, 2006, FC and the Office of the Administrator of Stool Lands (OASL) agreed to work out a framework for the implementation of the 50/50 stumpage sharing formula. This new formula was implemented with effect from January 1, 2007 (Ghana Forestry Commission, 2009b).

The existing scope of stumpage disbursement therefore stipulates that after the 10% administrative fee for the OASL has been deducted, the remaining stumpage payable shall be shared by a 50/50 ratio between the FC and the other stakeholders. FC's portion of the stumpage is to be applied to cover cost and expenses of staff remuneration, administration, operations and investment. The formula applies to both on-reserve and off-reserve situations. The 50% share for the other stakeholders is deemed 100% and then distributed based on the proportions spelt out as follows (Ghana Forestry Commission, 2009b):

- twenty-five percent (25%) to the stool through the traditional authority for the maintenance of the stool in keeping with its status;
- twenty-five percent (25%) to the traditional authority; and
- fifty percent (50%) to the District Assembly, within the area of authority of which the stool lands are situated.

1.5. Forest Products Trade

Ghana's forest products sector is characterised by very low lumber recovery factors that range from 20 to 40% of the log input (FAO, 2005). The total annual log requirement of sawmills in the country is about 1.3 million m^3 for lumber alone (Odoom and Vlosky, 2007). Over many decades, Ghana has been exporting a variety of wood products to about 80 different countries worldwide, spanning most continents, and countries from Angola to Yugoslavia. These included logs, air-dried and kiln-dried sawnwood, rotary and sliced veneer, plywood, machined timber (including mouldings and profiled boards), flooring and furniture parts. About 90% of all exports are made up of sawnwood, veneer and plywood.

Figures 2 and 3 show the trends in the quantities of wood production and exports for logs, sawnwood, plywood, and veneer from 1961 to 2010. Log production and exports reached their lowest levels in 1983, during a period of serious economic decline in Ghana. Exports of plywood and veneer have fluctuated since 1961, but showed a consistent upward trend after the log export ban on 14 timber species in 1979. The quantity of veneer exported over the study period has averaged 61% of production, representing the highest percentage of any forest product exported. Most of the veneer produced in Ghana is used for either decorative furnishing or for making plywood in importing countries.

The total exports of timber and wood products reached a volume of 466,155m^3 and value of €184.0 million in 2005, up by 2.4% and 7.9% respectively, as compared to 2004. The vast majority of the increase in Ghana's timber exports in 2005 was due to a substantial increase in quantity of plantation teak (*Tectona grandis*) that was made available to sawmills and the resultant rise in overseas teak shipments (Ghana Forestry Commission, 2006a).

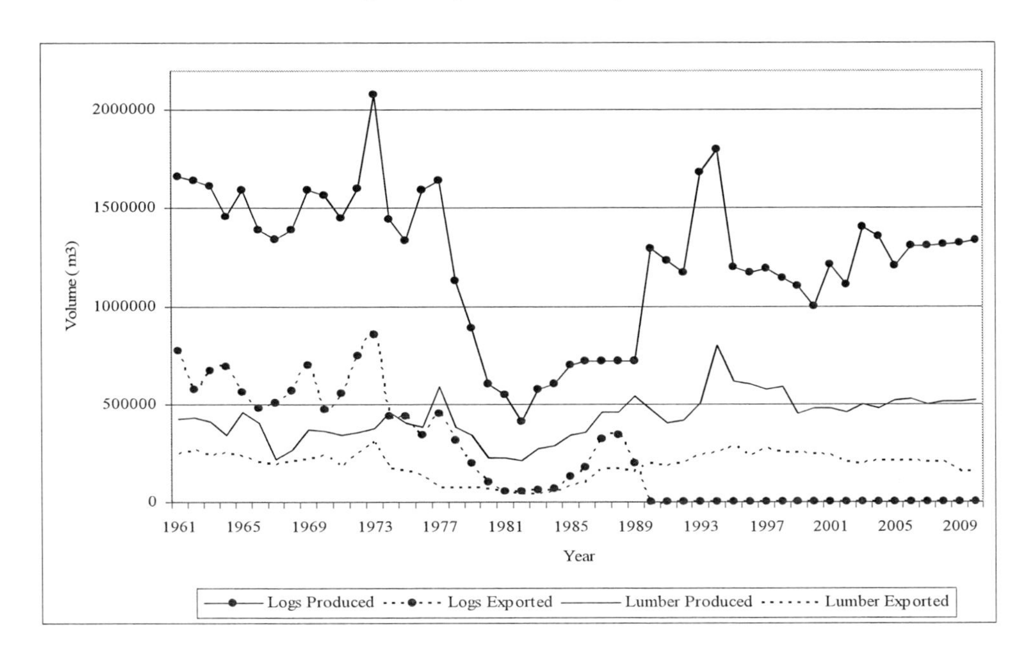

Figure 2. Quantities of logs and lumber produced and exported from 1961 – 2010.

Total volume exports and earnings fell in 2006 compared to 2005 mainly due to reductions in exports of air-dried sawnwood and rotary veneer. The Forestry Commission attributed the downward trend in exports to structural constraints within the forestry sector. However, exports increased again in 2007. A total volume of 528,570m^3 was exported in 2007, earning about €184.2m compared to 451,608 m^3 and €170.1m in 2006, representing increases of 17.0% in volume and 8.3% in value, respectively. Countries in the European Union (EU) continue to be Ghana's major trading partners in wood products. In Africa, the

Economic Community of West African States (ECOWAS) is the main importers of Ghana's forest products (mainly Nigeria, Senegal, Niger, Gambia, Mali, Burkina Faso and Togo). These countries imported 81.5% of all products exported to African countries in 2007, which increased to about 91% in 2008 (Ghana Forestry Commission, 2008a). In 2010, Ghana earned €137.9 million (US$189 million) from the exports of forest products compared to €128.2 m in 2009, an increase of about 7.8%. The volume of exports, however, decreased from 426,220m^3 to 403,250 m^3 (a decrease of 5.4%) from 2009 to 2010 (Ghana Forestry Commission, 2011).

Nanang (2010) examined the factors affecting the external demand for Ghana's forest products. The following conclusions are drawn from that study. External and domestic factors combine to influence the demand for Ghana's timber exports internationally. In the domestic forestry sector, initiatives that affect the timber supply (such as AAC restrictions), regulations that influence the domestic processing and export of forest products and the fiscal framework that determines the external value of the local currency (devaluation) all play important roles. On the international scene, income of importing countries and world prices of the wood products are significant determinants of export volume (Nanang, 2010).

Sawnwood and plywood face stiff competition in the international market, and this has revenue and tax policy implications for Ghana's forestry sector. The demand for Ghana's forest products is tightly linked to fluctuations in the economies of the importing countries (Nanang, 2010). This was confirmed by recent data from the Ghana Forestry Commission on forest products exports. Analysis of export figures for January to May 2009 showed decreases of 37.0% in revenue and 32.5% in volume of wood exports over the same period in 2008. The data showed considerable reductions in market demand for wood products especially in the major importing countries (Germany, Italy, Spain, U.K. and U.S.A) due to the credit crunch and the global economic downturn which generally affected the cash flow of most buyers of wood products (Ghana Forestry Commission, 2009a).

The reliance of Ghana's forestry sector on external markets further implies that revenues from forest products would be closely tied to the international business cycle, as economic downturns in importing countries would decrease revenues and vice versa. The price elasticities for sawnwood and plywood showed that a significant degree of substitution for these products could occur with small price increases in timber prices. Consumer responses to changes in the prices of forest products and the potential for substitution by other wood or non-wood products represent two fundamental elements of long-term forest resource planning and development.

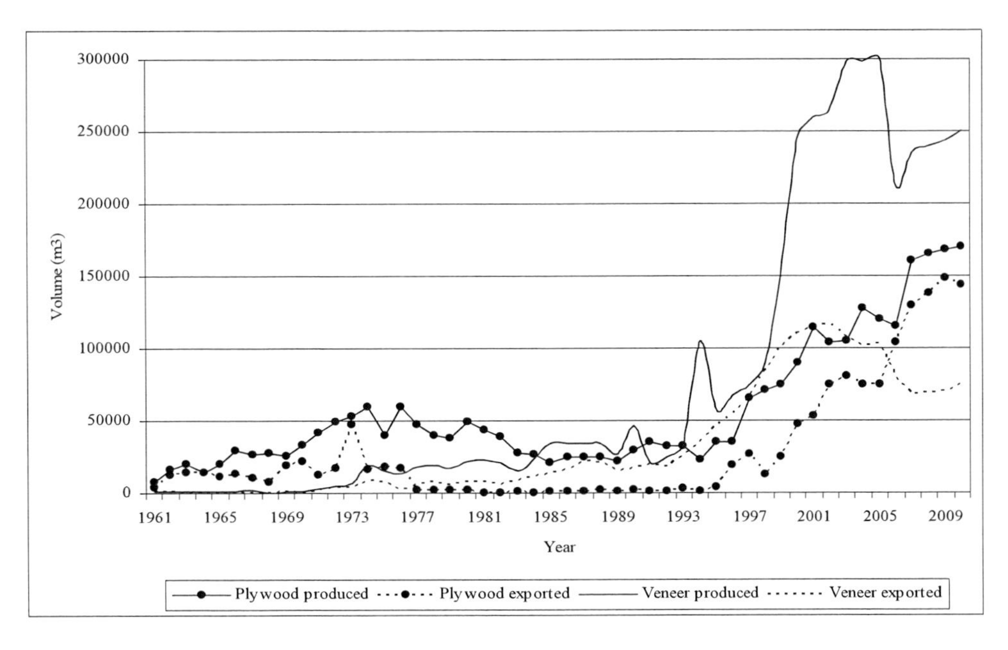

Figure 3. Quantities of plywood and veneer produced and exported from 1961 – 2010.

From a forest tax policy perspective, the elastic demand for sawnwood and plywood implies that Ghana's exporting firms may not be able to shift a large proportion of any domestic tax increases on these products to importing country consumers, without losing market share or profits. Export tariffs on wood products would in general increase the prices of the final products in importing countries. For sawnwood and plywood, most of the tariff would be paid by the producers in Ghana, with a marginal increase in final price. If the producers were to shift the tax to consumers in importing countries in higher prices, the demand for Ghana's sawnwood and plywood would fall. Veneer has inelastic demand, and hence more of the tax burden can be passed on to consumers, with relatively little or no impact on exports.

Nanang (2010) concludes that policies that encourage domestic processing and restrictions on both legal and illegal harvesting would work to ensure greater value-added benefits to Ghana. The focus on exported forest products at the expense of the domestic market is a major cause of illegal logging practices that are inconsistent with sustainable forest management principles. There is an urgent need to increase the supply of legally processed wood products to the domestic market.

1.6. Challenges Facing the Forestry Sector

Despite an encouraging and a strong indication of steps towards sustainable forest management through policy and legislative changes, and international involvement, the forestry sector continues to face several challenges.

- The most critical threat to Ghana's forestry sector is deforestation. Poor management of some forest reserves and off-reserve forests, increasing human population, uncontrolled wildfires, illegal logging, high dependence on wood energy, corruption, etc. all contribute to the high rates of forest cover loss in Ghana.

- Given that successive silvicultural systems since 1946 have been unsuccessful at achieving a desirable level of natural regeneration - including the current polycyclic felling system - there is a clear challenge for an innovative forest management system that would ensure sufficient regeneration of the natural forests. The most appropriate silvicultural system should be one based on biological, social and economic considerations to meet the goals of sustainable forest management.
- Thirdly, wood harvesting and processing continue to be inefficient, thereby leading to enormous waste in processing and destruction of residual forest stands. It is estimated that about 30 to 35% of the wood harvested are left in the forest as waste, while the conversion rate of logs into lumber is about 40% (Dauda, 2009). The combination of inefficient extraction methods and processing result in a final lumber volume that is only 25–40% of the total log volume extracted (Chachu, 1989). These high levels of inefficiency reduce the international competitiveness of Ghana's forest industry and contribute to excessive exploitation of the natural forest resource.
- The rising demand for wood products resulting from population growth will continue to pose challenges in terms of how to meet this demand from a diminishing forest resource base. In this case, improving efficiency of forest resource use, diversification towards value-added wood products, increasing wood supply through plantation development and the use of lesser-known tree species will be essential.
- Ghana's timber products face high international competition from other tropical forest products, due to their elastic demand (Nanang, 2010). The demand is also affected by the economic health of importing countries; hence fluctuations in the international business cycle partly drive export demand for Ghana's wood products.
- High poverty levels in Ghana in general and among forest fringe communities in particular constitute a major stumbling block to sustainable forest management by making it difficult to control illegal forest product exploitation activities. Forest management strategies that ensure effective community participation will mitigate this problem.
- Another major challenge to the forestry sector is the competition between forestry and other land uses. Ghana is mainly an agricultural country, and hence expansion of agricultural cultivation, mining and urbanisation into forests continue to contribute to deforestation and forest degradation.
- Finally, there are institutional constraints as well. Policy implementation failures are widespread in Ghana. Developing sound policies and legislations are good first steps in managing forest resources; however, without the capacity to implement and enforce those policies effectively, they amount to nothing. The FC seems to be ineffective against the more powerful timber industry in enforcing forestry legislation and policies. This puts into question the ability of the FC to protect the public interest when it comes to forestry issues.

1.7. A Forest Sector in Crisis

Wood production and consumption in Ghana have been increasing steadily since 1961, the earliest date for which reliable data are available. This section projects the production and supply of roundwood for Ghana from 2010 to 2100. The analyses are based on the following information and assumptions. The historical roundwood production data is taken from FAO (2011), and this production is assumed to meet the current demand for roundwood in Ghana (though in reality, the supply falls short of demand). The roundwood production includes all wood harvested from Ghana for fuelwood, charcoal, sawlogs, veneer logs and wood for plywood and other industrial wood requirements. In 2010, roundwood production was 36.755 million m^3 (FAO, 2011). Based on the historical data of roundwood production from 1961 to 2010, the average annual increase in roundwood production was about 3%, identical to the average population growth rate in Ghana over the same period. The growth in roundwood production was assumed to remain constant throughout the projection period.

It is assumed the supply of timber will come solely from the natural forest. Based on FAO (2011) estimations, the average growing stock of forests in Ghana was $49m^3$/ha, with a mean annual increment (MAI) of $4m^3$/ha/yr in forest reserves. Given that the MAI in off-reserve areas is likely less than that in the reserved areas, an average MAI of $3m^3$/ha/yr is used and applied to the total forest area in Ghana. It is further assumed that all the growing stock can be profitably recovered at a reasonable cost (though this is not true, it ensures the supply projections are over-estimated).

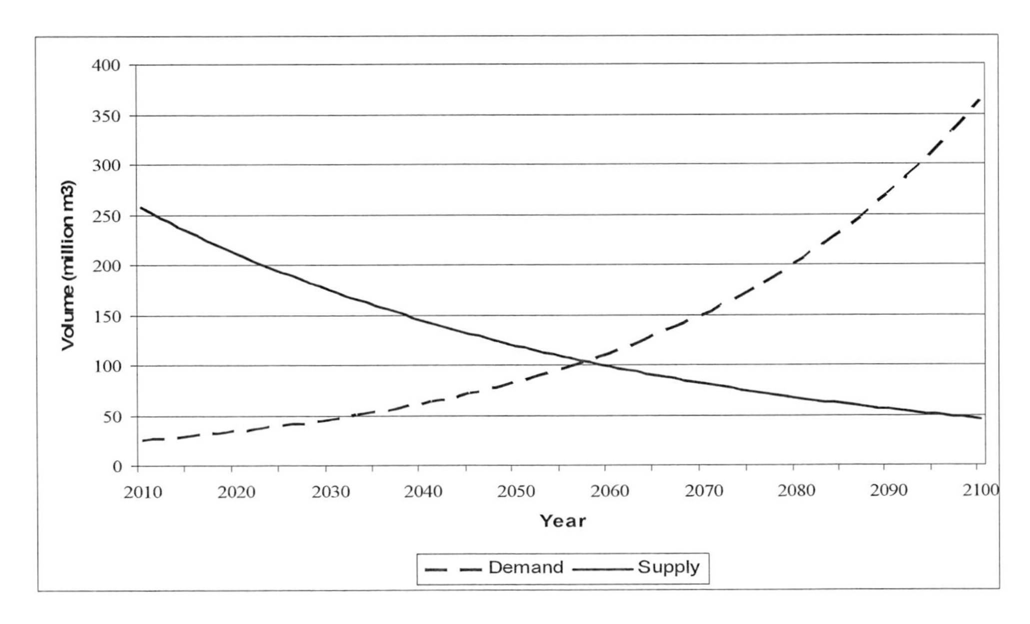

Figure 4. Projected demand and supply of roundwood in Ghana from 2010 – 2100.

The average deforestation rate in 2010 was about 1.9% of the forest area and this was assumed to be constant throughout the projection period as well. Therefore, supply in each of the projection years was estimated as the forest area remaining in that year multiplied by the combined growing stock per ha and the MAI. Figure 4 presents the demand and supply curves. At the current rates of deforestation and natural growth rates of the natural forests, the demand would outstrip supply of roundwood by 2058 even with these optimistic projections of supply. One of the key assumptions is that all timber stocks can be harvested, which, of course, is not the case. This, together with continuous forest degradation, means that Ghana would be unable to meet its demand from the natural forests much sooner than 2058. At the current deforestation rate, the total forest area would be reduced to about 0.90 million ha by 2100, from the 4.94 million ha in 2010. If we consider that the rate of deforestation is likely going to be higher than that estimated for 2010, then the timber supply outlook is even gloomier! Given that not all the wood in the forest can be recovered, Ghana can run out of wood, even if we do not run out of forests.

It is important to highlight that the gap theory applied in this analysis assumes a continuous constant (linear) decrease in wood resources as human populations continue to grow and assumes that nothing is done to increase wood supplies through plantation development and natural regeneration. If the decrease in wood supplies is at a faster rate than assumed here, the shortages would occur sooner than projected. However, if wood supplies increase through plantation development or increased natural regeneration, it would mitigate the wood scarcity, and shortages would occur later than projected, all other things held constant. On the demand side, the analysis does not recognise the phenomenon of substitution whereby, as wood becomes scarce, people will use wood more efficiently or substitute with other materials. Substitution and efficiency have the potential to decrease the demand for wood supplies and hence would be critical to reducing the severity of the wood crises.

The above analyses suggest that Ghana's forestry sector is already facing a crisis. The forest stocks are being depleted at a faster rate than they are being renewed. The depletion of the forests would lead to serious ecological, economic, social and environmental consequences. New forest resources must be created to offset greater losses of tree stocks and minimise the impacts of the crisis.

1.8. Overview of Plantation Forestry in Ghana

1.8.1. Brief History of Plantation Development

Plantation forestry development in Ghana is the most promising option to reduce deforestation and ensure the availability of forest products to meet social, environmental and economic objectives. The need to develop forest plantations to meet the wood products requirements of the inhabitants of the Guinea savannah zone was recognised as early as 1956. A national plantation project was launched in 1970, but during a decade of economic decline there was little activity. Another plantation programme was launched in the Northern and Upper Regions in 1976, but also slowed down due to lack of funds in the early 1980s (FAO, 1995). As of 1997, there were only 40,000 ha of forest plantations consisting of about 15,000 ha planted by the erstwhile Forestry Department and the rest by forest industry firms. There are also a large number of small holdings of forest plantations with teak as the main species (FAO, 1997).

In 1989, a Rural Forestry Division was established within the Ghana Forestry Department to encourage the establishment of plantations in order to mitigate the effects of projected wood shortages identified by the World Bank (1988). This programme was implemented under the Rural Afforestation Programme (RAP) from 1989 to 1995. To implement this programme, the new Rural Forestry Division was given a mandate to establish and expand existing nurseries, to initiate and expand community and individual plantations, and to provide technical advice to farmers on establishment, management and protection of the trees. In addition, the new Division was to provide extension services and education on rural forestry and agroforestry. Of the planting that did take place, woodlots accounted for about 90% of the seedlings while the remaining 10% were aimed at agroforestry, boundary planting, wind breaks and home gardens (FAO, 2002a).

The years between 1996 and 1999 were quiet on the plantation development front, with no major plantation forestry initiatives. The interest in plantation forestry development was renewed with the passing in 2000 of the *National Forest Plantation Development Fund Act*, 2000 followed by the launching in 2001 of the National Forest Plantation Development Programme (NFPDP) with a target planting of 20,000 ha/annum. Interestingly, while the NFPD Fund was empowered by legislation, the NFPDP was not. The Forestry Commission has enumerated a number of challenges that they face in the implementation of the NFPDP. These challenges are (Ghana Forestry Commission, 2008b):

- The FC acknowledges that the late release of funds towards programme implementation continues to hamper the success of the NFPDP. The lateness results in late payment to communities for services rendered as well as inability to acquire inputs in a timely fashion to implement the programme. In addition, the limited funding results further in inadequate logistics (such as vehicles, accommodation and equipment) and capacity of existing staff to execute their duties effectively;
- A serious problem that needs to be addressed is how to ensure that farmers under the modified taungya system (MTS) continue to tend the farms. Despite the increase in benefits to farmers under the MTS, the success of the MTS is being hindered by the inability of taungya farmers to tend their farms. Some planted sites have apparently been abandoned by some taungya farmers one or two years after planting the tree seedlings;
- Many of the participating taungya groups lack adequate capacity to cope with the cost of tending large areas developed so far;
- The activities of cattle herdsmen to the security of FC staff, their role in setting wildfires to the dry vegetation during the dry seasons, trampling of seedlings and/or browsing of the herds in young plantation resulting in damage to some of the growing trees has continued to affect the success of plantations in some areas of the country; and
- Finally, wildfires have continued to be a major concern for the survival of planted seedlings, yet the FC has limited fire management capacity to deal with annual wildfires that destroy forest plantations.

1.8.2. Governance and Legislative Framework

Forest plantation development in Ghana is currently led by the Plantations Department (PD) of the Forest Services Division (FSD) of the Forestry Commission (FC), which is responsible for the implementation, coordination and management of the NFPDP. The programme is currently being implemented under three main strategies and five components: the Modified Taungya System (MTS), Government Forest Plantation Development Programme (HIPC), and Private Plantation Development. A purely research-based Model Plantation component was added in 2007 to offer the FC plantation managers the opportunity to undertake mixed species trials, experiment various planting designs and tree spacing trials (Ghana Forestry Commission, 2007b).

The Modified Taungya System (MTS) involves the establishment of plantations by the FSD in partnership with farmers. The farmers, in addition to the food crops they harvest, have a 40% share in the returns from the investment. The Government also has a 40% share while the landowner and community have 15% and 5% shares, respectively. This is funded through the African Development Bank (AfDB) Community Forest Management Project (CFMP) and the government funded Modified Taungya Component. The CFMP, unlike the other components under the NFPDP, is being executed by the Ministry of Lands and Natural Resources (MLNR) through the FSD and the Ministry of Food and Agriculture field staff in four Forest Districts in three Regions of the country.

The Government Forest Plantation Development Programme (HIPC): This second strategy utilises hired labour and contract supervisors to establish industrial plantations. Plantation workers are hired and paid a monthly allowance to establish and maintain plantations while plantation supervisors are given one year renewable contract employment to supervise and offer technical direction (Ghana Forestry Commission, 2006b). This strategy is under the Government Plantation Development Programme (GPDP) which is funded through the Highly Indebted Poor Countries (HIPC) benefits. Under this scheme the plantations developed are owned by government and the respective landowners who are entitled to royalty payments. The Private Plantation Development component of the strategy involves the release of degraded forest reserve lands by the FC to private entities after vetting and endorsing their reforestation and business plans (Ghana Forestry Commission, 2006b).

Policy and legislative support for plantations are provided by the 1994 Forest and Wildlife policy and the *Forest Plantation Development Fund Act* (Act 583). The 1994 Forest Policy encourages the establishment of forest plantations and calls for the provision of incentives for their establishment and management. The main legislation governing forestry plantation development in Ghana is the *Forest Plantation Development Fund Act* (Act 583), which was passed in 2000 to offer financial support and other incentives to plantation developers in the private sector. The Act was subsequently amended in 2002 into the *Forest Plantation Development Fund* (*Amendment*) *Act* (Act 623), to cover plantation growers, both in the *public* and private sectors.

1.8.3. Extent of Plantation Forests in Ghana

In 1998, there were approximately 50,800 ha of forest plantations in Ghana inside and outside of forest reserves owned by the Government of Ghana, private timber companies, gold mining companies and tobacco companies. Table 1 presents the distribution of the plantations by species.

Table 1. Distribution of forest plantations in Ghana as at 1998

Species	Area (ha)
Teak	22,900
Rubber wood	11,200
Gmelina	6350
Miliaceae/mixed hardwoods	6070
Cedrela	3840
Terminalia	400
Eucalyptus	42
Total	50,802

Data source: FAO (2002a).

Table 2 presents the areas planted under the National Plantation Development Programme that was launched in 2001. The implementation of the programme however, begun in 2002. The planting target of 20,000 per annum was reduced to 16,250 ha in 2004 and further to 10,000 in 2005. The annual targets were reduced to allow the FSD field staff to cope with the large annual target while at the same time maintaining the large areas already established (Ghana Forestry Commission, 2006b). A wide range of tree species are planted in the plantations, including both indigenous and exotic economic tree species. The indigenous species include *Mansonia altissima* (Oprono), *Terminalia superba* (Ofram), *T. ivorensis* (Emire), Mahogany spp., *Ceiba pentandra* (Onyina), *Heritiera utilis* (Nyankom), *Entandrophragma angolense* (Edinam), and *Triplochiton scleroxylon* (Wawa) and the exotics are predominantly *Tectona grandis* (Teak), *Cedrela odorata* (Cedrela) and *Eucalyptus camaldulensis* (Eucalyptus). The exotic species form about 95% of the areas planted (Ghana Forestry Commission, 2008b).

Table 2. Areas planted under the National Forest Plantation Development Programme from 2002-2008

Year	Area planted (ha)					
	Modified Taungya system (MTS)	Community Forest Management Project	HIPC funded	Private Developers	Model Plantations	Total area
2002	17460	-	-	1609	-	19069
2003	17691	-	5650	1609	-	24950
2004	16090	-	5300	1609	-	22999
2005	9105	1136	6575	1609	-	18425
2006	9401	2298	6075	1609	-	19383
2007	8711	2731	5312	1613	69	18436
2008	111	2930	3740	5374	160	12314
Total	78,569	9,095	32,652	15,032	229	135,576

Data source: Ghana Forestry Commission (2008b).

The NFPDP has achieved some successes since 2002, but clearly the planting of 135, 576 ha is insufficient to reduce the current burden on the natural forests. If it is assumed that the plantations in Table 1 all exist in good condition and all the trees planted in Table 2 survived, the total area of plantations in Ghana would be about 216,000 ha. Assuming a mean annual increment of $14m^3$/ha/year (based on the average MAI in Chapter 2, Table 1), these plantations would be producing only 3.02 million m^3 of wood annually, which is more than the official annual allowable cut (AAC) of 1.0 million m^3, but probably equivalent to the unofficial quantities harvested in Ghana in a year.

In terms of exports, plantation wood products still play a very minor role in the export markets. The only plantation species that is currently exported in significant quantities is teak (Section 1.5). In 2010, a total of about $50{,}000m^3$ of teak wood in the form of air-dried lumber, poles and billet were exported, mainly to India (Ghana Forestry Commission, 2011).

1.9. Conclusions

Ghana's forestry sector is important for its environmental, social and economic contribution to the lives of Ghanaians. Analyses of the current state of the natural forest estate in this Chapter have shown that this natural resource base is not sustainable without a serious effort to stop the high rates of deforestation. Plantation forestry development is a key component of the effort to arrest deforestation and ensure the availability of forest resources to society.

The government of Ghana has an important role to play in this direction. The government needs to seriously implement measures to reduce deforestation. The forest industry and its supporters like to argue that reductions in harvesting levels will result in lay-offs of employees. Politicians like to use this argument to rationalise their own inaction. It seems clear that this focus on short-term economic gain for long-term pain is unwise; precisely because when the forest is completely depleted everyone in the industry will then be unemployed. The most sensible approach is to manage the resources sustainably, even if it will cost a few jobs in the short term.

Chapter 2

Plantation Forestry Development

Chapter 1 provided an overview of Ghana's forestry sector and revealed that the sector is already in a crisis. In Chapter 2, we will examine the reasons why Ghana should intensify the development of forest plantations and explain why the current state and management of the natural forest resources would be unable to meet the future wood and non-timber needs of Ghanaians and the export market. Several models of plantation development are examined and the policy as well as the technical and economic factors affecting plantation establishment and management discussed.

2.1. Defining a Forest Plantation

Definitions are by nature difficult and controversial, and defining a forest plantation is no different. Over several decades, there have been considerable attempts to define many concepts related to forestry and forest plantations (Carle and Holmgren, 2003). According to Ford-Robertson (1971), a forest plantation is "a forest crop or stand raised artificially, either by sowing or planting." This definition has been accepted and used for several decades until the FAO started work to harmonise forest definitions in order to ensure consistency, effective communication and data exchange on forest-related issues between countries and organisations. In 2003, the FAO proposed a comprehensive definition for plantation forests as *"planted forests that have been established and are (intensively) managed for commercial production of wood and non-wood forest products, or to provide a specific environmental service (e.g., erosion control, landslide stabilisation, windbreaks, etc.)."* Though intensive management is an important component of the definition, it should be noted that planted forests established for conservation, watershed or soil protection may be subject to little human intervention after their establishment (Carle and Holmgren, 2003).

Plantation forestry is usually associated with afforestation or reforestation activities. Both afforestation and reforestation involve tree planting; the only difference between them is the time span over which there was no trees on the land being replanted. Afforestation is usually used to describe tree planting on areas that have not had a forest for more than 50 years. Reforestation on the other hand, refers to land cover change into a forest of an area that has not carried a forest within the last 50 years.

2.2. The Need for Forest Plantations in Ghana

The reasons for pursuing plantation forestry differ across the various countries of the world. For Ghana, the overarching strategic reasons for promoting forest plantations are to ensure that the country is able to meet the demand for forest products for its growing population, reduce the pressure on the natural forests and contribute to their sustainability. Forest plantations, usually of fast growing exotic species[1], also provide forest products to communities that do not have natural forests. Plantation forestry in Ghana is one arm of the axes of policy tools to reduce deforestation, protect the natural forest resource base thereby arresting and /or reversing deforestation, and ensure the availability of forest products to meet social, environmental and economic objectives. Other reasons for pursuing forest plantations include: high productivity of plantations compared to natural forests, flexibility to site plantations, rural and economic development and for carbon sequestration to obtain carbon credits.

2.2.1. Decreasing Supply of Forest Products from Natural Forests

Ghana's population has traditionally relied on the natural forests to provide its forest products needs. However, analyses of demand and supply of the forest resources suggests that in future, demand would outstrip supply and hence new resources need to be created. The decreasing supply of forest products is a result of increasing destruction of the natural forest due to natural and anthropogenic factors, unsatisfactory and unreliable natural regeneration, poor management of existing forests, and exclusion of some natural forests from wood production.

A) Deforestation and Forest Degradation

Ghana has one of the highest rates of forest loss in Africa, raising concerns about the adverse consequences of diminishing forest cover and woodland areas in the country. The accelerating rate of deforestation shows that Ghana's reliance on its natural forests to supply its raw material needs cannot be sustained in the long term. The total forest area in Ghana was estimated at 4.94 million ha in 2010 (FAO, 2011) with 1.123 m ha classified as production forests, 0.352 m ha maintained for the protection of soil and water, 43,000 ha for the conservation of biodiversity, 59,000 ha for social functions and unclassified forest area of 3.361 m ha. There are 204 forest reserves in the high-forest zone covering an area of 1.58 million ha and 62 forest reserves in the savannah zone covering 600,000 ha.

Deforestation averaged about 1.8% per year between 1990 and 2000 and increased to 1.9% per year between 2000 and 2010[2], according data from the Global Forest Resource Assessment (FAO, 2011). Against this backdrop of a declining resource, roundwood production has been rising steadily over the years.

Figure 1 illustrates the continuous decline in the forest area vis-à-vis the drastic increases in the production of roundwood for wood fuel, sawlogs and veneer logs, and for other industrial

[1] The term exotic is used to describe a species introduced into a country from outside, as opposed to indigenous species that grow naturally within a country.

[2] The forest area reduced from 7.448 million ha in 1990 to 6.094 million ha in 2000 and to 4.94 million ha by 2010.

purposes. Roundwood production increased steadily from 1961 until 1990, but thereafter rose drastically. In 1961, the forest area was about 9.6 million ha and declined to 4.94 million ha by 2010. This represents a 49% reduction in forest area in only 50 years. Based on the estimated remaining forest area and the deforestation rate in 2010, it is obvious that with no effort at maintaining and/or increasing the forest area, Ghana's forests could be completely depleted by 2058 or degraded to the point where no commercial logging can take place (see Chapter 1).

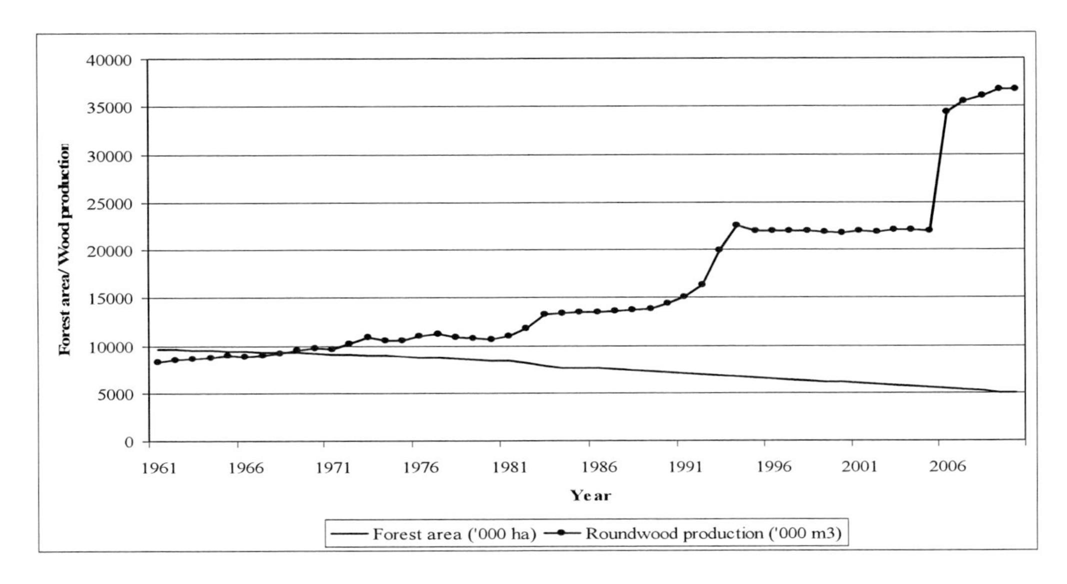

Figure 1. Forest area and roundwood production in Ghana from 1961 to 2010.

As a result of concerns over sustainability of the timber resource, the commercial species were classified into 'scarlet', 'red' and pink' groupings based on their sustainability. 'Scarlet' species were being over-cut at a rate greater than 200% the estimated sustainable yield and were thought to be under threat of economic extinction, 'red' species were being cut at a rate of 50–200% the sustainable yield, and 'pink' species were being harvested at less than 50% the sustained yield (Ministry of Lands and Forestry, 2004)

There is little consensus on the roles and, in particular, the actual causes of deforestation (Lugo and Brown, 1982), partly because some studies mistake causes for effects (Deacon, 1994), and because interactions among the causal factors vary from one country or region to another (Allen and Barnes, 1985; Saxena et al., 1997). The few rudimentary, primarily qualitative forestry studies for Ghana have generally concentrated on community-specific ethno botanical and anthropological issues (the only exceptions are Benhin and Barbier (2001), Benhin and Barbier (2004), and Owusu (1998), but all focused on effects of the structural adjustment programme on deforestation). For example, Dei (1992) and Dei (1990) qualitatively analysed the process of deforestation for a specific rural Ghanaian community, while Townson (1995) assessed various economic activities generated from forest products. Appiah et al. (2009) through a survey in southern Ghana, identified four most highly ranked causes of deforestation as poverty-driven agriculture, lack of alternative rural wage employment other than farming, household population levels, and conflict in traditional land practices. These studies for particular communities help provide more focused solutions to local or specific deforestation problems, but are inadequate in providing strategic policy direction for reducing deforestation at the national level.

There is potentially a long list of factors that cause deforestation in Ghana. The most common ones include: population growth, forest products exploitation, over-reliance on wood fuel for energy, uncontrolled forest fires, agricultural and mining expansion into forested areas, corruption, lack of enforceable property rights, etc. However, broad generalisations and qualitative statements about the factors that cause deforestation in Ghana do not contribute much to the search for specific policy prescriptions. Hence, quantitative assessments of the causes of deforestation in Ghana are required to analyse the complex relationships between deforestation and its hypothesised causes. This would provide the information needed to evaluate possible policy interventions, and provide insights especially on how limited resources could be prioritised in managing Ghana's forest resources.

Nanang and Yiridoe (2010) provided the most comprehensive analysis of the causes of deforestation in Ghana yet. The study modelled deforestation using two-stage regression methods: as an interaction of interlinked key sectors in the Ghanaian economy (including forest products exports, fuelwood energy consumption, cocoa production, infrastructure development and food crop production), which compete for forest land use or forest products. In the first stage, the four direct (or first-level) causes of deforestation were regressed on various second-level causes. In the second stage, deforestation was regressed on the estimated first-level causes of deforestation. The impacts of various causes of deforestation were quantified using elasticity estimates. Policies aimed at minimising deforestation in Ghana should be classified and prioritised, based on whether the effects on forest and woodland area loss are direct or indirect. The causes of deforestation are not linked to the forestry sector alone, but are also affected by agricultural, economic, demographic and political factors. The results indicate that fuelwood consumption and food crop production are two of the leading direct causes of forest and woodland area loss, with the elasticity of deforestation being highest with respect to fuelwood consumption (3.5), and then with respect to deforestation in previous years (0.7).

B) Natural Disturbances in Natural Forests

Natural disturbances such as fires, pests and diseases are part of the dynamics of every natural forest ecosystem, and Ghana's forests are no exception. Fire is an important part of the Ghanaian culture; as it is used frequently in land clearing for agricultural purposes, hunting, honey collecting, palm wine processing and ceremonial celebrations. Fire activity in forests is strongly influenced by weather and climate, fuels, ignition agents and human activity. There are social and ecological benefits of fire in terms of forest ecosystem renewal, composition, species diversity and importantly, carbon balance. However, wildfire in forests and savannah woodlands cause irreversible environmental damage by reducing the productive capacity of forests, damage water supplies, impacts on water quality, reduces soil fertility and hence agricultural productivity, and kills wildlife and biodiversity. Forest fires occur regularly and cause severe damage; the annual financial loss due to wildfires is estimated to be about US$24 million (Ministry of Lands and Forestry, 2004). FAO (2011) estimates that up to 80% of the forest area in Ghana were affected by wildland fire in 2005. The most devastating wildfire in recent memory was in 1983, which resulted in massive damage to forests, woodlands and cocoa farms across the country. It is estimated that about 30% of the forest in the moist semi-deciduous zone were either destroyed or degraded by these fires (Hawthorne and Abu-Juam, 1995) and the loss of about 4 million m^3 of high-quality timber. Due to the high relative humidity and short dry season experienced in the HFZ, the frequency and extent of wildfires are less than in the savannah zone (FAO, 2002a). In the savannah zone, wildfires are an annual occurrence, and hence the trees are adapted not only to survive recurrent fires, but also to withstand drought conditions as well.

All forest tree species in Ghana have at least some pests that can cause serious damage under some circumstances (Wagner et al., 2008). Some of the most valuable timber species in Ghana are susceptible to various kinds of pests and diseases. For example, the shoot borer- *Hypsipyla robusta,* has been the major obstacle to the propagation of the khaya spp; while the main constraint to *Milicia excelsa* is the gall *Phytolama lata,* which attacks the leaves of seedlings (FAO, 2002a). In savannah ecosystems, the *analeptes trifasciata*, which is a longicorn beetle attacks several species in the family *Bombaceae* such as *Adansonia digitata* (baobab), C*eiba pentandra* (silk cotton) and *Eucalyptus alba*.

There is no formal monitoring of forests to detect pests and diseases, and hence there are no statistics to quantify the economic damage to forests, though these costs are considered to be substantial. There is little, if any, information on forest diseases in Ghana. Pests and diseases tend to reduce natural forest productivity and therefore impact on their sustainability. Efforts to control forest and forest product pests in Ghana dates back to the colonial times when the West African Timber Borer Research Institute was establishment in Kumasi in 1953 (Nair, 2007). Attention was only paid to the entomology of other pests after the establishment of the Forest Products Research Institute in 1964 (now the Forest Research Institute of Ghana (FORIG) (Nair, 2007). Plantation forests of exotic species are less vulnerable to indigenous pests and diseases of their new environment, at least for the first rotation. Large amounts of wood residue from felling debris and the presence of stumps are favourable for colonisation by insect pests and as sources of infection (Evans, 2001). This can usually be reduced by modifying the silviculture or application of specific protection measures (Evans, 2001).

C) Unsatisfactory and Unreliable Natural Regeneration

Ghana has, for many decades, failed to obtain satisfactory and reliable natural regeneration from its natural forests and woodland areas. There could be several reasons for this: poor management, little understanding of the ecology of desired species, institutional and regulatory failures, the focus on promoting regeneration of only economic species, and the inability and /or high cost of controlling weeds. The ecology of tropical forests is more complex than that of temperate forests or plantations, and understanding of this ecology is still limited for many tree species. In addition, laws, regulations, policies and standards for timber harvesting that could promote natural regeneration are often not effectively enforced by the forestry authorities.

Since 1946, several attempts at the development of a silvicultural system for the indigenous forests of Ghana have been made, beginning with the tropical shelterwood system (TSS), which was based on the Malaysian uniform system (MUS). This was followed by the modified selection system (MSS), implemented between 1956 and 1970. Thirdly, enrichment plantings were carried out to improve the stocking of the poorly stocked wet evergreen forest reserves as well as to sustain the supply of the then desirable species (FAO, 2002a). All the silvicultural systems failed to achieve satisfactory natural regeneration and were abandoned. Despite these failures, the *polycyclic felling system* (PFS) used in Ghana today evolved from the modified selection system in the 1950s. The PFS is basically selective logging with diameter limits and uses advanced regeneration (i.e., residual stands below the cutting limit at the time of initial harvesting) to develop into the next crop on a 40-year cutting cycle. The key silvicultural considerations in implementing this system are to minimise damage to advanced regeneration during harvesting and defining an appropriate length of the cutting cycle (Buschbacher, 1990). A necessary condition for successful regeneration under the PFS is to ensure that logging operations have little to no impact on the residual trees through the use of proper felling directions, the right equipment and skilled workers (Buschbacher, 1990). Though no statistics exist of the level of damage to residual stands during harvesting, it seems clear that there is considerable damage to residual stands and soil disturbance, which is jeopardising any hopes of achieving sufficient natural regeneration under the current silvicultural system.

The *Timber Resource Management Act*, 1997 (Act 547) prescribes that a timber utilisation contract (TUC) holder shall submit an undertaking to execute a reforestation plan during the period of the contract to the satisfaction of the Chief Conservator of Forests (section 8(d)). TUC holders are also required under Section 11 (d) and (e) of LI1649 for the reforestation or afforestation in any area that the CEO of the Forestry Commission may approve and show evidence of capability to undertake reduced impact logging. These reforestation requirements, together with natural regeneration were expected to ensure the renewability of the forest resource. However, two main problems have been encountered in the implementation of these requirements over the years. First, forestry companies are reluctant to reforest because of the uncertainties of whether they would benefit from the reforestation activities, given the long rotation ages of the trees planted and the short duration and uncertainty of their harvesting rights to the contract area. Secondly, the forestry authorities have been unable to strictly enforce these regulations against the often powerful international forestry companies that operate in these concessions for fear that they might withdraw their investments. The best way to ensure reforestation occurs after logging is to levy the forest companies, and then use the revenue to hire a private company to undertake the reforestation.

Achieving natural regeneration in Ghana's forests is possible. Hawthorne (1995) showed that on a national level, the majority of the forest species in Ghana have sufficient natural regeneration. However, the right silvicultural system and forest management plans have to be developed and implemented in order to harness this potential for sustainable forestry practices to be realised. Compared to plantations, natural forest management is less intensive, with low yields and low capital investments (Buschbacher, 1990). Hence natural regeneration and growth in natural forest ecosystems continue to be poor, and puts into question the ability of Ghana's natural forest estate to meet the forest products needs of its growing population.

D) Exclusion of Some Natural Forest Resources

Some of the natural forest areas in Ghana are inaccessible to harvesting. These areas include forests that perform ecological functions such as erosion control and watershed protection. About 5% of Ghana's land area is under legal reservation. This includes 352,500 ha representing 22% of the permanent forest estate under permanent protection (areas in hills and swamps, sanctuaries, areas of significant biodiversity, fire protection, etc.) which are excluded from commercial timber harvesting. Approximately, another 397,000 ha of the permanent forest estate is excluded from timber harvesting due to the state of degradation and would have to be rehabilitated. These are found mainly in the savannah-forest transition zone.

Other factors that render forests inaccessible include forests with difficult terrain (hills or swamps); timber located in far away locations where the cost of harvesting is exorbitant compared to the benefits, low stocking, unavailability of roads, etc. These factors together reduce the amount of forest area that is potentially available for timber production in Ghana to only 762,400 ha.

2.2.2. Flexibility to Locate Plantations

Unlike plantations which can be sited by choice, natural forests in Ghana are endowed by nature and located in areas where we find them. This represents a major advantage of forest plantations over natural forests. Based on need, forest plantations can be located to serve environmental needs such as prevent soil erosion or protect watersheds and other vulnerable sites or close to where the products are in demand. Forest plantations can be used to promote and increase fuelwood supplies in areas that did not originally contain forests or where the forests have been depleted or degraded. In addition to the choice of location, plantations offer the flexibility to choose the species to be planted to achieve the social, economic or environmental objectives of the plantation.

2.2.3. High Productivity of Plantations

The natural growth rate of Ghana's natural forests is estimated at 4.6 million m^3 with an annual increment rate of $4m^3$/ha (FAO, 2011). This mean annual increment (MAI) falls well within the range of 0.4 to $7m^3$/ha for tropical forests, but still far below the 25-45 m^3/ha for tropical hardwood plantations reported by Evans and Turnbull (2004). Reasons for the high productivity of plantations include the ability to control inputs, match species and site characteristics, and intensive management. Unlike tropical natural forests that are complex in their species composition and structure, forest plantations are simpler and the ecology and silviculture of the species are better understood and hence they are easier to manage. With plantations, the initial spacing can be chosen to ensure full stocking of the site and pursue subsequent stand density management that optimise tree growth.

If plantations of exotic species are used, they are often immune from the pests and diseases associated with natural forest species, which would have otherwise impeded their productivity. The high productivity of plantations permits less land to be used in producing the same quantity of timber compared to natural forests, which leaves more land for other uses. However, it has to be noted that plantations may not be able to replicate the total goods and services provided by natural forests. Table 1 presents the MAIs of selected plantation species grown in Ghana by vegetation zone.

Table 1. Estimated mean annual increments (MAI) for selected species in Ghana

Species	Good site		Poor site	
	Zone	MAI (m^3/ha/yr	Zone	MAI (m^3/ha/yr
Tectona grandis	Moist semi-deciduous	12	Transition zone	6
Gmelina arborea	Moist evergreen	20	Moist semi-deciduous	12
Cedrela odorata	Moist semi-deciduous	18	Moist semi-deciduous	12
Azadirachta indica	Guinea savannah	12	Guinea savannah	4
Ceiba Pentandra	Moist semi-deciduous	18	Guinea savannah	12
Triplochiton scleroxylon	Moist evergreen	20	Moist semi-deciduous	12

Data sources: Nanang (1996), Nunifu (1997) and FAO (2002a).

2.2.4. Rural and Economic Development Tool

The ability to develop plantations in a location of the planter's choosing allows governments, non-governmental organisations (NGOs), and the private sector to use forest plantation initiatives as rural or economic development tools and to improve food security and alleviate poverty through payment for ecosystem and environmental services (e.g., developing eco-tourism). To achieve poverty reduction objectives, it is necessary to incorporate poverty alleviation as a major objective of both government- and private sector-led plantation development efforts. The benefits accruing from plantations must reach rural communities in order for their livelihoods to improve.

In many rural areas without natural forests, the major forest products need is often for wood energy and wood for building construction. Community and small-holder individual plantations can be used to serve these needs of rural folks. Some of the forest products can also be sold to supplement household incomes. In addition, government and NGO-sponsored plantation projects offer job opportunities to rural people, where there is usually limited access to alternative employment opportunities. This is particularly true if industrial plantations are established with wood processing facilities that can employ a diversity of professionals, and bring along development of much-needed infrastructure to the communities. These projects allow communities to participate meaningfully in the management of their resources and also build their capacity in project management. For example, the World Bank-sponsored Rural Afforestation Project that was undertaken in the savannah regions of Ghana from 1989 to 1995 employed several hundreds of technical and unskilled staff to produce, distribute and transplant seedlings for interested farmers, communities and organisations. With the closure of the project, many of these people were left unemployed.

In fact, the provision of employment to rural dwellers and re-stocking of commercial species in degraded forest reserves were the main objectives of the plantation development programmes by the then Forestry Department in the 1960s (FAO, 2002a). Again, in 2001, the Government of Ghana launched the National Forest Plantation Development Programme (NFPDP) with a target planting of 20,000 ha/annum and a secondary objective of employing the youth in rural areas. Job opportunities created under the NFPDP were either full time or casual by-day jobs. The full- time jobs were in the form of farming opportunities granted to peasant farmers from forest-fringe communities. The casual by-day jobs were offered mainly for activities such as site preparation, peg cutting, pegging, seedling production, and planting (Ghana Forestry Commission, 2006b). In 2005, the programme recorded an estimated 31,500 full time jobs, and 600,905 man-days of casual by-day jobs (Ghana Forestry Commission, 2006b). When the government re-launched the NFPDP in 2009, job creation for rural dwellers was again at the centre of its focus, promising to create 30,000 jobs by the end of 2011 (Ghana News Agency, 2009).

2.2.5. Environmental Benefits

Given the multi-purpose functions of forest plantations, they can play critical roles in protecting the environment, such as soil conservation, watershed protection, flood reduction, and shelter from wind and rainstorms. This is especially true in the savannah areas of Ghana which is more prone to these environmental problems due to lack of adequate tree cover. In degraded landscapes, plantation forestry can be a useful tool in restoring, rehabilitating or reclaiming these areas. In recent years, flooding has become an annual occurrence in the Guinea and Sudan savannah areas of Ghana. Such areas would benefit significantly from increased tree cover to reverse the degraded lands and reduce the impacts of these environmental hazards.

Increasing concerns about global warming and the role forests play in mitigating the concentrations of greenhouse gases in the atmosphere through carbon sequestration has provided an additional impetus to plantation development. For example, deforestation and forest degradation are estimated to account for almost 20% of the global greenhouse gas emissions (IPCC, 2007). Forestry sector mitigation measures are considered to be cost effective in combating global warming while providing a range of social, environmental and economic benefits to society. A growing tree can store as much as 45% of its stem dry weight as carbon (Evans, 1992) and this makes forest plantations particularly attractive in achieving climate change mitigation because of their fast growth. One of the flexible mechanisms of the Kyoto protocol is the Clean Development Mechanism (CDM), which allows Annex 1 countries (those countries that are signatories to the Protocol and have binding emission targets) to develop afforestation or reforestation projects in developing countries and use the resulting carbon credits to meet their own Kyoto targets. This provides an incentive for Annex 1 countries to meet their greenhouse gas obligations more cost effectively, and an opportunity for developing countries to attract the kind of resources required to establish plantations. It is important to point out that the benefit of plantations in reducing greenhouse gases is highest where trees are planted on degraded or grass lands, not when plantations replace natural forests.

2.3. Policy Considerations in Plantation Development

Developing forest plantations for any purpose involves several long-term decisions that have economic, environmental and social implications, and therefore requires that careful planning and due diligence be undertaken before embarking on the project. In this Chapter, I focus on the important considerations in plantation forestry development. There are two levels to this: the national level policy decisions; and the plantation level considerations. At the national level, policy decisions are based on the overall national requirements for forest resources and environmental protection and the role of plantations to meet these needs. From the analyses in Chapter 1 and the reasons given in Section 2.2, it is clear that at the national level, Ghana needs to implement forest plantation development initiatives in order to avoid acute wood shortages.

Once a decision is made to pursue forest plantations, there is a secondary question of what approach (model) or models of plantation development should be used. Approaches in the literature include government-sponsored plantation projects, community-based approaches, and private industrial plantation development initiatives (Enters et al. 2004). Each of these has their advantages and disadvantages and each country needs to decide which combination of these approaches would best suit its needs. At the plantation level, there are factors that investors have to take into consideration when establishing forest plantations. These factors are considered at the planning stage of the project to ensure that the investor is aware of the requirements surrounding the investment and also identify where technical help might be required by the investor.

2.3.1. Survey of Plantation Models

In an FAO study in the Asia-Pacific Region reported by Enters et al. (2004), there were notably three distinct models commonly applied for the establishment of large-scale forest plantations:

i. plantation establishment led by a central or state-government planting programmes;
ii. plantation establishment under community-based development; and
iii. private sector-led plantation establishment.

The first model is popular in China, Vietnam and Malaysia. For example, the Chinese Government had planned to increase the country's forest cover from 14% to about 20% by 2010 (Enters et al., 2004). In fact, the Chinese government achieved a 20.36% forest cover by 2009, two years ahead of schedule! The Chinese government attaches great importance to forestry development through various government-supported planting programmes, including the promotion of private sector participation with funding through foreign investment and joint ventures. Ghana has implemented a few state-led plantation development programmes in the past including the Rural Afforestation Programme (1989-1995) and the National Forest Plantation Programme started in 2001.

In the second model, participatory activities in plantation establishment at all levels form the key element for the successful implementation of the project. This model is adopted in a number of developing countries including India, Nepal, the Philippines, Bangladesh, Laos, Myanmar and Sri Lanka (Enters et al., 2004). The basic concept of this model recognises the importance of the local people in protecting, managing and developing forests and envisages mobilising the communities through community forestry programme. In this context, communities are empowered to manage degraded forests, including undertaking reforestation activities on a benefit sharing basis (Enters et al., 2004). This system is identical to the modified taungya system being applied in the HFZ of Ghana under the National Forest

Plantation Development Programme. The RAP also emphasised community participation in plantation forestry, and hence about 90% of all seedlings planted during the implementation of that project were in community plantations (FAO, 2002a). Recently, the FC has taken community involvement in forest plantation development more seriously and as a result, a Community Forestry Unit has been created within the FC.

The third model is through the involvement of the private sector. In this model, the government initially establishes a critical mass of plantations, but ownership and responsibility for the management of the plantation is gradually handed over to the private sector (Enters et al., 2004). Countries that have adopted this model are New Zealand and Australia. Plantation forestry is regarded as a business and afforestation investments are primarily assessed based on their potential profitability (Enters et al., 2004).

2.3.2. Plantation Models Used in Ghana

The development of forest plantations in Ghana has relied on different types of models. Generally, forest plantations are either undertaken by individuals at their own expense, or sponsored by the central government, the private sector, or non-governmental organisations. Private plantation forestry is limited to a few companies, whilst almost all plantation forestry since the 1950s, has been sponsored by the government in one way or the other. Historically, the most widely used models in Ghana are the small-holder individual and community plantations. The modified taungya system and agroforestry systems have also been applied on limited scales across the country. Most industrial plantations are owned by private companies to produce poles for electrification, sawnwood and pulpwood.

A) Individual Farm Plantations

Small-holder plantations in Ghana include tree planting on individual household farm properties and around homes. These are either private or government-sponsored. Small-scale plantations on farmlands often reflect land tenure constraints that limit land ownership sizes to small parcels of land per person. Farmers have known for centuries that incorporating trees among agricultural crops can produce valuable cover for shade-tolerant subsistence crops, recycle nutrients from deep in the soil and improve crop yields. Household trees can provide individual families with their own supply of nearby and protected fuelwood, thereby eliminating or reducing lengthy travel time to collection areas.

Traditionally, trees have been planted for shade, protection against wind, and for medicinal purposes. However, tree planting in plantations or on farmlands for poles and fuelwood has not been practised by rural and urban communities. For example, mature neem trees around homes in northern Ghana are usually managed by the coppice system for the continuous production of poles and rafters for building construction. With the advent of the government and World Bank sponsored Rural Afforestation Programme in 1989, neem was grown for poles and fuelwood in small to medium plantations (0.5 - 3.0 ha) adjacent to villages where they are easily maintained and readily accessible for harvesting (Nanang, 1996).

Under the NFPDP, about 37,000 ha of small-holder plantations were developed in the first three year period (2001-2003). To ensure success of smallholder tree planting it is important to identify the target group in need of this service, assess their needs, use local knowledge, solicit local participant's input into project constraints and potential, and recognise and accommodate the socio-cultural dynamics.

B) Community Plantations/Woodlots

Community plantations have the potential to provide large amounts of wood products for a community, and were the most popular plantation model under the RAP. Communities are heterogeneous in character, encompassing a variety of conflicting goals and expectations, and communities are embedded within a larger regional population (Skoupy, 1991). When community members are able to participate in all phases of the project planning, implementation and evaluation they are afforded a sense of ownership and control and will, as a result, be more committed to the project (Skoupy, 1991).

Although they may produce large quantities of wood, community plantations may be ineffective in areas where land tenure is uncertain or where land shortages occur. Moreover in such systems, the dietary subsistence needs of small-holder farmers are often ignored at the expense of wood production. Crush and Namasasu (1985) studied a woodlot project in Lesotho and concluded that its failure was due to uninterested communities because (a) the project did not recognise the realities of rural existence and (b) it did not meet the immediate needs of the people. The project was successful in growing trees and after 12 years of operation across 5000 ha in numerous localities, there was still limited farmer participation in the establishment and operation of the woodlots (Crush and Namasasu, 1985). Because community woodlots have relatively long rotation lengths they cannot provide immediate benefits. Furthermore, equity problems often arise in the distribution of benefits to poorer community members, those with lower status, or where benefits cannot be transferred across generations. Such people usually receive a significantly smaller proportion of the benefits (Crush and Namasasu, 1985).

In regions where community plantations may be an appropriate fuelwood conservation strategy, they may succeed in promoting self-sufficiency by allowing more wood to be produced, sold and used by a target group. Communal fuelwood plantations can be successful in locales where there is a strong sense of community and a history of community action (Munslow et al., 1988). Additionally, community woodlot initiatives are likely more appropriate in areas where secure communal land ownership is possible (Crush and Namasasu, 1985). With increasing population, it is likely that land that is truly communal, in both title and practice, will become scarce in Ghana and pose serious challenges to community forestry projects.

C) Agroforestry Strategies

Agroforestry is a collective name for land use systems and practices in which woody perennials are deliberately integrated with crops and/or animals on the same land management unit (World Agroforestry Centre (ICRAF), 1993). The integration can be either in a spatial mixture or in a temporal sequence. There are normally both ecological and economic interactions between woody and non-woody components in agroforestry (World Agroforestry Centre (ICRAF), 1993). The initial concept of agroforestry was based on its usefulness at the farm level, while more recent thinking conceives agroforestry within a landscape level, which is capable of providing wood products and environmental services to sustain the livelihoods of communities.

The benefits of agroforestry are often underestimated by development agencies for the benefits that it can provide to communities in terms of wood production and subsistence measures. If start-up costs are not available through outside donors or well-funded government programmes, agroforestry systems may be the best choice for peasant farmers. Agroforestry technologies are usually lower in cost because they require no new technology and are simply an extension of present land use systems where trees are incorporated into existing agricultural crop systems (Manshard, 1992). Such systems may be among the best suited for communities across Ghana where the fiscal resources to implement large-scale initiatives like state forest department strategies often rely on outside donor funding. Furthermore, land tenure constraints that restrict individual land holdings to a few hectares of land, and the need to produce enough food on the same piece of land means that for most farmers, it is not possible to acquire additional land solely for growing trees. For these reasons, agroforestry would be more attractive than conventional plantation forestry. It is well known that in Ghana, farmers traditionally practised agroforestry on their farmlands by protecting some indigenous tree species as they clear land for crop production and / or plant trees for shade on cocoa farms.

The Ministry of Food and Agriculture created an agroforestry unit within the ministry to promote agroforestry practices within the country. Research in agroforestry had also increased, with local universities establishing departments to teach and research into agroforestry. Ghana developed a National Agroforestry Policy in 1986 to promote agroforestry practices for sustainable land-use (MOFA/AFU, 1986). The National Agroforestry Policy recognised the fact that an organised and co-coordinated approach was required if agroforestry was to play a role in the promotion of sustainable agricultural development (Asare, 2004). In this light, the Government of Ghana, with assistance from the UNDP and FAO, initiated a national programme to support agroforestry. The three main areas for implementing this policy was research, training, and extension education in agroforestry.

Several NGOs such as the Ghana Rural Reconstruction Movement (GhRRM), Adventist Development and Relief Agency (ADRA), CARE-Denmark, and Conservation International have been influential in supporting government's effort in empowering farmers to engage in sustainable agriculture through agroforestry (Asare, 2004). For example, ADRA supported the government's effort in 1989 by launching the Collaborative Community Forestry Initiative (CCFI) programme that established nurseries and supported households with seedlings. Under this programme 20 nurseries were established within 10 years producing more than 4 million assorted tree seedlings including fruit trees like mangoes, cashew, orange and guava. Woody tree species including teak, eucalyptus spp., neem, and *Albizia lebbeck* were also produced (Djarbeng and Ameyaw, 2002).

D) Modified Taungya System

The Taungya system, which was developed in Myanmar (formerly called Burma) is a plantation development model in which farmers are given parcels of degraded forest reserves to produce food crops and to help establish and maintain timber trees (Agyemang et al., 2003). Ghana adopted this system in the 1930s to help produce a mature crop of commercial timber in a relatively short time, while also addressing the shortage of farmland in communities bordering forest reserves (Agyemang et al., 2003). This approach has merit because cocoa, the dominant cash crop grown by farmers in the reserve, is a shade tolerant plant especially during its initial stages of growth. The problem with the traditional taungya operations was that timber seedlings were simply ignored by the farmers. Farmers had no incentive to protect the forest trees. The modified taungya system (MTS), introduced in 2002, is a framework that allows farmers and their families to keep portions of the revenue from harvested trees on their farms as an incentive for them to tend the trees. This system allows degraded forest reserves to be forested with selected tree species, intercropped with food crops and designed to entitle all stakeholders to the plantations' benefit and give them a long-term interest in maintaining tress (Agyeman et al., 2003). The taungya system in Ghana relies on both indigenous and exotic tree species. The system offers 40% of proceeds from sales on the standing tree value to participating farmers. The farmers contribute their labour for their share of the benefits from trees and food crops. Additionally, the contribution of the taungya system to ensuring food security in the country since 2002 has been remarkable. For example, food production from the plantation coupes under the MTS in 2008 was estimated at 471,875 tonnes (Ghana Forestry Commission, 2008b).

E) Private Industrial Plantations

Apart from these Government and NGO-initiated programmes, several private companies and individuals have established plantations at various scales and for various purposes using various partnerships, programmes and strategies in the past decades (FAO, 2002a). However, these private industrial plantations are still limited in terms of the number of companies participating and the areas planted. Notable among these are the Pioneer Tobacco Company Ltd. (PTC) who own about 5,000 ha of teak plantations mainly in the Brong Ahafo Region, the Ashanti Goldfield Company Ltd., with over 1,400, 50, and 42 ha of teak, gmelina and eucalyptus species, respectively. The Bonsu Vonberg Farms Ltd., owns about 500 ha of teak (FAO, 2002a).

The Samreboi Timber and Plywood Company Ltd. (Samartex) has initiated a forest plantation programme at Oda Kotoamso near Asankrangwa in the Western Region. The area is an abandoned cocoa farm outside the reserved forests, while the Ghana Primewoods Products Ltd., Takoradi (GAP) has initiated a "Joint Forest Management Project" outside the reserved forests in late 1995 at Gwira Banso in the Western Region (FAO, 2002a). In addition, some other timber companies have also established plantations on limited scales, e.g., Sunstex Ltd., Western Hardwoods, WLVC, Ghana Prime Woods Products Ltd., Specialised Timber Products, John Bitar and Co., Asuo Bomosadu Timbers and Samartex Ltd (FAO, 2002a). The recent rural electrification programme in Ghana has created a market for electric transmission poles, which hitherto was supplied from thinnings of the Government owned plantations. This has generated unprecedented interest in teak plantations by individuals and organisations.

F) Timber Companies

As part of their timber utilisation contract (TUC) obligations and also to meet their future demands, timber companies are expected to develop and implement afforestation or reforestation plans in the areas in which they operate. These companies usually involve farmers in afforestation or reforestation and often provide them with training in nursery, plantation establishment and agroforestry techniques. Farmers are also provided with basic materials such as tree seedlings and nursery equipment. Examples of timber companies undertaking afforestation/reforestation under the TUC obligations include Suhuma Timber

Co. Ltd, Swiss Lumber Co. Ltd, Samartex Timber and Plywood Co. Ltd, Logs and Lumber Co Ltd, Bibiani Logs and Lumber Co Ltd, and AG Timbers.

G) Other Potential Models

FAO (2002a) has identified three other models used in rubber and oil palm plantations in Ghana, which may have potential for adoption to forestry plantations. These models normally operate between companies and individual farmers under the headings: small-holder scheme; out-grower scheme; and the lease-back system. Under the small-holder scheme, a company arranges for credit and provides extension services and planting material for the establishment and maintenance of the tree crops to an individual planter or farmer who does not own the land. The company also serves as the market for the produce. The out-grower scheme is identical to the small-holder scheme, except that the planter owns the land or has a lease, freehold or share-cropping arrangement. In the lease-back scheme, the farmer owns the land and leases it to a company for plantation development. The farmer usually takes a share of the final value of the harvest (FAO, 2002a).

2.4. Technical and Economic Considerations in Plantation Establishment

Establishing a forest plantation involves several technical steps and decisions. There are three distinct management phases that can be distinguished: seed collection and handling; nursery practices and plantation establishment; and plantation management. The initial plantation establishment phase is divided into the following activities: species selection, site preparation and planting operation (Camirand, 2002). The remaining decisions include where to plant trees, acquiring seeds, producing the planting stock, preparing the site, planting, weeding, pruning, thinning, final harvest, and regeneration. Depending on the purpose of the plantation, the frequency and timing of these operations would vary. The following factors need to be considered in making decisions related to the plantation enterprise.

2.4.1. Purpose

The first consideration in establishing a plantation is to clearly define the purpose for which the plantation is being established. This will include the investor asking the questions: "what are the outputs required and within what time period?" The answers to these questions will have tremendous impacts on many other decisions that will have to be made in respect of the plantation enterprise. For example, the purpose of the plantation will determine the species to be planted, where the plantation should be sited, the kinds of management regime to be used, etc.

2.4.2. Site Selection

A second important consideration is selecting the site for the plantation. Good site selection is one of the most important decisions that can lead to improved yields, reduced rotation lengths, and increased economic returns for plantations. In this case, the most important factor is to match species with the site in order to optimise productivity. Furthermore, sites are selected to minimise the risk of plantation failure due to poor soil drainage, drought or inability to control weed competition. The important components of site quality are: soil depth and drainage; soil physical and chemical composition (including pH); amount and pattern of yearly soil moisture availability; frequency and nature of common and occasional winds, storms and fires; and the general climate of the area (FAO, 2002b). A general approach to assessing site quality is to examine the presence and importance of competing vegetation, and of populations of animals, insects and microorganisms that are either damaging or beneficial to trees, which can affect management of forest plantation sites. The size and health of trees already present on the sites are good but imperfect indicators of site quality (FAO, 2002b).

Site selection may be limited by the availability of land for the investor. Most small-holder farmers would be restricted to planting trees on their own farmlands or family lands, which may not be the most suitable for forest plantations. This would reduce the productivity of such plantations, thereby requiring more land to achieve the same level of output. Community plantations rely on communal lands, while industrial plantation investors would acquire large tracks of land through purchasing or leasing. Industrial investors have the most flexibility in choosing the site, since they would be purchasing or leasing the land. Large-scale investors would also need to consider the location of the plantations vis-à-vis processing facilities, markets for their products, cost of land preparation, labour force availability and available infrastructure. The ability to match species with site is limited by the lack of technical information on many plantation species and the sites that optimise their productivity. Matching species with sites is complicated by the wide range of potential plantation species and infinite combinations of sites available to choose from. Consequently, a spatial map that shows the suitability of various plantation species for the different vegetation zones would be a useful tool for plantation forestry investors in Ghana.

2.4.3. Choice of Tree Species

In addition to choosing tree species to match their most productive sites, species are also chosen to fulfil the purpose of establishing the plantation (end-uses). There is also the question of whether the plantation should be a single species (monoculture) or multiple species (mixed). Each of these has their advantages and disadvantages. If the forest plantation is to provide only wood products, then most often, a single species suitable to the site may be sufficient. However, if the forest plantation is to provide aesthetic, wildlife-habitat, biological diversity, and other services in addition to wood production, multi-species forest plantations will often be favoured (FAO, 2002b). Extensive planting of one species, whether indigenous or exotic, inevitably results in some areas where trees are ill-suited to the site and suffer stress. This may occur where large mono-specific blocks are planted or where exotics are used extensively before sufficient experience has been gained over a whole rotation (Evans,

2001). Species choice is also related to whether an indigenous species or an exotic species should be used. In Ghana, most exotic species outperform indigenous ones in terms of productivity (Section 2.2.3), and would be more favoured in afforestation projects. In addition, exotic species offer a wider range of choices, are often free from local diseases and pests, and usually their silviculture is better understood than indigenous species (Evans and Turnbull, 2004). However, indigenous species offer advantages in terms of acclimatisation and would be preferred in re-stocking degraded forest reserves, for example. If two species are identical except that one is indigenous and the other is exotic, the former should be preferred. In most cases, however, the choice may be to use a mixture, rather than a monoculture.

Species are also chosen based on the outputs desired. If fuelwood is the major need, then species that produce large amounts of biomass within short periods, and also coppice well would be the appropriate choice. However, if the need is for building construction and electrification, then tree species that have straight boles and with natural durability would be more suitable. Tree species that are water intensive should neither be planted in the drier areas nor used as ecological restoration around rivers or streams. Species for restoring degraded lands and watershed protection would have characteristics that are consistent with their functions such as fast growth, provision of shade, litter production, deep-rooted, erosion resistance, etc. When it comes to ecological restoration, non-indigenous species can pose a major problem because they are often aggressive and can overwhelm native species, thus altering ecosystem structure (Berger, 2006).

Because most plantations in Ghana and other tropical countries use exotic species, it is often assumed that exotic species are better suited for forest plantations than native species. The literature identifies several advantages and disadvantages of using either species. The main advantages of native species are (after Evans and Turnbull, 2004):

- the growth of indigenous species in natural stands gives an indication of their potential, reducing the risk of complete failure when grown in plantations;
- indigenous species are naturally adapted to their environment and developed resistance to known diseases and pests;
- indigenous species play ecological roles, and may be critical habitats for local animals and plants; and
- plantations of native species conserve the native flora, especially in areas of deforested or degraded forests.

Advantages of exotic species include (Smith, 1986; Evans and Turnbull, 2004; Zobel and Talbert, 1984):

- there is a wide variety of exotics to choose from, and hence increases the chance of finding one that would be suitable for the purpose of the plantation;
- tree improvement programmes have identified some exotic species with preferable genetic and phenotypic characteristics;
- exotic species lend themselves to mass propagation, as required in large plantation programmes;

- most exotic species grow more rapidly, providing higher yields of wood and other products and higher economic value;
- exotics may have resistance to local diseases and pests;
- experience with the exotic species in other parts of the world may provide dependable silvicultural methods for managing the species; and
- uniform stands of exotic species are easier to manage intensively over short rotations

Despite these advantages, plantations of exotic species have been known to produce poor results in test plantations on sites unsuited for long-term plantation development, susceptibility to local pests and diseases, failure of single species stands due to slowly developing effects of sites, unacceptable stem form, undesirable wood qualities, poor adaptability to environmental conditions and development of wood characteristics, and unsuitable to local markets needs (Zobel and Talbert, 1984; Zobel et al., 1987). In order to minimise the risk of exotic plantation failure and optimise the benefits from exotic species, Zobel et al. (1987) recommend that species should be carefully chosen to match the site, use only resistant species and seed sources with attributes of local interest; use a variety of seeds or cuttings from several sources to ensure genetic diversity; ensure sanitation in raising nursery stock and transporting seedlings; and provide protection to the established trees.

2.4.4. Regeneration Method

The method of regenerating the plantation and subsequent management regime are critical to determining the success or otherwise of the plantation investment. Plantations can be regenerated from direct seeding, using container or bare-stock seedlings, or the use of cuttings. The best method will depend on the species and site characteristics. Another method of regeneration from an existing plantation is the coppice method. For example, coppicing is an alternative reforestation tool for teak in the tropics, substantially reducing regeneration time/costs and associated demands for labour and seed when available (Bailey and Harjanto, 2005). Growth rates of coppiced material are rapid in most situations, and make this method suited for managing woodlots in Ghana. Bailey and Harjanto (2005) compared coppiced teak plantations to paired seed-origin plantations, at ages 3, 8, and 13 years, on Forest State Corporation managed land located in Java, Indonesia. The plantations were evaluated for height, diameter, lower-bole straightness, and presence of disease in both plantation types and at three ages. Mean height and diameter of trees in coppiced plantations were both significantly greater than those in their paired seed-origin plantations at all three ages. Furthermore, heights and diameters in coppiced plantations were higher than expected based on established growth tables for Java. Coppiced plantations were less symptomatic of disease than seed-origin plantations, which promise better wood production and quality. Lower-bole stems in coppiced plantations developed less straight than those in seed-origin plantations, but these deviations faded with time and could likely become insignificant within a 60-year rotation. The authors concluded that coppiced plantations in Java can make a major contribution to teak production in Indonesia (Bailey and Harjanto, 2005).

2.4.5. Level of Investment

An investor in plantation forestry needs to consider the amount of investment required for the enterprise in terms of labour, capital, time and infrastructure. In industrial plantations, hired labour will be needed, and the investor would need to ensure that there is available labour force within a reasonable distance from the plantation site, as this will affect the cost of labour. In Ghana and other developing countries, labour costs are generally low and the labour force is usually available, especially in rural areas where unemployment rates are often high. For small-scale farmers, labour is usually not an issue, as they rely on their own labour and that of friends and family members. Unless subsidies are granted, capital investments in plantations can be high, especially in the first few years. And hence investors need to consider the costs of planting and managing the plantation until the final harvest. There is also investment of time in the plantation venture for managing the staff and ensuring that the plantation is protected from fires, insects and diseases.

Basic infrastructure such as roads, electricity and transport are required to get to a forest plantation site in order to plant, for later management and harvest. The relative availability of access does not directly affect productivity; however, it can indirectly affect it by its influence on the ability of management to exert leverage through such things as post-planting release, thinning and pruning, and other management activities (FAO, 2002b). Furthermore, accessibility affects the energy it requires to deliver forest products to their users, and thus the economic feasibility of using them (FAO, 2002b). If basic infrastructure such as roads, water, communications are not available at the chosen site, the investor would have to provide these at his own expense (either fully or partly) and this will impact on the profitability of the enterprise.

2.4.6. Economic Factors

Economic factors relate to first and foremost whether the investor would turn a profit on the venture. Profitability is affected by several other economic, social, and technical aspects of the plantation development enterprise. For example, economic viability of plantations is affected by the technical aspects of the plantation enterprise discussed in the preceding sections (e.g., matching species with sites), as some of these may impose economic costs on the plantation investment.

Another important economic consideration is the ability to finance the project at reasonable interest rates. Even if the investor can finance from his own resources, the issue of the rate of return on the plantation over the entire rotation would still need to be taken into account and compared with alternative investments. Benefits accruing to the investor are partly affected by the level of taxation imposed on the plantation and /or its outputs, and the property rights regime under which the investor operates. Therefore, investors need to be aware of the taxation regime in effect where the plantation is located. Furthermore, investors need to worry about whether there is a market for the outputs and the distance to those markets. In addition to these, the opportunity cost of the plantation would be an important consideration. That is, alternative uses of the inputs (e.g., land) would need to be compared with the use of the land for forestry. This would ensure that the land is put to its best use.

There are also a combination of taxes, levies and fees on forestry operations and forestry products in Ghana. From the legislative framework, there is still some confusion as to what levies and fees are applicable to privately established forest plantations. The potential fees and levies include: stumpage fees, fees for management services, rent for contract areas, property marks fees, export levies for selected species, and timber rights fees. With the exception of export levies which apply to teak (10% under the *Trees and Timber (Amendment) Act* (1994)), it is not clear which of the other charges plantation owners are subject to.

Consideration of profitability is not restricted to industrial investors in forest plantations. At first glance it may not be entirely obvious that forest plantations at the community level for the provision of fuelwood and other basic needs such as fodder, construction wood, fruits, etc. should also consider economic principles in their decision making. It is true that under these instances, the emphasis is not normally on the selling of commodities to maximise net returns based on commercial transactions, but communities try to maximise their well-being. Evans (1992) provides evidence that under certain conditions, smallholders do plant trees in anticipation of future markets.

2.4.7. Legal and Institutional Requirements

Plantation establishment is also affected by the legal and institutional framework within Ghana. Forestry practices are governed by various policies, laws and regulations and these have to be understood and respected when establishing plantations. For example, the Forest and Wildlife Policy stipulates that environmental impact assessment must be carried out as a prerequisite for resource development and utilisation projects, in compliance with approved standards (i.e., Environmental Assessment regulations). Secondly, investors need to be aware of the different incentives available to forest plantation developers, and the kinds of restrictions that exist on exports of various forest products from their plantations. The *Trees and Timber Management (Amendment) Act*, 1994 (Act 493) and the *Timber Resources Management Regulations*, 1998 (L.I. 1649) also specify various export levies and stumpage rates for timber products, including plantation species such as teak (export levy 10%) and ceiba (export levy 30%). Investors need to consider these in assessing the economic feasibility of the plantation enterprise.

Property rights form the bedrock of private decisions in resource use and allocation, and hence the property rights regime within the location of the plantation and how benefits are shared among property rights holders would also affect the kinds of management activities that would be undertaken and the benefits to be derived. There may also be land-use zoning restrictions, water and electricity rights-of-way, etc., that could interfere with plantation location and management. These need to be carefully considered.

2.4.8. International Environmental Concerns

Another factor that is likely to impact on the marketability of forest products from plantations that has to be considered, especially for large-scale industrial plantations, is the emerging issue of environmental concerns regarding forest plantations. These environmental considerations have led to the introduction of forest certification programmes around the world. Forest certification is a system for identifying forests that are managed to maintain ecological, economic and social components of the ecosystem. Certification is a market-based mechanism to reward sustainable forest management (SFM), and in a way resembles a command-and-control approach as those who do not certify their forests or products could be penalised by the loss of market share for their products.

Forest products and forest certification grew out of environmental concerns about the sustainability of natural forests, and is seen as a tool for reaching consensus on forest management and a means of addressing market failures in forest management resulting from multiple benefits and information asymmetry. Through certification, producers are able to prove to consumers that environmental concerns have been addressed, and hence certification acts as a link between the consumer and producers of forest products.

Although importing countries have not yet made forest certification mandatory for tropical plantation products, it is clear that countries that wish to have unfettered market access around the world would be compelled to show that their wood is coming from sustainably managed forests. Ghana has been developing the processes needed to certify its forests and this is a step in the right direction. The Ghana Forest Management Certification System Project was initiated in 1997 with the assistance of the European Union and the Netherlands to develop draft standards for certification. Field tests on the resulting chain-of-custody and log-tracking systems have been carried out since 2002, as has the development of standards for sustainably managed forests (ITTO, 2005a).

At the global level, the two competing certification schemes with different operating modalities are the Forest Stewardship Council (FSC), which provides all the necessary elements of certification through centralised decision-making on standards and accreditation and the Programme for the Endorsement of Forest Certification (PEFC), on the other hand, which operates as a system for mutual recognition between national certification systems. Almost two-thirds (65%) of the world's certified forests (in 22 countries) carry a PEFC certificate, while the FSC's share is 28% (in 78 countries); the remaining forests are certified solely under national systems. Most of the certified forests in the tropics are FSC-certified (ITTO, 2008).

Although originally intended for natural forests, certification schemes have recently been extended to cover plantation forests. The FSC has 10 principles and criteria that must be satisfied before a plantation is certified (after FSC, 2010):

1. Compliance with all applicable laws and international treaties
2. Demonstrated and uncontested, clearly defined, long–term land *tenure* and use rights
3. Recognition and respect of indigenous peoples' rights
4. Maintenance or enhancement of long-term social and economic well-being of forest workers and local communities and respect of worker's rights in *compliance* with International Labour Organisation (ILO) conventions
5. Equitable use and sharing of benefits derived from the forest
6. Reduction of environmental impact of logging activities and maintenance of the ecological functions and integrity of the forest
7. Appropriate and continuously updated management plan
8. Appropriate monitoring and assessment activities to assess the condition of the forest, management activities and their social and environmental impacts

9. Maintenance of High Conservation Value Forests (HCVFs) defined as environmental and social values that are considered to be of outstanding significance or critical importance
10. In addition to *compliance* with all of the above, plantations must contribute to reduce the pressures on, and promote the restoration and conservation of, natural forests.

Despite the general acceptance of certification schemes, there are some concerns regarding their implementation. These include: the lack of consensus on certification standards and definitions; the costs of certification and who should pay for them; whether certification will drive up the cost of wood and reduce demand for forest products; the impact of certification on small businesses such as loggers and sawmillers; and economic issues and welfare implications of certification. In addition, there are major problems associated with regulating markets with certification due to insufficient information for consumers in their decision making; unpredictable changes in consumer and producer response to certification, whether the goal of forest management is only for consumer values or societal values, the lack of coordination between certification schemes and government regulations, and how to deal with goods and services that are not traded in the market system.

In spite of these problems, it is accurate to say that certification schemes in whatever form they take are here to stay, and large-scale plantation developers who may be competing internationally for market share for their products, will be obliged or persuaded to certify their plantations. Even if plantation investors in Ghana do not follow any nationally or internationally recognised certification schemes, there may still be some pressure to ensure that environmental concerns of forest plantation development are addressed in one way or the other. These could increase plantation development and management costs that may be non-trivial, but which may be compensated for by increased market share of marketed products.

2.5. Sustainability and Environmental Concerns in Plantation Forestry

2.5.1. Sustainability Considerations

The word sustainability means different things to different people. In relation to plantations, there are two components to sustainability. The first component relates to the broad issues of whether using land and devoting resources to tree plantations is a sustainable activity from the economic, the environmental or the social sense (Evans, 2001). The second component focuses on the narrow issue of whether tree plantations can be grown indefinitely from one rotation to the next; i.e., whether their long-term productivity can be assured (Evans, 2001).

Economic and social sustainability have to be assessed on an individual basis, as they depend on several factors. Economic sustainability relates more to whether the enterprise is profitable enough to be self-sustaining, considering all the timber and non-timber benefits from the plantation as well as all direct and indirect costs. Establishment of forest plantations has economic implications beyond the forestry sector. For example, forest plantations may increase local and regional land and property prices and increase local inflation as well. But there are also positive economic benefits and spin-offs from plantation forestry including the generation of local employment opportunities at all phases of the plantation project, development of infrastructure or the development of local housing and accommodation to meet the needs of migrating populations.

Social sustainability has to do with how the plantations fit into the social structure of the community, social acceptability and social benefits that are derived by communities. The establishment of forest plantations may impact on rural populations (e.g., displacing them from their homes or farmlands). People may move into, or away from, locations where plantations are established, hence the community and population dynamics are likely to change in response to plantation development in a region. The remaining discussion in this section focuses on the narrow definition of sustainability, which has attracted more attention.

Because of the intensive management often associated with plantations and the high nutrient demands of most plantation species, there is the question of sustainability of forest plantations as a result of depletion of the soils following continuous biomass removals. The issue of how much biomass can be removed without negatively affecting the sustainability of plantations has gained greater importance with current emphasis in international climate change discussions to use bioenergy as an environmentally friendly energy supply alternative to fossil fuels.

Hartemink (2003) has extensively studied nutrient losses from plantations in the tropics and notes that harvesting techniques for logs affect the soil carbon and nitrogen stocks in forest plantations. Forest plantations mimic the natural forest in which nutrient cycling is fairly closed, and unless thinning of trees has occurred during the first years after planting, the major drain of nutrients is at harvest. Nutrient losses can be minimised if only the stem wood is removed from the field, but soil disturbance and complete removal of the above ground biomass is likely to induce considerable nutrient losses as well (Hartemink, 2003). For example, Chacko (1995) observed site deterioration under teak in India with yields from plantations below expectation and a decline of site quality with age and attributes this to poor supervision of establishment, over-intensive taungya (intercropping) cultivation; delayed planting; and poor after-care. Chundamannii (1998) similarly reports declines in site quality of teak plantations over time and blames this on poor site management. Site deterioration has also been identified as a problem in Indonesia and according to Perhutani (1992), is caused by repeated planting of teak on the same sites.

Comparisons of the physio-chemical properties of soils under two distinct forest covers (logged native forest and teak plantations) at three locations in the forest-savannah transition zone of Ghana by Salifu (1997) showed that in two of the locations, nitrogen (N), magnesium (Mg) concentrations and organic matter (OM) contents in the soil horizons were significantly higher under logged forest than under teak plantations. In the third location, phosphorus (P) and potassium (K) concentrations were significantly higher under the logged forests. In general, total nutrients were higher in soils under adjacent logged forest compared to teak plantations. The higher nutrient concentrations and contents in soils under logged forest were due to more undergrowth, litter and organic matter under logged forest and a lesser demand for these nutrients by tree species in the logged forest. Lower soil macro-nutrient concentration and contents in soils under teak was due to lower organic matter content under teak cover and /or associated with higher nutrient demand and nutrient immobilisation by teak (Salifu, 1997).

Evans (2001) acknowledges that plantations and plantation forestry operations do impact the sites on which they occur and under certain conditions nutrient export may threaten sustainability. However, site quality can be maintained through care with harvesting operations, conservation of organic matter, and management of the weed environment (Evans, 2001). Evans (2001) concludes that plantation forestry appears entirely sustainable under conditions of good husbandry, but not where wasteful and damaging practices are permitted.

There are several approaches proposed by Evans (2001) to help sustain the productivity of forest plantation sites through subsequent rotations. These include: (1) genetic improvement of planting stock, which includes change in species, seed origin, use of new clones, use of genetically improved seed and, in the future, genetically modified trees all offer the prospect of better yields in later rotations; (2) better understanding of the silviculture of plantation species through manipulation of stocking levels to achieve greater output of fibre or a particular product, matching rotation length to optimise yield, and use of mixed crops on a site to aid tree stability, lower pest and disease threats; (3) fertiliser application to compensate for nutrient losses on those sites where plantation forestry practice does cause net nutrient export to the detriment of plant growth; (4) site preparation and establishment practices to alleviate soil compaction after harvesting or weed control strategies to reduce competitive vegetation; and (5) conserve organic matter through preventing systematic litter raking or gathering during the rotation and conserving organic matter at harvesting.

2.5.2. Environmental Considerations

Despite the advantages of forest plantations over natural forests described in the preceding sections, it is important to note that due to their simpler structure and most often monoculture nature, forest plantations may not be able to provide the same kinds of environmental and social goods and services provided by natural forests. As a result, there are some environmental issues to keep in mind when considering, establishing and managing plantations.

Sites for plantation establishment have to be chosen carefully to exclude areas that are critical habitats for species needed to conserve biological diversity. In this case, replacing natural forests with a forest plantation is a bad idea. Forest plantations usually have less total biological diversity than do indigenous forests, and their associated biota are also different in composition from those of indigenous forests in the same area (FAO, 2002b). It is also possible for exotic plantation species to introduce diseases and pests to their new environment, which could infest native tree species. Moreover, some exotic species are invasive by nature, and could become difficult to control once they establish themselves within a location. An example is neem in some areas of the coastal savannah areas of Ghana. Therefore, tree species for introduction must be chosen carefully. Despite these potential problems with plantations, it is fair to say that the different suites of biota provided by species planted in forest plantations would contribute to regional biological diversity. In addition, tree planting can be used to improve stocking of a natural forest where necessary.

Some forest plantations such as teak hardly support any undergrowth. This has often led to excessive soil erosion especially when teak is planted on a slope. Some soil degradation under teak in Ghana has been observed in parts of the plantations at the Ho Hills Forest Reserve in the Volta Region and the Yendi Town Plantation (FAO, 2002a). There are two schools of thought about the lack of undergrowth in pure plantations of teak. The first attributes the lack of undergrowth under teak plantations to allelopathic effects of leachates from teak that tend to limit the amount of seedlings that can survive and grow under teak plantations. This claim has however, been more circumstantial with little or no empirical studies to support it (Healey and Gara, 2003). Although studies by Murugan and Kumar (1996) found significant concentrations of phenolic acids which have been implicated in regeneration failures in some forest types (e.g., Pelissier and Souto, 1999; Mitzutani, 1999) in the foliage of teak, there has not been any strong documented evidence that directly links the release of these substances at concentrations high enough to impede the establishment of undergrowth in teak plantations. Studies by Jadhav and Gaynar (1994) and Tripathi et al. (1999) on the effects of teak leachates on rice, soybean and cowpea have demonstrated potential allelopathic effects.

Healey and Gara (2003) investigated the effects of teak plantation on the establishment of native species and observed significantly lower abundance, diversity and size of the native species growing under teak plantations as compared to adjacent land. Their conclusion on the potential allelopathic cause of these differences was speculative. Empirical investigations of potential allelopathic effects of teak are often confounded by other factors. As alluded to by some researchers (e.g., Pelissier and Souto, 1999), allelopathy is not about the concentrations of allelopathic substances; other factors including climate, target plant physiology and the potential interaction with other soil compounds are also important (Healey and Gara, 2003).

The more popular school of thought attributes lack of undergrowth to the excessive shading of the broad leaves of teak trees particularly after crown closure, which makes it impossible for the development of any competing vegetation in the understory. This, according to FAO (2002a), is made worse by fires in the dry season which burn the leaf litter, further reducing the possibility of any undergrowth and leaving the soil surface unprotected from heavy rains, particularly at the beginning of the rainy season before the overstory has developed new leaves. This issue appears to be much more legitimate and may relate well with concerns about the sustainable management of multiple rotations of teak. While teak is a high nutrient demander, erosion of top soils under teak plantations coupled with high concentration of nutrients in teak leaves (Weaver, 1993) which are burnt annually may lead to a net nutrient loss between rotations. While this may be known from experience there is generally a lack of information about the effect on yields of the second rotation (Hall et al., 1999). The use of mixtures of different species especially the nitrogen fixers with teak has been widely promoted to help prevent soil nutrient loss and improve productivity. An extensive literature review, carried out as part of a general study on mixed tree species plantations in the tropics (FAO, 1992) showed many examples of teak grown in species mixtures in the 1930s largely from India but also Indonesia, Nigeria, Benin and Sri Lanka (Hall et al., 1999).

Some environmental organisations strongly oppose the use of exotic monoculture plantations, arguing that they are not environmentally sound. According to AFORNET (2008), these claims about the ecological fragility and damaging effects of such plantations are only partly justified because (AFORNET, 2008):

- there are few cases where monitoring and investigation of soil decline affecting long-term production have shown that this actually is a problem;

- fast growing plantations, whether exotic or indigenous, will have an impact on water balances in direct proportion to their growth rates;
- biodiversity in monoculture plantations will inevitably be less than in a natural forest but not necessarily less than in agricultural lands on which these plantations are established. In addition, biodiversity increases with increasing age of plantations;
- incidences of pest and disease damage occur but there are only half a dozen or so examples where such attacks have acquired serious and permanent proportions (e.g. *Dothistroma pini* on *Pinus radiata*); and
- plantations can also play positive ecological roles in terms of enhancing the natural regeneration of indigenous species on reforested degraded sites.

2.6. Conclusions

The main reasons for promoting forest plantations in Ghana are to ensure the availability of forest products for present and future generations and reduce the pressure on the natural forest resource, thereby contributing to the sustainability of the latter. The concept of forest plantations is not new to Ghana, neither is the need to develop plantations a recent invention. In fact, the need for plantations to provide wood products to the savannah regions of Ghana has been recognised since 1956 (FAO, 2002a). The slow rate of plantation development has more to do with the lack of will to follow up and meet the targets set by previous programmes. There is opposition in certain NGO communities to plantation development, but a fair assessment of plantations has to take into account the actual land use prior to planting and the positive benefits plantations provide. Some environmental groups have argued that plantations are detrimental to biodiversity, but this is only so if they replace natural ecosystems or traditionally managed forests. Even though plantations may not replicate all of the functions of a natural forest, they do play positive environmental roles in protecting sensitive areas and prevent or even to some extent reverse land degradation. The benefits of plantations in most cases would outweigh their economic, social and environmental costs. There is a cost to doing nothing about land and environmental degradation resulting from the lack of tree cover, which those who criticise plantations underestimate. The costs of environmental disasters that can be avoided with the presence of forest cover can be very high.

Plantation establishment and management involve a series of complicated processes, as they are affected by technical, social, economic, environmental, institutional and legal factors. Investors at the smallholder, community or industrial plantation levels would need to weigh these factors and make decisions that will determine the success or failure of the plantation venture. These decisions are further complicated by high levels of uncertainty in future costs and prices, legal and institutional operating environments. Feasibility of the plantation establishment would rely on sound and technical expertise from professionals in each of these aspects to minimise risk and optimise the benefits to be derived. Historically, Ghana has relied more on promoting individual and community forestry plantations, often sponsored by the central government and international donors. However, it would serve the country well to look at other models that support industrial plantations more seriously than is being done presently.

Although certification programmes are already in place, there are some sticky issues that still need to be resolved. Despite these concerns, it is likely that forest plantations that are targeted to produce timber for the international markets would be obliged either by the national government or international environmental movements to certify their plantations.

Chapter 3

CONSTRAINTS, OPPORTUNITIES AND INCENTIVES FOR PLANTATION DEVELOPMENT

3.1. INTRODUCTION

Historically, Ghana has not done much by way of developing forest plantations, probably because of large tracts of natural forests that were available. As a result, the rate of plantation development has often lagged behind the rate of deforestation. The limited success of plantation development may be due to barriers and lack of opportunities to promote plantation forestry.

This Chapter analyses the barriers to, and the opportunities and incentives for, plantation development in Ghana. In general, there are four kinds of barriers to plantation development: economic, technical, institutional and biophysical. Conversely, there are clearly some opportunities at the national and international levels which Ghana can seize to achieve its plantation development goals. These opportunities include Ghana's forest legal and policy framework, availability of land and a favourable climate, available labour force, the Clean Development Mechanism under the Kyoto Protocol, etc. The final section of this Chapter examines the issues around incentive packages to promote plantation development. It is recognised that the characteristics of forest plantations as an investment are unique, compared to other alternative land-use investments. The reasons for providing incentives are explored, followed by a description of the different kinds of incentives for plantation development in Ghana, and the problems associated with incentive programmes.

3.2. CONSTRAINTS TO PLANTATION DEVELOPMENT

Constraints or barriers to plantation development refer to the factors that prevent or discourage individuals, private firms, communities, NGOs and other public sector organisations from engaging in plantation establishment and management. It makes sense to assume that because Ghana is a tropical country that has a favourable climate that supports fast tree growth, it should do better than temperate countries in plantation development. However, growth rates are only one aspect of economic competitiveness in forestry. The fact that Ghana has not done well in this area, whilst countries such as Finland and Sweden continue to develop forest plantations and compete with tropical countries such as Brazil, suggests that there are other factors more important than a favourable climate that determine the success of plantation forestry. These other success factors may be lacking in Ghana, which together would constitute barriers or constraints. Even with a favourable climate it still takes considerable time for investors to recover investments from trees. Private and public individuals and firms will not invest in plantations unless the political, institutional and economic environments not only permit, but actually encourage, people to make money from the plantations. The constraints to plantation development have been categorised into economic, technical, institutional, and site and biophysical factors depending on their origin and impact.

3.2.1. Economic Constraints

The net financial returns from wood production is determined by a complex interaction of factors such as climate, soil type, land use regulation and those aspects of market structure that influence input and wood prices (Bhati et al., 1991). The main economic barriers to plantation development in Ghana are:

A. Uncertain Economic Profitability

A major economic barrier is the fact that plantations have long rotation periods (the time interval between planting and harvesting of the plantation) and may not be the most profitable investment option compared to other, more short-term, investments. Most native species in Ghana have very long rotations of up to 80 or more years, while exotic species may have shorter rotations of about 30-50 years. Economic profitability is highly dependent on the length of the investment: the longer the rotation, the less profitable the investment is likely to be, all things being equal. Questions regarding economic profitability also reflect increased risk of fires, pests and diseases and accumulation of interest on the capital that is tied up in the plantation venture. Another issue is that entrepreneurs are usually interested in investments that would yield profits within their lifetime, and hence would tend to shy away from long-term investments as encountered in plantation forestry. Unless thinning is undertaken, most trees do not generate any revenues between planting and harvesting; therefore landowners who do not have other sources of income to live on while the trees reach maturity will be reluctant to invest in plantations compared with other land uses such as agricultural and cash crops.

Available published analyses of profitability of forest plantations in Ghana are limited. FAO (2002a) compared the benefit/cost ratios of several native and plantations of exotic species on good sites in the HFZ using a discount rate of 10%. The study concludes that among the exotic species, teak and *Cedrela odorata* were the most profitable, with benefit/cost ratios of 2.4 and 1.7 respectively; while *Ceiba pentandra* and wawa (*Triplochiton scleroxylon*) were the most profitable among the native species with 1.4 and 1.3 benefit/cost ratios, respectively. It is not possible to make a general case as to whether *all* forest plantations in Ghana are profitable investments or not. Several factors influencing plantation profitability are: choice of species and site, rotation length, prices of outputs, input costs, and the discount rate. Therefore economic profitability has to be analysed on a case-by-case basis.

B. Lack of Financing

Timber plantation establishment requires huge investments of financial resources to be successful. Given the risky nature of such long-term investments, commercial banks are usually reluctant to provide credit for plantation investors. Furthermore, the land titling system in Ghana does not provide enough security for forestland to be accepted as collateral by the commercial banks. Lack of financing at reasonable interest rates therefore constitutes an economic barrier to plantation development. Interest is the fee for borrowing money, and usually this amount is paid to the lender over and above the amount borrowed (principal). In Ghana, commercial interest rates are habitually high (between 20-40%). With such high rates, it is almost impossible to return a profit from investments in forest plantations that span over several decades. The higher the interest rates, the higher the interest that a borrower would pay on the initial capital. For example, if one borrows Gh¢100, 000 to establish a plantation at a 5% interest rate over 50 years, the amount of interest at the end of the 50 years will be Gh¢1, 046,740. However, if the interest rate was 20%, the interest on the same capital will be Gh¢909, 943,815 over the same period.

The main goal of the Forest Plantation Development Fund (FPDF) was to support the development of plantations in both the public and private sectors in Ghana and thus, reduce the financial barrier for investors. Despite the good intentions of the FPDF, management of the fund has been weak and ineffective, with allegations of corruption and misapplication of funds by some beneficiaries. For example, in August 2009, the Minister of Lands and Natural Resources lamented that there was nothing to show for the Gh¢226 million that had accrued to the FPDF since 2000 (Business News, 2009). The Minister indicated that the former Board of the FPDF had contravened the FPDF Act by investing a large part of the resources accruing to the fund in money market instruments for interest, rather than investing in actual plantations, which was the core objective of the fund. The Minister probably forgot that the problem arose from the ambiguity in Section 7(b) of the FPDF Act itself that empowers the Board to invest the funds. Consequently, members of the Board were replaced in August, 2009 (Business News, 2009). There are also institutional problems plaguing the FPDF. Since its inception, no loans have been given for plantations under 3 ha and to farmers who intercropped with cocoa or palm oil (Boni, 2006). The land rights documentation required to obtain a loan is costly and the bureaucratic procedure lengthy, tiresome and inaccessible to small-scale farmers (Boni, 2006).

C. Marketing Challenges

The installed milling capacity of the timber industry in Ghana over the years has been geared towards processing large-diameter trees that are harvested from natural forests. Plantation grown trees generally tend to have smaller diameters, either because they are harvested much sooner than their natural forest counterparts, or because some exotic trees are generally smaller in size than native species. The local milling sector is not equipped to process small-diameter plantation logs and may therefore be unable to utilise some of the plantation grown wood in their mills. This could reduce the demand for plantation grown logs. While there may be alternative markets for this wood, eliminating the milling sector from this market has a potential to depress log prices of plantation wood, and hence have a negative impact on investments in plantation development. This is exacerbated by the complete ban on export of logs in 1995, which has already depressed the prices of logs sold in the domestic market (FAO, 2002a). As a result of this ban, private companies without timber processing plants located in Ghana will be obliged to sell their logs in alternative markets or to the few millers that can utilise the products at depressed log prices. Another issue is related to marketing of lesser-known plantation species. While teak is well-known internationally for its superior wood qualities and properties, the properties and uses of other popular plantation species in Ghana such as gmelina and cedrela are still relatively unknown and have therefore only been marketed on a limited scale within Africa (FAO, 2002a). This is so even though some of these plantation species have properties that are identical to some widely marketed native species.

Getting forest products to markets requires a certain minimum level of infrastructural development, such as roads that can be used to transport inputs to, and outputs from, the plantation and processing sites. Most areas in Ghana still lack basic feeder and trunk roads, technology and skilled labour to support industrial plantation development. Therefore, investors in these areas may be required to develop the infrastructure themselves, which will ultimately decrease the profitability of their investments. The lack of infrastructure that can support integrated wood-processing industries is therefore a barrier to forest plantations, given that integrated wood production, processing and marketing offer more advantages than non-integrated production systems. For example, integrated forest companies (i.e., those that own both forests and processing facilities) can ensure constant supply of raw material at predictable prices and have lower transaction costs compared to those who own only one of these, and hence would be able to reduce their production costs and increase competitiveness.

3.2.2. Technical Constraints

Technical barriers are related to the lack of materials, limited information on matching species to sites, limited fire management capacity, limited knowledge and technology required for successful plantation development, and the transfer of that information to those who need it most.

A. Limited Technical Information and Dissemination

Successful plantation development is underpinned by a good understanding of the silviculture and ecology of the tree species used. Before using a tree species in plantations, it is important to understand the site and climatic requirements of the species, the pests and diseases that affect the species, growth characteristics, management requirements, uses of the species, and market potential. In Ghana, this technical information is available for a limited number of tree species. Hence, this has limited the number of species that have been planted in plantations so far to only a few exotic and native species such as *Tectona grandis, Gmelina arborea, Cedrela ordorata, Eucalyptus spp.*, *Terminalia superba* and *Triplochiton scleroxylon*.

A critical piece of information for those interested in plantation development is to know what species to plant on what type of land. At the national level, there seems to be no scarcity of potential land for forest plantations. However, by determining and providing information to potential clients that helps them to correctly match species to their most productive sites, the profitability of plantations can be greatly enhanced. The Forestry Department provided some guidance on species to be used in the HFZ for plantations and improvement planting of degraded natural forests (line planting and enrichment planting) (FAO, 2002a). This information should be expanded to include the savannah and transition zones, and cover more tree species.

Research is needed to develop plantation-related technical information over time and provide these to the end-users. Although some work has been done by local research organisations such as the Forestry Research Institute of Ghana (FORIG), much still needs to be done. There has been lack of a consistent long-term research into plantation forestry issues in Ghana, which is the main reason for the current state of insufficient information. Research should be a key component of any national plantation development strategy. Such information needs to get to the target investors through effective extension services in the form of technical packages. Historically, extension services are well developed for the agricultural sector; the forestry sector can benefit from a similar model.

B. Limited Fire and Pest Management

Fire is a major part of Ghana's forest ecosystems, and hence is important for the management of plantations. Fire hazards are less in the HFZ as a result of high relative humidity and a short dry season, which together reduce the amount of combustible material (fuel load) in the forest. However, the savannah zones are subjected to annual recurring fires, which put all plantations at risk of burning. Unfortunately, there is limited forest fire management strategy in place across Ghana to eliminate or minimise the impacts of these fires on forest plantations. Even more serious than that, unlike other countries, there is no significant active research into forest fire management in Ghana. It is common for trees planted in the rainy season in the savannah zone to be completely burnt down the next dry season due to the lack of fire management. Research that develops management techniques to protect plantations at the individual, community or industrial level will be indispensible to successful plantation development. The current approaches to managing fires in plantations by the Forest Services Division of the FC include wildfire awareness campaigns, fire ride construction, ground patrol, fire suppression and enforcement of legislation relating to forest fires (Ghana Forestry Commission, 2008b). The FSD has acquired 12 fire tenders for its Wildfire Management Project to help in fire fighting operations in the fire prone zones of Ghana, especially in forest plantations. Despite these efforts, there is still a need for additional resources, training, research and capacity in fire management. In addition to fire, there is also limited information on how to manage the pests and diseases that affect forest plantations, especially, the exotic species. Research will need to be intensified in these areas to ensure the survival and sustainability of forest plantations.

C. Lack of Materials

Another technical barrier to plantation development is the lack of materials required for raising seedlings, transporting, transplanting, and managing the plantations over the rotation. This is particularly the case for individual plantation owners, who lack the financial capacity to purchase these materials on their own. To optimise the benefits from forest plantations, it is critical that seedlings be of superior phenotypes, or of genetic quality that are resistant to common local diseases and pests, be able to survive fire and drought conditions, have fast growth and high-wood quality. These characteristics suggest that a national organisation be responsible for providing such seedlings to end-users. During the implementation of the Rural Afforestation Programme (RAP), the then Forestry Department was responsible for providing all planting materials and technical assistance to individuals, communities and other organisations.

3.2.3. Institutional Constraints

A) Policy and Regulatory Framework

As discussed in Chapter 1, the forest sector is governed by a hierarchy of policies and regulations which have been developed and modified over time. The overarching document that provides the strategic direction for forest sector activities is the Forest and Wildlife Policy adopted in 1994. These policy and regulatory frameworks determine the operating environment, provide incentives, and in some cases, act as barriers against plantation development.

The 1992 Constitution, the 1994 Forest and Wildlife policy, the *Timber Resources Management Regulations*, 1998 (LI 1649) and *Timber Resources Management (Amendment) Regulations*, 2003 (LI 1721), the *Trees and Timber Management (Amendment) Act*, 1994 (Act 493), the *Administration of Stool lands Act*, 1962 (Act 123) and the *Forest Plantation Development Fund (Amendment) Act*, 2002 (Act 623) contain provisions that potentially pose barriers to developing forest plantations. The barriers posed by the policy and regulatory framework relate to ambiguities in some provisions of the policies/regulations, lack of coordination of the policies, and poor implementation of existing policies and regulations.

The *Timber Resources Management (Amendment) Regulations*, 2003 (LI 1721) impose stumpage fees on natural forest timber including some native species which are planted in plantations. According to these regulations, the stumpage rate shall be determined by the Minister responsible for forestry in consultation with the FC and the Administrator of Stool Lands, having regard to the market demand and inventory levels of timber species. This provision in particular is very problematic as it provides no clarity and could become a major disincentive for investment in plantations. Who would want to invest in a system when the investor does not know how much stumpage fees they will be charged? It should be pointed out that the previous LI 1649 of 1998 included a formula for calculating the stumpage fees for the various species, which was eliminated when the Regulations were amended in 2003 under LI 1721. In fact, this power for the Minister to determine stumpage rates has resulted in a court action by the Ghana Timber Association (GTA) against the Forestry Commission.

Ayine (2008) narrates the lawsuit against the FC by the GTA in relation to the determination of stumpage fees. In April 2005, the GTA sued the FC before the High Court, challenging its formula for computing stumpage fees. As noted above, under the LI 1721 (2003), the Minister responsible for forestry has discretionary power to fix stumpage fees, in consultation with the FC and the Administrator of Stool Lands. The GTA challenged the stumpage fees determined by the Minister on several grounds. It claimed that the FC's determination was illegal, as no guidelines regulating the exercise of the Minister's discretion had been formulated or published; and that the Minister did not consult the FC and the Administrator of Stool Lands as required. In addition, as the fees were also charged on trees that were not commercial logs, GTA claimed that its members had to pay higher-than-acceptable stumpage fees, which significantly affected their business operations. Therefore, the GTA sought a court declaration that the FC's system for the calculation of stumpage fees is defective and has resulted in the GTA's members being overcharged. The FC denied that the stumpage fees were illegal and threatened to publish in the media the names of the timber companies that were not paying stumpage fees as required. In retaliation, the GTA also threatened to expose collusion between forestry officials and its members. Eventually, the FC settled the case out of court (Ayine, 2008).

Another ambiguity in the legislation is whether export levies apply to timber harvested from plantations and if so, how much. There are export levies on processed and unprocessed wood from natural forests under the *Trees and Timber Management Act*, 1994 (Act 493). The list of species for which these levies apply includes only one exotic plantation species (teak) and leaves unanswered the question as to whether all timber harvested from plantations will be subject to similar export levies and royalties applicable to natural forests.

The *Timber Resources Management (Amendment) Act*, 2002 (Act 617) specifies that timber harvesting rights are transferred from the State to companies through a timber utilisation contract (TUC).The Act further specifies that no timber rights shall be granted in respect of land with private forest plantation or land with any timber grown or owned by any individual or group of individuals. Although it is clear that the government has no interest in the ownership of the trees planted by individuals or in private plantations, and will not allocate such trees to timber firms for harvesting, it is still unclear whether the grower of the trees is accorded ownership rights regardless of their rights to the land on which such trees are planted. It can be interpreted from Act 617 that a private plantation owner does not require a TUC to harvest his/her trees for private and commercial purposes. However, there are still some ambiguities regarding whether a plantation owner needs a registered property mark to harvest his/her own trees for individual and commercial purposes as required under the *Timber Resources Management Act*, 1997 and *Trees and Timber Management Act*, 1994 (Act 493) for natural forests.

With the competitive bidding used to allocate forest harvesting rights, companies who win bids are expected to pay a fee for the timber rights. Given that plantations do not qualify to be given out under a TUC, it is safe to assume that no such timber rights fees are applicable, although this is not explicitly spelled out in the legislative framework. These legal and regulatory ambiguities in Ghana's forestry sector show the lack of strategic direction and coordination, which tend to confuse private investors. It is not obvious that potential investors in plantations clearly understand the nuances and technicalities involved in navigating through these legal and policy confusion.

Another institutional bottleneck is the ineffective implementation of existing policies and regulations. Under the *Timber Resources Management Act*, 1997 (Act 547) and the *Timber Resource Management Regulations*, 1998 (LI 1649), TUC holders are required to develop and implement reforestation, afforestation and plantations in any area that the Chief Conservator of Forests[1] may approve. TUC holders make undertakings (including a performance bond) to execute the reforestation plan during the period of the contract to the satisfaction of the Chief Conservator of Forests. The FC and its predecessor Forestry Department have lacked the capacity to enforce these regeneration and plantation establishment regulations. This inability to enforce legislated plantation initiatives contributes to ineffective plantation development in Ghana.

[1] Although Ghana's legislations still refer to the Chief Conservator of Forests, this position no longer exists. The equivalent position is the Chief Executive Officer of the Forestry Commission.

A potential cause for concern with the FPDF is Section 7(b) of the *Forest Plantation Development Fund Act* (2000), which specifies one of the functions of the FPDF Board as "*attracting contributions into the Fund and investing the moneys of the Fund.*" This legal provision of "investing" the money has resulted in a situation where the money for plantation development was being invested in money markets for interest, rather than in plantations (Business News, 2009). Therefore, the FDPF has not achieved much since 2001, and the financial difficulties faced by investors in plantation forestry persist. This Act needs to be amended to eliminate this problematic clause.

B) Customary Practices and Tree Tenure

Apart from the legal and policy barriers, there are cultural practices that also pose challenges to forest plantation development in Ghana. Traditionally, most subsistent farmers have planted, tended and conserved only fruit trees with economic value around their homes and farmlands. In the northern regions, these include shea, dawadawa and mango (*Mangifera indica*), which can be observed on most farmlands and around homes. In southern Ghana, the most common fruit trees include citrus, banana, plantain and cash crops such as cocoa. However, there is no traditional culture of planting and tending trees for purposes of fuelwood or timber production in any part of Ghana. The concept of forest plantations to provide wood products is therefore alien to the culture and traditional practices of most farmers in Ghana.

Insecure tree tenure, resulting from the different configurations of ownership of planted trees, and the potential benefits that accrue to farmers/tenants can influence incentives for plantation development (Owubah et al., 2001). The forms of tenure that have longer terms, are more clearly defined, provide more of the economic benefits to their holders, are likely to simulate tree planting (Zhang and Pearse, 1996). In most cultures, a tenant cannot exercise ownership rights over trees on the land for which he was only granted farming rights. Under the traditional land-use system, the planting of trees by tenant farmers is generally considered as an attempt to perpetuate their stay, which may in turn indirectly imply ownership of the land (FAO, 2002a). Except in the Upper West Region, a tenant farmer does not own the natural or planted trees on the land issued to him (FAO, 2002a). These cultural barriers work together to impede the development of individual and community plantations.

C) Land Tenure

The importance of land tenure to forestry is related to the fact that tenures define the types of property rights held by users of the land. Property rights define the extent to which the holder can enjoy the benefits accruing from the asset; hence property does not have to be a tangle good, but rather a defined set of rights over something (Pearse, 1992). Therefore, efficient land tenure systems that offer appropriate property rights to holders contribute to effective resource management and socio-economic development.

The configuration of land ownerships across Ghana is complex. This is a result of a mixture of cultural practices, traditions and legal and constitutional frameworks, which produces a complex series of rights and interests in land. There are generally two main categories of land ownership in Ghana: public and customary lands (Stool/Skin lands, clan lands, family lands, etc.). Under the 1992 Constitution, all public lands in Ghana are vested in the President on behalf of, and in trust for, the people of Ghana. The constitution also recognises stool/skin lands, which are vested in the appropriate stool/skin on behalf of, and in trust for, the subjects of the stool/skin in accordance with customary law and usage. In Ghana, customary lands constitutes about 78% of land ownership, State lands constitute 20%, while vested lands (split ownership) is 2%. These complex land ownership arrangements lead to inadequate security of tenure, conflicts, difficult accessibility to land, confusion over ownership of some lands and tension between the state and customary authorities. These tenures confer different kinds of ownership and benefits and hence may imply incentives or barriers to plantation development.

Generally, three kinds of tenure arrangements are available: leasehold, freehold interest, and communal or family land-use rights. Under indigenous land-use rights, the user has no discretionary land transfer rights because the land belongs to a corporate body (Benneh, 1989). Individuals can sometimes enhance their rights in such holdings by making some long-term investments (Aidoo, 1996). Freehold interests in lands confer an absolute and secured land rights to the user (Benneh, 1989; Aido, 1996). Leaseholds take two forms: sharecropping and annual land rental payment. Sometimes land rental payment is small or merely symbiotic (Zhang and Owiredu, 2007). Leaseholds are restrictive and do not offer any security of land tenure to the farmers (Benneh, 1989). The period of leasehold ranges from a year for the cultivation of annuals to 50 years for the establishment of tree plantations (Zhang and Owiredu, 2007).

Several studies in Ghana have examined land tenure systems, methods of land acquisition, and management models for forest plantations (Odoom, 1999); farmers' willingness to establish plantations (Owubah et al., 2001) and the impact of land tenure and market incentives on farmers' actual forest plantation activities (Zhang and Owiredu, 2007). These studies conclude that the type of ownership is an important determinant of the willingness of people to invest in plantations. In particular, farmers who own lands outright are more likely to invest in plantations (Zhang and Owiredu, 2007) compared to those who do not have such security to title.

The key constraint imposed by the land tenure arrangements in Ghana related to forest plantation development are summarised by FAO (2002a) as: a) multiplicity of interests and rights in land which may vary in different parts of the country leading to conflicting claims to ownership; b) lack of reliable maps indicating stool/skin land boundaries which can give rise to disputes; c) cumbersome land disposal and documentation procedures; and d) potential conflict concerning tenurial and management arrangements for plantations within forest reserves where parties other than the FC are involved.

In addition, communal and family ownership of lands leads to fragmentation of the land base into small parcels, which poses a challenge for large-scale plantation projects that require lands that will cross ownership boundaries. Conflicts over land ownership and land fragmentation together act as serious barriers to plantation development in Ghana. An investor will have to negotiate land titles with numerous owners to acquire land to meet his/her needs. These negotiations can be long, frustrating and expensive for investors either at the negotiation stage or after the project has begun.

3.2.4. Site and Biophysical Constraints

In general, most vegetation, climatic and soil zones in Ghana can support tree growth; however, not all soils can support economically profitable forest plantations. The risk to economic sustainability of plantation forestry depends on the ecological capability of the site to support the planted trees. Site characteristics such as low nutrient reserves, poor nutrient retention ability and susceptibility to drought are therefore major limiting factors to using tropical soils for short rotation tree crops (Tiarks et al., 1998). Because of the importance of the litter layer to nutrient supply and soil structure, disturbance of the surface should be kept to a minimum to maintain productivity (Tiarks et al., 1998). Maintenance of the litter layer or other vegetative cover is necessary in limiting erosion and keeping a suitable soil moisture balance (Spaargaren and Deckers 1998). Despite the high rainfall received in parts of the tropics, soil water deficit can be a recurring constraint on productivity on many sites in subtropics and tropics (Gonçalves et al., 1997; Landsberg, 1997). In the semiarid and arid tropics, water stress may limit rates of growth below commercially viable levels; soil management for conserving available water is a critical consideration (Tiarks et al., 1998).

Forest and savannah fires contribute to dry conditions and may limit tree growth as well. The soils of the savannah zone tend to be eroded, have only a small capacity to keep water and are deficient in nutrients, notably phosphorous and nitrogen (Boateng, 1966). The soils in the HFZ, due to the high rainfall are highly leached and acidic (pH 4.0 -5.5), with low cation exchange capacity, available phosphorus, nitrogen and organic matter (FAO, 2002a). These differences in soil characteristics mostly account for the low productivity of the savannah zones compared to the high forest zones in Ghana.

3.3. Opportunities for Plantation Development

3.3.1. Legal and Policy Framework

Even with the imperfections of the regulatory and policy framework governing Ghana's forestry sector identified above, these frameworks do provide some opportunities for developing forest plantations. The major opportunities are contained in the 1994 Forest and Wildlife Policy, the *Timber Resource Management Regulations*, 1998 (LI 1649), the *Forest Plantation Development Fund Act* (2000), the *Forest Plantation Development Fund (Amendment) Act* (2002), and the National Forest Plantation Development Programme (NFPDP).

Section 5.2 of the Forest and Wildlife Policy (1994) emphasises reforestation initiatives towards restoring a significant proportion of the country's original forest cover, while Section 5.4.6 calls for incentives to promote investments in forest plantations to ensure sustainable supplies of marketable products. The support to plantations provided by the Policy is significant, given that this is the document that provides strategic direction for Ghana's forestry sector. It is a sign of commitment from the highest levels of government to promote plantation development and provides the enabling policy for subsidiary policies and regulatory initiatives to be developed to promote plantation forestry.

The *Timber Resource Management Act*, (1997) provides for a TUC holder to establish and manage forest plantations in contract areas and reforest or afforest in any area that the Chief Conservator of Forests may approve. These provisions provide opportunities for the Chief Executive of the FC (who is equivalent to the Chief Conservator of Forests under the previous Forestry Department) to direct plantation development through the timber utilisation contract process.

Since 2000, governments of Ghana have taken steps towards promoting forest plantation development. This began with the passing of the *Forest Plantation Development Fund Act* (Act 583) in 2000, which was subsequently amended in 2002 to cover plantation growers, both in the *public* and private sectors. If managed effectively, the funding provided by the Fund would help to reduce the financial barriers faced by plantation developers. It also offers opportunities for the Fund to meaningfully support research into plantation development, given the limited research funding from alternative sources.

In 2001, the government launched the National Forest Plantation Development Programme (NFPDP). Under the programme both indigenous and exotic economic tree species have been planted. These efforts should be continued and intensified especially at the community level. In the savannah regions, the greatest need is for fuelwood and other wood products for building construction. Therefore, the financial assistance under the FPDF should be provided to individuals and communities who show the willingness and capacity to develop community plantations.

3.3.2. Land Availability and a Favourable Climate

Data from the Ministry of Lands and Forestry on land use patterns in Ghana show that there are about 10.7 million ha of savannah woodlands and unimproved pasture, comprising 45% of the total land area of Ghana (FAO, 2002a). These lands are potentially available for forest plantation development. This area is in addition to degraded reserved and unreserved forest areas that can be replanted into plantations or improvement planting. The actual area that is available for forest plantations will need to be determined, taking into consideration other land use needs and whether each parcel of land is suitable for plantations.

Ghana is also blessed with a favourable climate without extremes of temperatures, rainstorms and other weather-related hazards. Rainfall is adequate to support tree growth in almost all areas of Ghana. Not only that, as shown in Table 1, Section 2.3, some plantation species can reach MAIs of up to $20m^3$/ha/yr on good sites in the HFZ. Even on poor sites in the savannah zone, plantation growth is as good as that of natural forests on good sites in the HFZ (section 2.3). As with the case of land, there is a need to develop maps of Ghana that match tree species to soils and climate for each ecozone and forest type.

3.3.3. Labour Force Availability

Reliable official statistics on unemployment are difficult to find, and so estimates of the unemployment rate varies significantly among agencies reporting these statistics. Currently, the unemployment rate is estimated to be between 11 and 20%. Unemployment in Ghana is a function of over-reliance on the government as the main employer in the country and limited job opportunities from the private sector. The agricultural sector employs about 50% of Ghana's population. There are many junior and high school drop-outs who are walking the streets in cities, selling everything from dog chains to candies to survive. Many others are under-employed, especially in the savannah areas of Ghana, who have little to do during the long, dry season. Plantation development initiatives will have access to this large pool of labour at reasonable costs to perform skilled and unskilled tasks from tree nurseries to harvesting and processing wood products. This will serve as an opportunity for investors to reduce their investment costs, and will also be an economic development tool for rural areas with few alternative employment opportunities. The similarity between the operations in the agricultural and forestry sectors is an advantage because many job seekers are already familiar with the basic tasks associated with agricultural practices.

3.3.4. Available Market Demand

Current and future shortages of wood products from the natural forests for domestic use and export provide an opportunity for forest plantations to fill the gap between demand and supply. As shown in Chapter 1, the demand for wood products will continue to rise and hence there is a ready demand for plantation grown wood. For example, with increasing rural electrification, the demand for teak for use as electric transmission poles will continue to increase. Secondly, increased demand for teak products by India offers an exciting market opportunity that Ghana is already exploiting. The majority of Ghanaians depend on wood as their main energy source (Benhin and Barbier, 2001), and hence the demand for roundwood in Ghana is driven mainly by the demand for fuelwood. Although it is fair to note that not all plantation species will have as much demand as teak, species that yield large quantities of biomass within short rotations will have ready markets as fuelwood.

3.3.5. International Climate Change Initiatives

The Kyoto Protocol (KP), which was negotiated in December 1997, requires developed countries as a whole to reduce their greenhouse gas (GHG) emissions by 5.2% compared to 1990 levels, during the first commitment period from 2008 to 2012. The KP recognises that forests, forest soils and forest products all play important roles in mitigating climate change. The Clean Development Mechanism (CDM) is one of the flexible mechanisms that was included in the Kyoto Protocol to help developed countries meet their reduction targets in a cost-effective way. Specifically, the KP recognises afforestation and reforestation as the only eligible land uses under the CDM. Since CDM projects can only be undertaken in developing countries, this offers interesting opportunities for the establishment of plantation forests for sequestering carbon in Ghana.

While developed country investors will benefit from carbon credits, Ghana will benefit from forest products and environmental services (e.g., reduced levels of GHG in the atmosphere) provided by such plantations. This is an opportunity for Ghana to incorporate CDM projects into its plantation development agenda, and draw foreign investors and institutions into the country.

3.3.6. Environmental Consciousness and Wood Shortages

Ghana's population is becoming increasingly educated in all spheres of life including environmental issues. Both governmental (e.g., the Environmental Protection Agency (EPA)) and local and international non-governmental organisations have been carrying out environmental campaigns across the country on the need to protect the environment. These environmental agencies continue to use their existing networks to further educate people on the benefits of increasing tree cover through plantation development. The need for additional forest resources to meet the needs of society is generally recognised, especially in the savannah areas where tree cover is disappearing at a very fast rate. Rural dwellers now travel longer distances to gather fuelwood and other forest products; while city dwellers now pay more for charcoal because it has to be transported over longer distances to markets than previously. Soil fertility is declining, and rainfall patterns have become more erratic and unreliable as well. In short, many people are aware of the need to increase tree cover in many parts of the country. This general awareness of the populace about the effects of deforestation and environmental degradation and the increasing scarcity of forest resources within their lifetimes will reduce their resistance to plantation establishment.

3.4. Incentives for Plantation Development

3.4.1. Definitions of Incentives

Several definitions of incentives abound in the literature. For example, Enters (2001) refers to incentives as the "incitement and inducement of action." Gregersen (1984) defines incentives as "public subsidies given to private sector to encourage socially desirable actions by private entities." For the purposes of this chapter, incentives refer to anything that motivates people to do something they would otherwise not have done. In line with this definition, incentives for forest plantations include economic and policy instruments that motivate investments in forest plantations.

3.4.2. Reasons for Incentives in Plantation Forestry

Most of the world's forest plantations have been established with incentives in the form of a subsidy of one sort or another at some time, either directly or indirectly (Whiteman, 2003; Cossalter and Pye-Smith, 2003). For example, in many parts of Latin America, Oceania and Asia, plantation programmes paid more than 75% of the establishment cost with additional allowances made for land, maintenance and many other costs (Brown, 2000). Given the pervasive nature of incentives in plantation forestry, the natural question is, why incentives?

The long-term nature of growing trees, whereby establishment expenditures are incurred early in the life of the plantation, and benefits accrue many years down the road, is the most important reason that makes promoting plantation forestry difficult. The long gestation period for trees makes the investment riskier than many other alternatives. The question is: why would anyone want to invest in such a risky enterprise? Incentives are appropriate when the private net returns are lower, but the returns including externalities are greater than the returns from alternative land uses. In this situation, incentives could be effective by creating a more socially desirable land-use pattern (Haltia and Keipi, 1997; Williams, 2001). If not, incentives will represent a misallocation of public-sector resources, merely helping investors to earn higher returns (Enters et al. 2003). Incentives may not be needed if the private returns from forestry exceed those from other land uses or if the addition of incentives will still not provide an attractive private return. In selecting an appropriate level of assistance, rates of return not only have to be compared with those from alternate land uses but also with investments in other sectors (Williams, 2001). Incentives may also be used to help investors overcome barriers such as the high capital cost of establishment and the relatively long waiting period for a return. If incentives are implemented effectively, they could provide the initial impetus to develop national forest industries for either foreign exchange or ecological and environmental benefits. Furthermore, incentive programmes can be used to encourage economic development and generate employment in specific, less developed regions and diversify the economy away from areas with limited economic potential, such as agriculture (Williams, 2001).

Most governments realise that they cannot undertake plantation development programmes successfully without the participation of the private sector, hence incentives are meant to reduce barriers to investments and remove structural impediments and operational constraints to maintain private sector interest and investment in plantations. Incentive programmes in Ghana should be targeted at reducing or removing entirely some of the barriers to plantation development identified earlier in this Chapter.

3.4.3. Types of Incentives

In a comprehensive study of incentives for forest plantation development in the Asia-Pacific region, Enters et al. (2004) categorise the kinds of incentives into direct and indirect incentives. Direct incentives are provided directly by governments, development agencies, non-governmental organisations and the private sector and include the following: goods and materials (e.g., seedlings, fertilisers etc.); specific provision of local infrastructure; grants; tax relief or concessions; differential fees and access to resources; subsidised loans; cost-sharing arrangements and price guarantees. For example, under the RAF in the late 1980s in Ghana, the government and World Bank-funded project provided direct incentives to individuals and communities interested in plantation establishment. These were mainly in the form of free seedlings, transportation and technical assistance in the planting and managing the plantations.

Enters et al. (2004) further divide indirect incentives into *variable incentives* and *enabling incentives.* Variable incentives are economic factors that affect the net returns that producers earn from plantation activities such as input and output prices, exchange rates, general taxes, specific taxes, interest rates, trade restrictions (e.g., tariffs) and fiscal and monetary measures (Enters et al. 2004). Enabling incentives on the other hand mediate an investor's potential response to variable incentives and help to determine land use and management (FAO, 1999). These include: land tenure and resource security, accessibility and availability of basic infrastructure (ports, roads, electricity etc.), producer support services, market development, credit facilities, political and macro-economic stability, national security, research and development, and extension services. Enabling incentives can also be viewed as elements in the investment environment that affect decision making (Enters et al., 2004).

3.3.4. Incentives for Plantation Forestry in Ghana

Incentives for plantation development in Ghana can be broadly classified into two categories: institutional and policy incentives, and economic incentives. Institutional and policy incentives provide the long-term enabling environment for investments in plantations. The most notable reform in this direction is provided within the context of land management under the National Land Policy reform. Economic incentives are more wide ranging and are intended to influence the benefit/cost structure of the plantation enterprise to make it more worthwhile for investors and/or support rural farmers with basic inputs into plantation establishment. The main policy and legislative frameworks that provide economic incentives are contained in the *Timber Resources Management Act 617 (Amendment) Act,* (2002), the *Forest Plantation Development Fund Act,* 2000 and the *Forest Plantation Development Fund (Amendment) Act,* (2002).

A) Institutional Incentives

One of the main constraints to plantation forestry development in Ghana identified above is the land tenure system. The National Land Policy (NLP) was the first comprehensive approach at dealing with the land tenure constraints in Ghana. It was formulated from 1994-1999 through participatory processes that included consultations with traditional authorities, farmers' organisations, academia, public sector institutions, researchers, government authorities, etc. The overall long-term goal of the NLP is to stimulate economic development, reduce poverty and promote social stability by improving security of land tenure, simplifying the process for accessing land and making it fair, transparent and efficient, developing the land market and fostering prudent land management practices.

The specific objectives were to facilitate equitable access to land, protect land owners and their descendants from becoming landless, ensure prompt payment of compensation for compulsorily acquired lands, minimise and where possible eliminate land boundary disputes, create and maintain effective institutional capacity at national, regional, district and community levels, and promote community participation and public awareness in sustainable land management. The NLP is expected to provide the appropriate institutional structures and clear, coherent, and consistent polices to support land development activities. The NLP is implemented under the Land Administration Programme in five-year phases over 15-25 years (starting from 2004). When these institutional changes are completed, they will minimise the barriers posed by the hitherto antiquated and confusing land administration system in Ghana and provide incentives for investments in forestry and other land-related projects.

B) Economic Incentives

The *Timber Resources Management (Amendment) Act, 2002* (Act 617) provides the framework for a wide range of incentives for forestry and wildlife sectors, including forest plantation developers. Section 14A prescribes that an investor in any forestry or wildlife enterprise is entitled to such benefits and incentives as are applicable to its enterprise under the *Internal Revenue Act*, 2000 (Act 592) and under Chapters 82, 84, 85 and 98 of the Customs Harmonised Commodity and Tariff Code scheduled to the *Customs, Excise and Preventive Service Law*, 1993 (PNDCL 330). The chapters of the Customs Harmonised Commodity and Tariff Code referred to above specify items that have been zero-rated under the law. In addition, an investor whose plant, machinery, equipment or parts of machinery are not zero-rated under the Customs Harmonised Commodity and Tariff Code scheduled to the *Customs Excise and Preventive Service Law*, 1993 (PNDCL 330), may submit an application for exemption of import duties, VAT or excise duties on the plant, machinery, equipment or parts thereof to the Forestry Commission which shall submit it to the appropriate tax authority. For the purpose of promoting strategic or major investments in the forestry and wildlife sector, the Minister responsible for forestry may determine or negotiate specific incentives in addition to the incentives provided above for such period as may be specified in the relevant timber utilisation contract.

Act 617 also provides for investment guarantees, transfer of capital, profit and dividends. It gives investors guaranteed unconditional transferability through any authorised dealer bank in freely convertible currency of dividends or net profits attributable to the investment; the remittance of proceeds, net of all taxes and other obligations, in the event of sales or liquidation of the operations of the investor or any interest attributable to the investment. There are also guarantees against expropriation. The operations of an investor shall not be nationalised or expropriated by Government; and no person who owns, whether wholly or in part, the capital of any forestry or wildlife investment shall be compelled by law to cede the interest in the capital to any other person; and there shall not be any acquisition of the operations of an investor by the State unless the acquisition is in the national interest or for a public purpose and under a law which makes provision for payment of fair and adequate compensation.

The *Forest Plantation Development Fund (Amendment) Act,* 2002 provides for financial assistance for the development of forest plantations and for research and technical advice to persons involved in plantation forestry on specified conditions. The Act encourages investment in forest plantation development through incentives and other benefits to be determined by the Fund Board. Incentives under the Act are defined as loans, rebates, grants and insurance to qualified beneficiaries. A beneficiary under this Act is entitled to tax rebates and such other benefits that are applicable to it under the *Ghana Investment Promotion Centre Act*, 1994 (Act 493). The Forest Plantation Development Fund is the main funding source for the National Forest Plantation Development Programme (NFPDP), which was launched by the government in September 2001. The programme was aimed at encouraging the development of a sustainable forest resource base that will satisfy future demand for industrial timber and enhance environmental quality. The programme was also expected to generate jobs and significantly increase food production in the country thereby contributing to wealth creation and reduction in rural poverty.

The NFPDP provides direct incentives to farmers participating in the programme in the form of paying for hired labour for seed collection, peg cutting, ploughing and lopping. The government also provides free seeds and seedlings, technical supervision, technical information on matching species with site, market for products, and the benefit from the food crops planted under the system.

3.4.5. The Problems with Incentives

Despite the great success of forest plantation incentives in countries such as Brazil, Chile, New Zealand and Australia, direct subsidies to industrial plantations have been widely criticised as inefficient and inequitable because of market distortions that can lead to economically incorrect allocation of productive factors (Evans and Turnbull, 2004; Enters et al., 2004). Inefficiency results from two reasons: a) the difficulty of proving a linkage between the incentive and environmental benefits resulting from the investment; and b) when incentives are provided to plantation growers who would have planted trees without them or when a higher rate of incentive is paid than would have been necessary to induce a grower to plant trees (Enters et al., 2004). Indirect incentives provided by policy and land tenure reforms are considered essential to industrial and socio-economic development. Most countries are re-directing their incentives at individual small-scale plantations and community forestry programmes ((Evans and Turnbull, 2004).

Direct incentives offered in the early stages of a forestry project may simply buy participation and may not lead to long-term interest in the project (Enters at al., 2004). Subsidies have often succeeded in stimulating the adoption of conservation measures that were abandoned or even actively destroyed once payments ceased (Lutz et al., 1994). The same has been observed for plantations (Sawyer, 1993). In fact, during the RAP in Ghana, the author observed that some farmers did not bother to protect their seedlings from fire in the dry season following planting, especially those who received food aid from the Adventist Development and Relief Agency (ADRA) as incentives. These same farmers requested for seedlings the following year to replant the same sites that were planted in the previous year! If an incentive is the primary cause for behavioural change, the discontinuation of that incentive is likely to become a cause for reversal (Enters et al., 2004).

3.5. Conclusions

Plantation development in Ghana continues to face economic, technical, institutional and biophysical constraints. The importance of these barriers differs across the various categories of plantation developers. The needs of individual and community small-scale farmers are different from those of large-scale industrial plantation investors. Clearly, there are opportunities as well, offered by local, national, and international demands for goods and services produced by forest plantations. The implementation of the Land Administration Programme is critical to streamlining land administration and minimising land-related barriers to plantation development in Ghana. However, such institutional changes should not be limited to legislation, judicial decisions, records management, titling, community-based land use planning, monitoring and evaluation, but should be expanded to include much needed reforms of the land tenure system itself as well.

With the exception of the incentives provided directly to farmers under the NFPDP, the most extensive economic incentive programmes provided by legislation are intended to create an enabling environment for commercial timber producers and foreign investors. There is no published information on the success of the incentives provided to industrial plantation developers in Ghana, and so it is difficult to assess their effectiveness. As shown above, response to the incentives provided to farmers under the MTS is encouraging.

Given the land tenure constraints discussed above, it seems that majority of the forest plantations in Ghana will continue to be small-scale individual and community plantations, and hence incentives targeted at this group should be encouraged and improved, in addition to those for industrial plantations. In the meantime, the debate as to whether incentives for forest plantations actually work and their impact on plantation development and national economies will continue to rage on.

Part II. Plantation Silviculture and Management

Chapter 4

CRITICAL SILVICS OF SELECTED PLANTATION SPECIES

4.1. INTRODUCTION

The term *critical silvics* is defined as the vital information on the biological behaviour of a tree species that is needed to grow it as a managed crop (Day, 1996). According to Day (1996), critical silvics information usually includes the genetic behaviour of the species when grown as a crop, the most suitable and dependable reproduction methods, the types of stand management required and any difficulties in growing and managing the species. The information provided by the critical silvics of potential tree species for plantation forestry is indispensible for successful plantation establishment and management. While this information is available for some plantation species in Ghana, it is absent for others. Secondly, there is no single repository of such information for landowners, plantation developers and forestry practitioners. Through a review of the literature, this chapter provides the critical silvics of selected tree species that are currently used in plantation forestry in Ghana. These species were selected based on their current or potential importance in national plantation development efforts. Based on the critical silvics, the suitability of each of the species for the major vegetation zones in Ghana are suggested.

4.2. CRITICAL SILVICS OF TEAK (*TECTONA GRANDIS*)

4.2.1. Teak in Ghana

Presently, teak is the most important plantation species in Ghana in terms of the areas planted and the value of its wood products. In 2010, almost 50,000m^3 of teak wood was exported in the form of air-and kiln-dried lumber, poles and billet (Ghana Forestry Commission, 2011). Teak was introduced into Ghana between 1900 and 1910 (FAO and UNEP, 1981). In fact, trials of teak in Ghana date back to 1905 under the German administration in the Volta Region (Kadambi, 1972). Teak has since acclimatised well and has been widely grown in both industrial plantations and small community woodlots. But large-scale plantations of teak in Ghana only started in the late 1960's, under a plantation programme that was initiated with the help of the Food and Agricultural Organisation (FAO) of the United Nations to supplement the supply of wood products from the indigenous natural forests (Prah, 1994). By 1987, these plantations were estimated to cover over 45,000 ha (Drechsel and Zech, 1994; FAO, 2002a) with over 30,000 ha planted to teak, of which an estimated 10,000 ha survived (FAO, 2002a).

A further increase in teak plantations occurred following a five-year Rural Afforestation Programme in 1989 under the erstwhile Ghana Forestry Department, which saw a boost in teak planting through the establishment of new plantations and small-scale community woodlots in Northern Ghana. Apart from electric and telephone transmission poles, teak is also valued by small-scale farmers and local communities particularly in Northern Ghana as poles for construction, fencing, rafters, fuelwood, stakes and wind breaks. It has also become an important source of income for private organisations and individuals especially farmers who plant the species on their farms.

Several private companies and firms and even individuals have established teak plantations at various scales and for various purposes using various partnerships, programmes and strategies in Ghana in the past decades (FAO, 2002a). Notable among these are the Pioneer Tobacco Company Ltd. (PTC) who own about 5,000 ha of teak plantations mainly in the Brong Ahafo Region, the Ashanti Goldfield Company Ltd., with over 1,400 ha of teak and the Bonsu Vonberg Farms Ltd., who own about 500 ha of teak (FAO, 2002a). Of the more than 50,000 ha of newly planted forests that were established between 2000 and 2004, 60% was teak (ITTO, 2005a).

4.2.2. General Description

Teak, also known commercially as *teek* or *teca* (Spanish) belongs to the family *Verbenaceae*. Teak varies in size according to locality and conditions of growth. On favourable sites, it may reach a height of about 40 to 45 m, with a clear bole of up to 25 or 27 m, and a diameter of between 1.8 and 2.4 m (Farmer, 1972). Generally a drought and heat resistant tree species, teak can survive and grow in a wide range of climatic and edaphic conditions (Hedegart, 1976). According to Kadambi (1972), records from Thailand reported a teak tree, claimed to be the world's largest tree (in 1965), with approximately 6.6 m diameter at breast height (dbh) and 45 m total height. In drier regions, trees are generally smaller. The boles are generally straight, cylindrical and clear when young, but tend to be fluted and buttressed at the base when mature (Kadambi, 1972). They tend to fork when grown in isolation, but are generally shade intolerant.

4.2.3. Natural Distribution

Teak grows naturally in Southern Asia, from the Indian Subcontinent through Myanmar (Burma) and Thailand to Laos, approximately 9° and 25° latitude and 73° to 103° E longitude (Troup, 1921). As an exotic species, teak grows in several parts of the world. First naturalised outside its natural range in Java some 400 to 600 years ago (Kadambi, 1972), the earliest plantation of teak has been traced back to 1680 by Perera (1962), when it was successfully introduced to Sri Lanka. Since then, teak has been introduced and acclimatised well in other parts of Asia. Horne (1966) stated that Nigeria was the first place outside of Asia where teak was introduced in 1902 and it quickly spread to other parts of tropical Africa, reaching Ghana in 1905.

4.2.4. Site Requirements

Teak grows on a variety of geological formations and soils (Kadambi, 1972; Seth and Yadav, 1959), but the quality of growth depends on the depth, structure, porosity, drainage and moisture holding capacity of the soil (Kadambi, 1972). Teak grows best on deep, well drained and fertile soils with a neutral or acid pH (Kadambi, 1972; Watterson, 1971), generally on elevations between 200 and 700 m, but exceptionally on elevations of up to 1300 m above sea level (Troup, 1921). Warm tropical, moderately moist climate is best for teak growth. Optimum annual rainfall for teak is 1200 to 1600 mm, but it endures rainfall as low as 500 mm and as high as 5000 mm (FAO, 1983; Hedegart, 1976; Kadambi, 1972; Troup, 1921).

4.2.5. Propagation

Teak is relatively easy to establish in plantations and because of the enduring global demand for products from teak it has good prospects as a plantation species (Krishnapillay, 2000). Plantation grown teak is established using stump plants rather than direct sowing of teak seeds which does not always give satisfactory results (Borota, 1991). Depending on desired product (fuelwood, poles, lumber or a mixture of products) and the site quality, the initial planting spacing generally range from 1.8 by 1.8 m to about 3 by 3 m (Kadambi, 1972). When planted in a taungya system, spacing could be as wide as 4.5 m between rows. Generally, on good soils, wider spacing is used. This results in better diameter and height growth, and also reduces nursery, planting and early thinning costs (Kadambi, 1972). On sloping terrain, wider spacing has been suggested to encourage ground cover and to avoid erosion (Weaver, 1993).

4.2.6. Management

Teak is generally shade intolerant but needs training for improved form. Accordingly, plantations must be thinned regularly and heavily, particularly in the first half of the rotation. Initial planting density is generally between 1 200 and 1 600 plants per hectare (Krishnapillay, 2000). Closer planting spacing is sometimes adapted to ensure quick canopy closure, thereby achieving training and reducing weeding cost (Adegbeihn, 1982; Kadambi, 1972). However, this practice necessitates early thinning. The time of the first thinning is largely determined by site quality. Lowe (1976) noted that although thinning may be delayed for 10 to 15 years after planting without unduly affecting the growth potential of the final crop, very heavy thinning becomes necessary if the growth of the final tree crop is to be maintained at satisfactory levels. Teak also coppices quite vigorously, making post-harvest re-establishment much easier than from seed.

4.2.7. Growth and Yield

Teak is generally fast growing when young, but its overall growth rates on rotation basis are not outstanding (FAO, 1956). It is considered moderate to fast growing (Briscoe and Ybarra-Conorodo, 1971). A study of the standing biomass of teak in India, showed height growth to be most rapid between 10 and 50 years after which it declined (Weaver, 1993).

The rotation of teak in India is a function of forest type and management systems (Ghosh and Singh, 1981). Plantation crops have rotations between 50 and 80 years, whereas in areas where teak occurs in mixed stands, rotation is about 70 to 80 years. Coppice systems or coppice with standards have rotations of between 40 and 60 years (Weaver, 1993). FAO (1985) quotes the peak ages for the mean annual volume increment at 50 and 75 years, respectively, for site classes I and II in Kerala, India, based on stemwood volume. In Indian yield tables for teak (Laurie and Ram, 1940), the maximum total volume growth occurs at ages between 5 and 15 years depending on site class. Similar estimates in Trinidad (Miller, 1969) are between 7 and 12 years. At Mtibwa, Tanzania, Malende and Temu (1990) estimated the peak ages of mean and current annual increments for teak to be at 42 and 55 years respectively. At base age 20, the site index for teak was estimated by Malende and Temu (1990) to be between 16 and 25 m. In Miller (1969), the estimate is between 15 and 23 m. Akindele's (1991) estimate for North-western Nigeria was between 10 and 29 m. At the same base age, figures from Laurie and Ram (1940) ranged from 28 m for site class I to 12 m for site class V. Similar results have been reported by Keogh (1982), Friday (1987) and Drechsel and Zeck (1994). In Ghana, a similar study for teak in the high forest zone reported indices ranging from 17 to 26 m (Anon., 1992).

Logu et al. (1988) estimated the aboveground biomass production for teak to be between 2.1 and 273 t/ha for ages 5 and 97 years, respectively. The mean annual biomass increment was estimated to peak at between 10 and 40 years depending on site conditions. Detailed information on the growth and yield of teak in Ghana are provided in later chapters of this book.

4.2.8. Uses

Teak has gained importance worldwide as a top quality timber species with attractive physical properties such as high natural durability, ease of seasoning without splits and cracks, and attractive grain and colour (Pandey and Brown, 2000). These physical properties make teak suitable for various purposes including; electric transmission poles, furniture, shipbuilding and decorative panelling. The growth characteristics of the species make it particularly attractive as a plantation species.

4.2.9. Pests and Diseases

Many kinds of pests and diseases have been identified for teak in other regions of the world. For example, the teak defoliator and skeletoniser (*Hyblaea puera* and *Eutectona machaeralis*) cause extensive damage to young plantations, while root rot due to *Polyporous zonalis* is also common in plantation grown teak (Anon., 2009). Pink disease fungus causes cankers and bark flaking, and a powdery mildew caused by *Olivea tectonae* and *Uncinula tectonae* leads to premature defoliation of teak trees (Anon., 2009). Fresh leaf extracts of *Calotropis procera, Datura metal* and *Azadirachta indica* have been found to be most effective against teak skeletoniser. This method is of immense importance in insect pest control considering its harmless and pollution free implications on the environment further avoiding the operational and residual hazards that are involved in the use of organic and inorganic insecticides (Anon., 2009).

4.3. Critical Silvics of Neem (Azadirachta indica)

4.3.1. Neem in Ghana

According to Streets (1962), neem was introduced into Ghana *circa.* 1915. It was first planted in small plots scattered throughout the country along roadsides and in amenity belts in towns and villages. Neem has acclimatised well throughout Ghana, and is popular as a source of fuelwood and as poles and rafters for building construction. Neem extracts are also valued for their medicinal properties in treating malaria and as an insecticide for storing grain. As neem is not browsed by animals and grows rapidly, it is easy to grow and maintain even within villages. Local processing of the many products from neem are essential to meeting basic needs at the rural household and community level and can give rise to the establishment of thriving village industries and thus generate employment and additional income (Radwanski and Wickens, 1981). Based on their research of various aspects of neem in Nigeria, Radwanski and Wickens (1981) concluded that there was enough scientific evidence on the subject to warrant the launching of a well coordinated multidisciplinary research and development programme leading to agricultural, industrial and commercial exploitation of neem. A survey of community woodlots in some districts of Northern Ghana in 1997 showed that over 60% of all tree species planted under the Rural Afforestation Programme (RAP) were neem (Ayamga, 1997).

4.3.2. General Description

Neem or nim (*Azadirachta indica* A. Juss.) , which is synonymous with *Melia indica* (A. Juss.) Brand., and *Melia azadirachta* L. belongs to the family *Miliaceae*. It is a deep-rooted, small to medium sized tree, broad-leaved and evergreen, except in periods of extreme drought (National Academy of Sciences, 1980). Mature trees attain heights of 7 – 20 m with a spread of 5 – 10 m and may live for more than 200 years (Ketkar, 1976). With its widely extended branches, the tree forms an ovate to round crown on a straight stem (Maydell, 1990). The bark is brown-grey, of medium thickness and longitudinally and obliquely fissured; the slash is reddish brown (Maydell, 1990). The heartwood is hard and durable with a specific gravity varying from 0.56-0.85 with an average of 0.68. The wood is resistant to termites and other wood-destroying insects even in exposed areas (Maydell, 1990).

The leaves are imparinpinate, alternate, 20-40 cm long and 1 – 3 cm wide on slim petioles; 6- 10 cm long. The flowers are white, yellowish or cream-coloured, small, numerous and honey-scented (Maydell, 1990). The fruit is an ellipsoidal drupe, with one, rarely two, seeds, 1.2 -1.8 cm long, green-yellow when ripe, with a thin cuticle and juicy fruit pulp (National Academy of Sciences, 1980).

Figure 1. A nine-year old neem plantation near the Business Secondary School, Tamale (in the picture is Dubik Moisob).

4.3.3. Distribution

In its native Indo-Pakistan subcontinent, neem is found in a large belt extending southwards from Delhi and Lahore to Cape Comorin. In South Asia, it is also found in Bangladesh, upper Burma, and in the drier parts of Sri Lanka (Ahmed and Grainge, 1986). In Southeast Asia it occurs scattered in Thailand, southern Malaysia, and in the drier Indonesian islands east of Java (Ahmed and Grainge, 1986). In Africa, neem is particularly widespread in Nigeria and Sudan; it is also found along the East African coastal plains stretching from Ethiopia across Somalia, Kenya and Tanzania to Mozambique and in West Africa in the sub-Sahelian region of Mauritania, Togo, Ivory Coast and Ghana (National Academy of Sciences, 1980).

4.3.4. Site Requirements

Neem is very drought resistant and grows with as little as 150 mm of annual rainfall. The optimum annual rainfall however, is 450 – 750 mm (Maydell, 1990). The tree has been successfully grown in regions with up to 2,000 mm annual rainfall and temperatures of between 0°C and 44°C though the tree is frost tender in the seedling and sapling stages (National Academy of Sciences, 1980; Radwanski and Wickens, 1981). Neem also grows better on dry, stony, shallow and nutrient-deficient soils than other species. Neem is salt-tolerant and can be grown on marginal soils with low fertility (Ahmed and Grainge, 1986).The tree requires neutral to alkaline soils and will not grow well where the pH is less than 6 (Laurie, 1974). It tolerates altitudes of 50 – 1,500 m above sea level. Generally, lateral roots may extend radially to 15 m. The tree however is intolerant to frequent inundation and lateritic outcrops (Maydell, 1990). Poor soil drainage may retard the growth of neem.

4.3.5. Propagation

Propagation is generally by seeds which should be sown immediately after maturity, i.e., December to the end of February (Maydell, 1990). This is because the seeds are short lived and do not retain their viability for long periods (Troup, 1921), with a two week upper limit. The seeds begin to germinate as soon as they fall from the trees (Evans, 1992). Loss of viability appears to be due to and is accompanied by the fermentation of the unopened cotyledons inside the inner seed case. If the cotyledons are green, the seeds are good and will germinate, but if the cotyledons have turned brown or yellowish, they are not likely to germinate (Smith, 1939). Nagaveni et al. (1987) recommend collection of neem fruits when they are greenish-yellow and still on the tree, as opposed to the usual practice of collecting fallen fruit. If this is followed by depulping or drying, immediate germination will be delayed and therefore permit longer storage (Evans, 1992). It is advisable to use only swollen seeds, and to transplant when seedlings are 30 – 50 cm high. Neem trees start bearing fruits from the fifth year onwards and a mature tree produces more than 20 kg of fruit, corresponding to 10 – 15 kg of seeds per year. There are about 4,000 – 6,500 seeds/kg (Evans, 1992). The bare-root method has been the traditional way of raising planting stock in the Sahel-Sudan zone of Africa though in dry years survival is often poor (Evans, 1992).

In a study by Oboho et al. (1985) in northern Nigeria, germination time did not vary with seed weight, but seedling height at time of appearance of leaflets did, being tallest for the high weight and lowest for the low weight class. The number of leaflets, biomass production and growth rate varied with seed weight. It was realised in this study that the biomass production of the high weight class was three times that of the medium weight class, due to the appearance of twin seedlings from the high weight class.

4.3.6. Management

Neem plantations can be grown successfully in all parts of Ghana. In Northern Ghana, it can be grown on biologically optimum rotations that range from 5 to 11 years. It is probable that neem may be grown to larger size on longer, less productive rotations (Nanang, 1996). Weeds do not affect growth as neem is very resistant to competition and may become a noxious weed under favourable site conditions since the seeds are widely distributed by birds (Maydell, 1990). However, in Nigeria, a study revealed that freeing young neem plants of grass had a striking effect on planted seedlings. The study showed that strips that had been hoed to rid them of grass, with little breaking of the soil, showed a spectacular difference in health and survival. The results appeared to be due entirely to the eradication of grass and not to disturbance of soil, for cultivation around plants had no better effect than light surface hoeing (Anon., 1952). The tree coppices freely and early growth from coppice is faster than growth from seedlings (Maydell, 1990).

Given the invasive potential of neem, it is important that management of neem includes some practical advice on how to reduce its population when so desired. This will ensure that the benefits of neem are maximised whilst the negative impacts are minimised. Neem is a prolific seeder, characteristic of invasive species. The fruit ripen at the onset of the rainy season. They then germinate and establish themselves while there is available soil moisture (CAB International, 2005). This means that neem has the potential to occupy the most fertile lands and displace native vegetation. Neem has a large root system and when the tree is felled, it re-sprouts easily and difficult to kill. The best way to kill neem seedlings is to uproot them, while they are still in their early stages. If the understory of the mature tree is cultivated or cleared, neem seedling survival is much lower than when it is left uncultivated. This shows that neem can compete with the other native grasses. It also indicates that with timely weeding, neem can be reduced in the field (Judd, 2004). The threat of neem to become an invasive species and spread uncontrollably has to be balanced with the potential benefits that can be derived from its sustainable and controlled management on degraded, low-fertility soils.

4.3.7. Growth and Yield

Neem is a fast growing tree: two-thirds of the height may be reached after 3 to 5 years (Maydell, 1990). The rate of development of young neem plants after the first season is fairly rapid. As a rule, the trees put on an average annual girth increment of 2.3 – 3.0 cm (0.73 – 0.96 cm in diameter), though more rapid growth is easily attained (National Academy of Sciences, 1980). In four different test plots in West Africa, the height of neem trees varied from 4 to 7 m after the first 3 years and from 5 to 11 m in 8-year-old stands. According to the National Academy of Sciences (1980), in West Africa, cropping is usually done on an 8-year rotation, with original spacing between the plantation trees most commonly 2.4 m x 2.4 m. In Ghana, first rotation yield (at 8 years) was 108 – 137 m^3 of fuelwood/ha and in Samaru (northern Nigeria), the yield at the same rotation was 19-169 m^3/ha (National Academy of Sciences, 1980; Maydell, 1990). Streets (1962) reported mean heights for Northern Ghana as 3.6 and 7.5 m after 2 and 5 years respectively. In Cuba, a small stand of neem trees planted on a fertile soil reached a mean height of 14.2 m and mean dbh of 27 cm eight years after planting (Betancourt, 1972). Mean tree diameter after four years of growth in Nigeria was reported as 5.14 cm and survival was more than 75% (Verinumbe, 1991). In the semi-arid Sahel of Africa, neem typically achieves growth rates of 5m^3/ha/yr (Evans, 1992). Under favourable Sahel conditions, the annual leaf biomass production may reach more than 10 tonnes/ha (Maydell, 1990).

4.3.8. Uses

Due to the multipurpose use of neem, it has been hailed as a wonder tree. Its uses span from improving soil fertility to curing various forms of ailments, giving it the name, the "pharmacy tree". Some of the major uses of neem are outlined below.

Energy and Fuelwood

Neem seeds contain up to 40% oil, which is used as fuel in lamps and as a lubricant for machinery (National Academy of Sciences, 1980; Maydell, 1990). Neem has long been used as fuel in India and Africa. It has become the most important plantation species in northern Nigeria and is planted for fuelwood and poles around the large towns (National Academy of Sciences, 1980). The high wood production capacity of neem, coupled with its high calorific value of 51.1 MJ/tree and fuelwood value index of 3.9 at 56 months after planting, make the species an ideal choice for this purpose (Lamers et al., 1994).

Construction, Shade and Windbreaks

The natural durability of neem makes it a good choice for use in building construction and for making furniture in rural areas. The wood is tougher than teak and very similar in characteristics to mahogany (National Academy of Sciences, 1980). In rural areas, supply of poles for building is frequently the most pressing need and for many purposes, pole diameters at breast height (dbh) of 5-10 cm are required along with a smooth surface, freedom from snags and resins and good natural durability (Evans, 1992). Neem possesses most of these qualities. Neem has been used successfully as a windbreak and as a source of shade for humans and cattle. It is a splendid street tree for the arid tropics. In Niger, shelter belts of neem interplanted with other species reduce wind speeds by up to 65% (National Academy of Sciences, 1980).

Pest Control

A pool of biologically active constituents, including the triterpenoids azadirachtin, margosan-o, salanin, and meliantroil, are found in neem leaf, fruit, bark, and seed (Schmutterer, 1982; Warthen, 1979; Evans, 1992). These compounds reportedly control more than 100 species of insects, mites, and nematodes – including such economically important pests as the desert and migratory locusts, rice and maize borers, pulse beetle and rice weevil, root knot and reniform nematodes and citrus red mice (Grainge et al., 1985; Warthen, 1979; Jacobson, 1958, 1975). Modes of control include antifeedant, growth regulatory, repellent, hormonal, or pesticidal action in larval and/or adult stages of these pests.

Traditionally, Indo-Pakistani farmers simply mixed 2 – 5 kg of dried neem leaves/100 kg of grain in order to control stored-grain pests (Ahmed and Koppel, 1985). Alternatively, empty sacks were soaked overnight in water containing 2 – 10 kg of neem leaves/100 litres water and then dried these sacks before filling them with grain (Ahmed, 1984). Evans (1992) indicated that mulches of neem foliage inhibit termite attack of newly planted trees. It is reported that the insecticidal effect of azadirachtin is as good as DDT, and is not toxic to man (Maydell, 1990).

Soil Improvement

In a case study in northwest Nigeria by Radwanski and Wickens (1981), soil analysis showed that the average pH value of red sands without neem was 5.4 in the first and second layer, and the average pH in the corresponding layers under neem was 6.8. A substantial increase in the soil pH values under neem was due to the accumulation and decomposition of leaf litter and is a surface phenomenon. The content of organic carbon had risen from 0.12% under fallow to 0.57% under neem, and the total nitrogen content increased from 0.013% in the top layer of the fallow soil to 0.047% in the corresponding layer of the neem phase. Neem is a nonleguminous tree and there is no symbiotic nitrogen fixation in the soil. However, free-living, nitrogen-fixing bacteria are known to be C heterotrophic and thus dependent on the supply of organic carbon for their energy (Mengel and Kirkby, 1978). It is possible therefore that a substantial increase of the organic carbon content in the neem soil may create conditions favourable for the proliferation of those bacteria and a consequent increase in the supply of nitrogen (Radwanski and Wickens, 1981).

Studies in the Gambia report a steady rise in the mean values of phosphorous content as neem matures at the site. Young neem trees do not drop much litter when establishing at a site, and no other plants vegetate the site and hence the organic matter is less for those locations with young stands. It is not surprising that neem increases phosphorus. As the neem grows older it changes the site, producing more and more organic matter, adding litter to the soil, thereby increasing the phosphorous and organic matter (Judd, 2004).

In order to evaluate the effects neem plantations have on the yield of food crops, surface soil under 12-year-old plantations of neem was used to grow food crops in north-eastern Nigeria. The results indicated that two months after planting, the crops produced five times higher biomass on the neem plantation soil than on the control. The trees had favourable effects on soil fertility and therefore improved crop yield (Verinumbe, 1991).

The extensive root system of neem can extract nutrients from deep subsoils and enrich surface soils through litter. Thus, in northwest Nigeria, significantly higher total cations, cation exchange capacity, base saturation and pH were observed in soils under neem than on similar soils under fallow (Radwanski and Wickens, 1981). Mulching sorghum (*Sorghum bicolor*) with neem leaves in Burkina Faso improved sorghum yields by up to 422% of the unmulched control (Tilander, 1993).

The oil cake of neem is a good fertiliser and is effective in reducing attack of agricultural crops by termites and diverse insects after fertilisation (Datta, 1978). In northwest Nigeria, neem is used on degraded agricultural lands for soil amelioration in order to improve the pH value and to make available soil nutrients for commercial crops (Maydell, 1990).

Medicinal Values

Many medicinal uses have been reported for neem. The bark, leaves, fruit, oil and sap reportedly cure various skin diseases, venereal diseases (syphilis), tuberculosis etc. (Maydell, 1990). Ahmed and Grainge (1985) and Hepburn (1989) also report that neem oil has contraceptive qualities. Undiluted neem oil showed strong spermicidal action and was 100% effective in preventing pregnancies in rhesus monkeys and human subjects (Sinha et al., 1984).

4.3.9. Pests and Diseases

Neem trees are generally pest-free, due perhaps to the presence of azadirachtin and other insecticidal compounds (Csurhes, 2008). However, neem plantations have been badly damaged by a scale insect, *Aonidiella orientalis*, in Africa, and to a lesser extent in India (NRC, 1992). Certain species of ants, moths and bugs are also known pests of neem (NRC, 1992). Live neem trees are susceptible to borers and termites (Hearne, 1975). Because neem has not been extensively studied in Ghana, there is no record of the pests and diseases that affect the trees in this environment.

4.3.10. Case Study: Assessing the Potential of Neem Plantations for Use in Agroforestry

An on-farm experiment to evaluate soil fertility status and crop productivity under neem plantations in an agroforestry setting was undertaken in Northern Ghana by Nanang and Asante (2000). Three neem plantations of different ages *viz*: 9, 7 and 5 years were selected. Soil nutrient status under these plantations was determined and sorghum and groundnuts were planted in the 7 year-old plantation. The treatments were: pruned, unpruned and control (adjacent plot without neem trees). Soil samples were collected from each plantation site at 0-20 cm and 20-40 cm depths. The experimental design was a randomized complete block with two replications.

Soil Analyses

Initial soil analysis before cropping indicated that pH under the neem plantation was significantly higher than pH on the control plot. Though differences in soil pH under plantation of different ages was not significant, a clear pattern of pH was observed i.e., the older the plantation, the higher the pH. Generally pH below 20cm depth was lower than pH at 0-20cm depth. Soil organic carbon ranged between 0.340 – 0.550%. The older the plantation, the higher the organic carbon content obtained.

Total nitrogen followed the same trend as soil organic carbon and ranged from 0.019-0.046%. The control plot had lower soil total nitrogen than soils under the neem plantations. The difference between total N level at 0 – 20 cm and 20 – 40 cm was not significant. Soil available phosphorus ranged from 1.16 -5.55 mg/kg. No clear trend was observed in relation to either the depth of sampling or the age of the plantation. Soil exchangeable potassium was clearly higher at 20 – 40 cm than at 0 – 20 cm. It ranged between 37.88 – 177.7 mg/kg. The age of the plantation did not influence soil exchangeable potassium content.

Soil pH after harvesting food crops from the treatments with neem trees ranged from 5.29 to 5.53. The difference between control and soil under neem plantation was not significant. The higher soil pH under the plantation could be attributed to higher amount of organic matter in the plantation produced by leaf fall as compared to the control plot with low vegetative cover.

Soil available phosphorus and exchangeable potassium may be more related to the soil than to the plant material addition. Generally, there was reduction in soil organic matter, carbon, total nitrogen at harvest compared with initial levels. This could be attributed to mineralisation and uptake. Organic carbon after harvest was between 0.34 and 0.38%, total nitrogen ranged between 0.015 – 0.023%, soil available phosphorus 2.55 to 5.22 mg/kg and exchangeable potassium between 22.87 and 29.76 mg/kg. These soil chemical properties were not influenced by plantation, either pruned or unpruned.

Crop Yields

Sorghum stover yield ranged from 4.5 – 9.7 t/ha for all treatments. Both pruned and control treatments gave significantly higher stover yield than the unpruned treatment. The difference between the pruned and the control plot was however not significant. Grain yield ranged from 2.2 – 3.7 t/ha. Groundnuts vine yield was significantly influenced by treatment. Vine yield on control was more than twice that of the pruned area and more than six times that of the unpruned plot. Pod yields followed a similar trend as the vine ranging from 0.72 to 1.24 t/ha. The differences in yield were significant at 5% significance level. Grain yield ranged from 0.067 to 0.107 t/ha (67 to 107kg/ha). The pruned plot had higher grain yield than the control, though the difference was not statistically significant. Both yields on the control and pruned treatments were significantly higher than the unpruned treatment.

The lower yields from the unpruned treatments for both crops were probably due to shading effect. The canopy of the plantation may have intercepted sun radiation of the intercrop reducing photosynthetic activities and hence reduction in dry matter production. The lower yield observed from the unpruned plot could also be attributed to moisture stress. The tree with its broad canopy could have large surface area for transpiration. This could result in competing for soil moisture with the intercrop. Competition for moisture was probably lower on the pruned plot compared to the unpruned. This could have resulted in less moisture stress on the intercrop in the pruned plots, hence its better performance.

These results suggest that neem plantations can positively influence soil fertility and hence have a potential for use in agroforestry practices in Northern Ghana. However, adequate pruning of the trees at certain stages of growth to reduce shading of the companion crops and minimise moisture stress of this evergreen tree is required. The pruning should be incorporated into the soil to increase soil nutrients. This approach has been used in Burkina Faso, where neem branches and suckers are trimmed back at the beginning of the rainy season, and left to enrich the soil (Bationo et al., 2004). This illustrates an integrated approach to manage neem on the farm: the crops benefit from the trees (organic matter and nutrients for the soil) and negative effects (shade) are minimised through management (Judd, 2004). Further studies in Northern Ghana can quantify the influence of shading and moisture stress on the yield of companion food crops.

4.4. Critical Silvics of Gmelina Arborea

4.4.1. Gmelina in Ghana

According to FAO (2002a), there are about 6350 ha of gmelina plantations in Ghana. The most extensive plantation of gmelina in Ghana is owned by the Subri Industrial Plantation Ltd. (SIPL), which is a parastatal organisation that was formed under the previous Forestry Department (FAO, 2002a). An estimated 4000 ha of gmelina was planted for the production of fibre for a proposed paper mill which was intended to be established at Daboase on the eastern banks of River Pra in the Western Region of Ghana (FAO, 2002a). Plantation establishment at the SIPL site started in 1971 and ended in 1996. The pulp mill was never built due to the lack of financing, and this delay in the establishment of the pulp mill resulted in the addition of the production of sawlogs to the management objective in 1990. Exports of gmelina wood products in 2010 were about 9,000 m^3 in the form of air-dried lumber, billet and poles.

4.4.2. General Description

Gmelina arborea Roxb. is a tree species belonging to the family *Verbenaceae*. It is a fast growing tree frequently planted in plantations to produce wood for light construction, crafts, decorative veneers, pulp, fuel, and charcoal (Hossain, 1999). The species is also planted in taungya systems with short-rotation crops and as a shade tree for coffee and cocoa (Hossain, 1999). Gmelina is a deciduous, medium-sized tree that grows up to 40 m tall and 140 cm in diameter, but usually smaller than this (Jensen, 1995). The tree form is fair to good, with 6–9 m of branchless, often crooked trunk and a large, low-branched crown (Hossain, 1999). The leaves are simple, opposite, more or less heart-shaped, 10–25 cm long, and 5– 18 cm wide (Hossain, 1999). Evans (1992) estimated that the number of seeds per kilogram varies from 700 to 2500. Gmelina begins to flower and set fruit at about 6 to 8 years of plantation age. The first flowers are borne 3-4 years after planting and, in nature, self-pollination is discouraged by the floral morphology (Orwa et al., 2009).

4.4.3. Distribution

Gmelina is native to the South and Southeast Asia from Pakistan and Sri Lanka to Myanmar (Hossain, 1999). According to Jensen (1995), the species has been widely planted in Southeast Asian countries such as Bangladesh, Myanmar and Thailand. However, it has not been as widely planted in tropical African and Latin American countries (Evans, 1992), although it has been introduced into many tropical countries, including the Philippines, Malaysia, Brazil, Gambia, Costa Rica, Burkina Faso, Ivory Coast, Nigeria, and Malawi (National Academy of Sciences 1980).

4.4.4. Site Requirements

Gmelina is found in rainforest as well as dry deciduous forests and tolerates a wide range of conditions from sea level to 1200 m elevation and annual rainfall from 750 to 4500 mm (Orwa et al., 2009). It grows best in climates with mean annual temperature of 21–28°C (Jensen, 1995). The best growth of gmelina is achieved on deep, well drained, base-rich soils with pH between 5.0 and 8.0. The species can tolerate a 6–7-month dry season and grows on many soils, acidic laterites to calcareous loams, doing poorly on thin or poor soils with hardpan, dry sands, or heavily leached acidic soils, well-drained basic alluviums (Duke, 1983).

4.4.5. Propagation

Gmelina can be propagated by seeds, cuttings, or stumps (Alam et al., 1985). Under natural conditions germination takes place in the rainy season soon after fruits fall from the tree. The germination rate for fresh seed is 65–80% (Hossain, 1999). Fresh seed can be stored at room temperature for about 6 months. Seed stored at 4°C will remain viable for about three years. The seed should be soaked in cold water for 24 hours before planting. Seeds should be planted in germination beds with a mixture of sand and loam and covered with a thin layer of sand or compost (Hossain, 1999). Seeds germinate in 2–3 weeks and are ready for transplanting to polybags when the first pair of leaves appear. Root pruning and hardening off of the seedlings are beneficial for maximum field survival. Within 6 months, seedlings reach a height of 30–45 cm, and are ready for planting in the field (Hossain, 1999). Seedlings are usually ready for stump preparation in 7–8 months and should have a root collar diameter of at least 2.5 cm (Hossain, 1999). The stem and roots of seedlings should be pruned back to 5 cm and 20 cm, respectively. Mortality rates for stump planting are usually as high as 50% (Hossain, 1999).

4.4.6. Management

In plantations, an initial spacing of 2 x 2 m is commonly used, while a spacing of 4.5 x 4.5 m is used for agroforestry (Hossain, 1999). Under favourable conditions the growth of the seedlings is rapid, particularly from the second year onward (Duke, 1983). Because gmelina is shade-intolerant and sensitive to competition, 3–4 weedings are required during the first two years of growth (Hossain, 1999). If gmelina is grown for pulpwood and sawnwood, rotations of 6 and 10 years, respectively are commonly used, while for fuelwood, between 5–10 years are common. If 10-year rotations are used, thinning 50% of the stand at about half-way through the rotation is carried out, and another 50% at seven years (Hossain, 1999). Gmelina lends itself to coppicing and the second rotation is usually produced using this method. If a third rotation is desired, seedlings and stumps are used (Duke, 1983). In Ghana rotations of 6 years are used for pulpwood purposes, while for sawnwood, the usual rotation is 10 years.

Gmelina has suitable characteristics for use in agroforestry systems, with its fast growth, ease of establishment, and relative freedom from pests outside its natural range (it can be browsed repeatedly without damage) (Orwa et al., 2009). It is an especially promising fuelwood species because it can be established easily, regenerates well from both sprouts and seeds, and is fast growing (Orwa et al., 2009). Although able to compete with weeds more successfully than many other species, it responds positively to weeding and also benefits from irrigation (Orwa et al., 2009). Planting gmelina with crops like maize and cassava has been found beneficial in increasing the simultaneous production of wood and food.

At the SIPL plantations, an initial spacing of 4 x 2 m (1250 trees/ha.) has been observed to be the best for gmelina for the production of pulpwood and sawlogs (FAO, 2002a). According to Cabaret and Nguessan (1988), due to difficulties of obtaining a straight bole and frequent problems of forking and multiple stems, a planting density of about 1,100 stems/ha should be used to ensure natural pruning and a sufficiently large stock for a good selection at the time of thinning.

4.4.7. Growth and Yield

Gmelina is a fast growing species, and on good sites, can reach 20 m height in 5 years, and can attain more than 30 m in height with about 60 cm dbh at maturity (Hossain, 1999). The form of the tree is fair to good, with 6–9 m of clear bole (Duke, 1983). Some trees can reach 3 m after a year from planting and 20 m after 4.5 years. In Nigeria, the yield of gmelina is 84 m^3/ha at age 12 in poor sandy soils, 210 m^3/ha at age 12 in clay or lateritic soils, and 252 m^3/ha at age 10 in favourable alluvial soils—all volumes are under bark to 7.5 cm top diameter (Adegbehin et al., 1988). The National Academy of Science (1980) reports annual increments greater than 30 m^3/ha on fertile sites. On the SIPL plantations in Ghana, MAI of about 19m^3/ha/yr have been recorded, with current annual increments of up to 22m^3/ha/yr (FAO, 2002a). With proper management of the plantations, it is possible to obtain higher yields than these.

4.4.8 Uses of Gmelina

Davidson (1985) describes the wood as yellowish or greyish-white, even grained, and very useful for planking, panelling, carriages, furniture, and carpentry of all kinds. The specific gravity of gmelina wood is 0.42–0.64 (Davidson, 1985). It is easily worked, readily takes paint or varnish, and is very durable under water. The wood is used for light construction and pulp as well as for fuelwood and charcoal. Fuelwood derived from gmelina provides 4400–4800 Kcal/kg (Davidson, 1985). Graveyard tests indicate that the untreated timber may last 15 years in contact with the soil, with pulping properties superior to most hardwood pulps (Duke, 1983). The wood makes a fairly good charcoal and according to Little (1983), the leaves are harvested for fodder for animals and silkworms; the bittersweet fruits were once consumed by humans.

4.4.9. Pests and Diseases

Numerous pests and diseases affect gmelina in its native range, but it is unknown what effect they may have on crop/production yields of native plant populations. Some fungal pathogens have been introduced into areas where the trees have been established as exotics Browne (1968). Among these, leaf spot caused by *Pseudocercospora ranjita* is most widespread although it has not caused any substantial damage (Wingfield and Robison, 2004). Wingfield and Robison (2004) note that a serious vascular wilt disease caused by *Ceratocystis fimbriata* in Brazil resulted in the most significant failure of gmelina in plantations. Among the insect pests, the defoliator *Calopepla leayana* (Chrysomelidae) appears to be most important, although no serious insect pest problems have been recorded where gmelina is grown as an exotic (Wingfield and Robison, 2004). A bark disease (worm disease) that can girdle the base of the tree and cause dieback of branches in 2-year-old plantations is spread by *Griphosphaeria gmelinae* (Orwa et al. 2009). In Indonesia, one of the insects consistently associated with the species is a carpenter worm *Prionoxystus sp.*, which bores into stems of saplings, feeds from within and weakens them (Orwa et al. 2009). In Ghana, heart-rot in gmelina plantations at the age of 20 years has been observed at the SIPL plantations, which resulted in reduced coppicing ability (FAO, 2002a). In addition, grasshopper attacks on the leaves of gmelina have also been reported (FAO, 2002a). There are opportunities to reduce insect and disease problems through biological control of insects and integrated disease and pest management, especially using vegetative propagation breeding and selection for insect and pathogen tolerance will facilitate the propagation of healthy trees (Wingfield and Robison, 2004).

4.5. Critical Silvics of Cedrela Odorata

4.5.1. Cedrela Odorata in Ghana

It is on record that *Cedrela odorata* was introduced into Ghana in 1898 (FAO, 2002a). Known plantations of cedrela in Ghana are: about 3800 ha owned by the Forestry Commission within forest reserves, 40 ha planted under the Gwira Banso Project (FAO, 2002a), 1000 ha at the SIPL site in Daboase and another 641 ha at the Fure River Forest Reserve near Prestea in the Western Region where a cedrela plantation was established in 1971. Due to its desirable wood characteristics, cedrela is one of the priority exotic species that is being promoted for plantation development programmes in Ghana. The Forestry Research Institute of Ghana has established a seed orchard for cedrela in order to produce good quality seeds for stakeholders. In 2009, about 2,000m^3 of cedrela lumber and 30,000m^2 of sliced veneer were exported from Ghana. Exports of cedrela wood products increased in 2010 to about 6,000m^3 in the form of lumber and veneer (Ghana Forestry Commission, 2010).

4.5.2. General Description

Cedrela odorata belongs to the family *Meliaceae.* It is a deciduous tropical tree that grows to a maximum height of about 30 – 40 m. About two-thirds of the bole is clear of branches. The wood has a strong aromatic odour and the fruit is a woody capsule, while the very small seeds are winged and spread when the ripen fruit splits open (Lamb, 1968). The species is insect pollinated and has wind-dispersed seed (Cavers et al., 2004). Trees bear fruit from the age of 10 years according to Lamb (1968) or 15 years according to Lamprecht (1989). Flowering is annual, but good seed crops occur every 1-2 years (Orwa et al., 2009). Flowers appear early in the rainy season and fruits mature during the dry season when the leaves become deciduous. Seeds are wind-dispersed (James *et al.,* 1998). Fruit development takes about 9 or 10 months and fruits ripen during the next dry season. The fruit, a large woody capsule, is borne near branch tips. Fruits ripen, split, and shed seeds while still attached to the parent tree (Lamb, 1968). Cedrela is listed in the International Union for Conservation of Nature (IUCN) category of species that face a high risk of extinction in the wild in the medium-term due to over-exploitation (CITES, 2007).

4.5.3. Distribution

The native range of cedrela is the forests of moist and seasonally dry subtropical or tropical life zones (Holdridge, 1976) from latitude 26° N on the Pacific coast of Mexico, throughout Central America and the West Indies, to the lowlands and foothills of most of South America up to 1200 m altitude, finding its southern limit at about latitude 28°S in Argentina (Chaplin, 1980; Tosi, 1960). Plantations of cedrela have been established both within and outside of its native range.

4.5.4. Site Requirements

Cedrela needs a plentiful supply of nutrients and is very intolerant of water logging (Cintron, 1990; Lamb, 1968). The species tolerates soils high in calcium and prefers fertile, free draining, weakly acidic soil (Orwa et al., 2009). It tolerates a long dry season but does not flourish in areas of rainfall greater than about 3000 mm or on sites with heavy or waterlogged soils (Beard, 1942; Malimbwi, 1978). Cedrela develops best in seasonally dry climates, as reflected in its deciduous habit. It grows at altitudes from 0-1900 m, with a mean annual temperature range of 22-26 °C. It reaches greatest prominence under an annual rainfall of 1200 to 2400 mm with a dry season 2 to 5 months long. The species can also survive in lower rainfall areas, to as low as about 1000 mm of annual rainfall, although under these situations, it will grow slowly and become stunted (Miller et al., 1957; Wadsworth, 1960). The common denominator appears to be drainage and aeration of the soil, not soil pH (Styles, 1972; Holdridge, 1972; Whitmore, 1976).

4.5.5. Propagation

Transplanting of naturally regenerated seedlings or establishing branch and stem cuttings are the most common propagation methods. Grafting and budding methods have also been successful (World Agroforestry Centre, 2009). The World Agroforestry Centre (2009) database on tropical forest species describes the propagation of cedrela as follows. Seeds may be broadcast or sown in lines in level nursery beds and lightly covered with soil, sand, sawdust or charcoal. Where there is adequate moisture, shade is not necessary; shade increases the risk of damping-off. Germination takes 2-4 weeks. It is fastest at temperatures 30-35°C, but seed also germinates at 15°C. Seedlings grow very quickly and may attain 40-50 cm height after 3 months and 130-150 cm after 12 months. Stumps, striplings and container seedlings are used for planting; occasionally wildings may be used. Direct seeding is feasible, as the young plants develop very quickly; as trees seem to experience a rather severe planting shock this method is recommended when there is no shortage of seed. Cedrela does not coppice readily nor produce root suckers; and is not fire resistant (Beard, 1942).

4.5.6. Management

Although cedrela tolerates weed competition during the seedling stage (Whitmore, 1976), it is classified as intolerant of weeds and shade at the sapling stage and beyond (Malimbwi, 1978). Pruning is not required when cedrela is grown as a stand, but trees affected by *Hypsipyla* attack may need pruning to remove multiple leaders formed (Orwa et al., 2009). The root system of cedrela is superficial, and this makes the tree subject to wind damage, especially after thinning. In mixed stands, it is realistic to raise only 10-20 high-quality trees/ha. Well-formed, straight stems are usual, except in trees grown in open places (Orwa et al., 2009).

4.5.7. Growth and Yield

Generally, cedrela is a very fast growing tree, once it passes the vulnerable early sapling stage, adding 2.5 cm or more in diameter and 2 m in height a year under good conditions (Omioyola, 1972). During the first 9 years in trial plantations of cedrela in Java, the mean annual increment was 17 m^3/ha at 650 m altitude and 28 m^3/ha at 800 m altitude. A 40-year-old plantation in Nigeria yielded a timber volume of 445 m^3/ha (Omioyola, 1972). Cedrela shows potential for plantations, as it is fast growing and produces multipurpose timber (Orwa et al., 2009).

4.5.8. Uses of Cedrela

Cedrela is widely harvested for use as timber by virtue of its durability, excellent working qualities and appearance. It was reported to be perhaps the most important local timber for domestic use in tropical America (Rendle, 1969). The wood is also in high demand in the American tropics because of its natural resistance to termites and rot (Cintron, 1990). The aromatic wood is also used to make cigar boxes. The wood is suitable for making non-structural elements for exteriors and interiors, quality furniture and novelty and craft items (Anon., 2004; Echenique-Marique and Plumptre, 1990). Cedrela trees have many low branches and a spreading crown, and this makes them suitable for use to provide shade and as a windbreak in courtyard gardens and in cocoa and coffee plantations (Orwa et al., 2009). The species can also be planted for ornamental purposes along roads and in parks.

4.5.9. Pests and Diseases

Plantations of cedrela have suffered snail damage in Malaysia and Africa, while slugs killed some nursery stock of an exotic provenance in the Virgin Islands. Beetle damage is a problem in some plantations in Africa (Malimbwi, 1978, Omoyiola, 1973). According to Holdridge (1976), the most serious insect pest of cedrela is the mahogany shoot borer *Hypsipyla grandella*. The larvae of this moth eat the pith just behind the growing tip of fast-growing shoots, causing death of the apical meristem, which slows seedling and sapling growth and may ruin tree form, since multiple leaders or bushiness often result (Holdridge,

1976). Shoot borer attack may also contribute to seedling mortality, especially in already stressed populations (Allan et al., 1973; Grijpma, 1976). Chaplin (1980) reports that provenance trials of cedrelas from a wide geographic range have shown that they may vary in response to attack and therefore careful selection may allow future development of tolerant strains. Because cedrela is highly susceptible to *Hypsipyla* attack, it is recommended that trees be planted in mixed plantations, for example with *Leucaena leucocephala*, or under the light shade of trees such as *Eucalyptus delgupta* (Orwa et al., 2009). Even though there is *Hypsilla alcate* in Ghana that attacks the *Khaya* species it has not been observed to attack cedrela yet (FAO, 2002a).

4.6. Critical Silvics of Kapok (*Ceiba Pentandra*)

4.6.1. Ceiba Pentandra in Ghana

Ceiba pentandra (kapok or silk cotton tree) is an important tree species occurring in the natural forests of Ghana. It is exploited as a commercial timber species from the wild and exported in the form of lumber, plywood, sliced veneer or mouldings. In 2010, more than 23,000 m^3 of ceiba was exported, mostly as sliced veneer and plywood (Ghana Forestry Commission, 2011). The main problem with ceiba exports is the relatively low value of its wood in international markets. Local and national markets for ceiba timber are weak; the average log price in Ghana's domestic market was US$31–55/$m^3$ in 2005, while the average sawn wood price was US$ 53/m^3 (ITTO, 2005c).

Although the potential of ceiba to be planted in plantations has been recognised and FORIG has been carrying out research on the artificial regeneration of the species including vegetative propagation, there are currently no large-scale plantations of ceiba in Ghana. The only known plantations of ceiba are two small ones in the savannah zone. These plantations are located near Tamale in the Northern Region and were planted to produce kapok and edible seeds (FAO, 2002a). The first plantation was established in 1993 and is only about 0.8 ha in size and was planted at a spacing of 8 x 8 m (FAO, 2002a). The second ceiba plantation is located at Bogunayili and was established on a 5 ha plot. This plantation is located on a sandy soil that is lying fallow, at a spacing of 10 x 10 m. The planting is said to have been done at about 1960 (FAO, 2002a). The measured average diameter of the trees was about 70 cm in 1997 (FAO, 2002a).

In addition to its use as timber, ceiba is widely planted in Ghana as an ornamental tree along avenues and around buildings. Its seeds and leaves are popular as sources of food to prepare local soups. The fibre is a main source of material for stuffing pillows and mattresses, and for insulation by local people across Ghana. Recent studies suggest that there is a potential to use new processing techniques to process the fibre in textile applications, as well as a biodegradable alternative to synthetic oil-sorbent materials, due to its hydrophobic-oleophilic properties (Duvall, 2009). The hydrophobic-oleophilic characteristics of the kapok fibre could be attributed to its waxy surface, while its large lumen contributes to its excellent oil absorbency and retention capacity (Lim and Haung, 2006).

4.6.2. General Description

Ceiba pentandra belongs to the family *Bombacaceae* and is a tall, deciduous tree bearing short, sharp prickles all along the trunk and branches; supported by pronounced buttresses at the base (Orwa et al., 2009). The trunk is cylindrical to slightly convex. The crown spherical to round, with bright green, open foliage; branches verticillate and abundant, sloping upwards; the bark is smooth to slightly fissured, pale grey, with horizontal rings, protruding lenticels and sharp prickles that are irregularly distributed on the upper part of the trunk (Jøker and Salazar, 2000). It has a light crown and is leafless for a long period. The ceiba cultivated in West Africa and Asia is a medium-sized tree up to 30 m tall, bole unbranched, usually spineless, buttresses small or absent, branches horizontal or ascending, leaves intermediate, flowering annually after leaf-shedding.

The tree is obligately deciduous, losing its leaves for 10–14 weeks in the dry season, and it usually flowers annually in the leafless period. Leafing and flowering periods are more regular in drier parts of the distribution area; in moister areas, leafing and flowering periods are highly irregular (Duvall, 2009). The flowers open at night and are senescent by midday; they are pollinated by bats, but are also visited by moths and bees. The fruits ripen 80–100 days after flowering, the dehiscent types releasing kapok with loosely embedded seeds that are wind-dispersed (Duvall, 2009).

Leaves are alternate with slender green petioles. There are usually 5 leaflets in a mature form. The leaflets hang down on short stalks; short pointed at the base and apex, not toothed on edges, thin, bright to dark green above and dull green beneath (Orwa et al., 2009). Fruit are leathery, ellipsoid, pendulous capsule, 10-30 cm long, usually tapering at both ends, rarely dehiscing on the tree. White, pale yellow or grey floss originates from the inside wall of the fruit. Seed capsules split open along 5 lines. Each capsule releases 120-175 seeds rounded black seeds embedded in a mass of grey woolly hairs. Seeds are usually dark brown (Orwa et al., 2009).

The wood of ceiba is creamy white to yellow, often with greyish veins with a coarse texture and interlocked grain. The wood is light and not durable unless treated with appropriate preservatives. It is amongst the most vulnerable African timbers to termites, *Lyctus* beetles, and other boring insects (Duvall, 2009). It is susceptible to white-rot and blue-stain fungi, but resistant to very resistant against brown rot. The wood is extremely vulnerable to decay when in contact with the soil; however, it readily absorbs preservatives: both vacuum-pressure and open tank systems give good penetration and absorption (Duvall, 2009).

4.6.3. Distribution

Ceiba originated in the American tropics. Natural occurrence from 16°N in the United States, through Central America to 16°S in South America, but is cultivated widely in the tropics between 16°N and 16°S (Jøker and Salazar, 2000). Its natural distribution has been obscured by its widespread introduction after about 1500 (Duvall, 2009). Duvall (2009) argues that although the literature has often indicated that ceiba was introduced by humans into tropical Africa, there is no historical evidence of such introduction, and there is strong ecological, botanical and cytological evidence that the tree is native to western and central Africa as well. Ceiba is now cultivated all over the tropics, but mainly in South-East Asia, especially in Indonesia and Thailand. In addition, there are records of the species in 13 other countries in East and southern Africa (including South Africa) and the Indian Ocean islands, but the tree has probably been planted in all other tropical African countries as well (Duvall, 2009).

4.6.4. Site Requirements

According to Orwa et al. (2009) ceiba can be found in various types of moist evergreen and deciduous forests, as well as in dry forests and gallery forests. As a pioneer species, it mostly occurs in secondary forests. The preferred altitude is from sea level to 900m above sea level with mean annual temperature range of 18-38°C. Ceiba requires abundant rainfall during the vegetative period and a drier period for flowering and fruiting (Duvall, 2009). Optimal mean annual rainfall is 1500 mm, but does well in its area of natural distribution with mean annual rainfall of between 750-3,000 m. Ceiba can grow on a variety of soils, from sand to clay soils provided they are well drained; but prefers alluvial soils, slightly acidic to neutral, with no water logging (Jøker and Salazar, 2000). Assessments of soils in the dry semi-deciduous forests of Ghana showed that most of the upland soils with the exception of those with shallow, eroded, ironstone concretions, boulders and outcrops were suitable for the growth of ceiba (Adjei and Kyereh, 1999).

4.6.5. Propagation

Ceiba is usually propagated by seed, although it can also be grown from cuttings (Orwa et al., 2009). Without any pre-treatment seeds germinate slowly (less than 10% one month after sowing) and germination may continue for 3–4 months. The seeds contain large amounts of oil that tend to go rancid quickly and the viability diminishes rapidly (Jøker and Salazar, 2000). Bushfires may cause simultaneous germination of seeds (Duvall, 2009). The following description of propagation is adopted from Duvall (2009). The seeds of ceiba may be stored up to one year in glass or plastic containers at 4°C and 60% relative humidity. Fresh seeds have germination rates of 90–100% within 3–5 days after sowing, when pre-treated by scoring lightly and soaking in water for 24 hours or by soaking in boiling water for 5 minutes. Germination is good in sandy soil with temperatures of 20–30°C (Duvall, 2009). When the young plants are 12–15 cm tall, they can be exposed to full sunlight. Young plants can be grown in a nursery and be transplanted into the field when they are 4–10 months old. It is recommended, however, to sow directly on land which has been properly cleared for planting. *Ceiba pentandra* can also be easily propagated from cuttings, which should be taken from orthotropic branches. Trees raised from seeds root deeper than those raised from cuttings, but develop slower (Duvall, 2009).

4.6.6. Management

Ceiba is light-demanding, and growth is spindly and poor and mortality is high for seedlings and saplings in shaded locations, including small canopy gaps that close relatively quickly (Duvall, 2009). The recommended initial planting distances should be 5 x 5 m. Plantations can be thinned to remove every other second row after 6 years of growth, to reach an optimal spacing of 10 m between rows (Duvall, 2009). In the first 2 years after planting, vegetation must be cleared periodically around saplings. General tending may be necessary in the following years, by cutting climbers and lianas, and removing dead and diseased trees. Plantations need not be thinned if planted with 7 x 7m spacing unless intercropped with smaller tree crops (Duvall, 2009).

4.6.7. Growth and Yield

Growth of ceiba is relatively fast. Seedlings planted in Ghana were 29 cm tall 6 weeks after germination and 63 cm after 51 weeks (Duvall, 2009). The annual increases in height and diameter during the first 10 years are about 1.2 m and 3–4 cm, respectively. In forest gaps height growth may be 2 m/year (Duvall, 2009). A tree 70 cm in diameter above the buttresses yields on average 4 m^3 of timber, and trees 100 and 150 cm in diameter above the buttresses yield 9.3 and 23 m^3 of timber, respectively (Duvall, 2009). Under optimum conditions a full-grown plantation tree may yield 330–400 fruits per year, giving 15–18 kg fibre and about 30 kg seed. A satisfactory average annual fibre yield is about 450 kg/ha, whereas about 700 kg/ha will be considered very good yield (Duvall, 2009).

4.6.8. Uses

Ceiba has many important uses around the world. The leaves, which contain about 26% protein is used as fodder to feed sheep, goats and cattle (Orwa et al. 2009). Supplementing goat feed with ceiba foliage increased growth rates of the animals and appears to be a viable option to improve the nutritional status of goats during periods of the year when grazing is restricted (Kouch et at., 2006). In addition, the fibre from the inner wall of the fruit is unique in that it combines springiness and resilience and is resistant to vermin, to make it ideal for stuffing pillows, mattresses and cushions (Orwa et al. 2009). It is light, water repellent and buoyant, making it ideal for life jackets, lifeboats and other naval safety apparatus. It is an excellent material for insulating iceboxes, refrigerators, cold-storage plants, offices, theatres and aeroplanes (Orwa et al. 2009). It is a good sound absorber and is widely used for acoustic insulation; it is indispensable in hospitals, since mattresses can be dry sterilised without losing original quality (Orwa et al. 2009). The tree is an important source of honey and also suitable for soil erosion control and watershed protection. In agroforestry it is grown with coffee, cocoa, and in Java as support for pepper trees. In India it is used in taungya systems (Jøker and Salazar, 2000).

Traditionally, ceiba has been used to make canoes by hallowing out the entire trunks and for lightweight furniture, utensils, containers, musical instruments, mortars and carvings (Duvall, 2009).The timber from ceiba, though not naturally durable can easily be machine planned and sanded and resistant to splitting when screwed. The wood is easy to peel for veneer. Reported uses of wood include plywood, packaging, lumber core stock, light construction, pulp and paper products, canoes and rafts, farm implements, furniture and matches. Ceiba seed contains 20-25% non-drying oil, similar to cottonseed oil, used as a lubricant, in soap manufacturing and in cooking (Orwa et al., 2009). Orwa et al. (2009) report some medical properties of ceiba. For example, the compressed fresh leaves of ceiba are used as medicine against dizziness; decoction of the boiled roots is used to treat edema; gum is eaten to relieve stomach upset; tender shoot decoction is a contraceptive and leaf infusion is taken orally against cough and hoarse throat. In Tamilnadu, India, the leaves are pounded together with fermented boiled rice water and the extract is administered to cows orally as a remedy for reproductive problems. The dose is approximately 500 millilitres applied three times a day for three consecutive days (Orwa et al., 2009).

4.6.9. Pests and Diseases

Insect defoliators include *Ephyriades arcas, Eulepidotis modestula, Oiketicus kirbiyi* and *Pericalia ricini*. The tree is also a host to parasitic plants such as *Dendropthoe falcata* and *Loranthus spp*. (Orwa et al., 2009). High seedling and sapling mortality may occur in humid climates as a result of leaf spot, dieback, damping off and anthracnose. These infections are caused by various fungal pathogens (Duvall, 2009). Ceiba is a host tree of Cocoa Swollen Shoot Virus (CSSV) causing swollen shoot disease in cocoa, a disease which has had a devastating effect on cocoa production in Ghana and neighbouring countries (Duvall, 2009). Ceiba itself shows considerable resistance to this disease.

In Ghana, dieback of seedlings and saplings has been observed at the nursery and plantation. Studies by Apetorgbor et al. (2003) to identify possible pathogenic causes of damping off, leaf spot, anthracnose and dieback in ceiba seedlings and saplings showed that leaf spot and anthracnose were caused by *Colletotrichum capsici*, whereas under favourable conditions, *F. solani* and *L. theobromae* were found to be associated with dieback of stem in both the nursery and field. The authors found that Kocide at 6.6g/l and Aliette at 5g/l were the most effective fungicides to treat these infections but retarded growth of ceiba seedlings. The fungicides were found to be most effective when applied as soon as seedlings emerged and continued fortnightly (Apetorgbor et al., 2003).

4.7. Critical Silvics of Wawa (*Triplochiton Scleroxylon*)

4.7.1. Wawa in Ghana

Triplochiton scleroxylon (known in Ghana as wawa) is a major commercial timber tree of West and Central Africa. In Ghana, it is found in the natural forests of the moist evergreen, moist semi-deciduous and the forest-savannah transition zones. The potential of wawa to be used in plantation development has been recognised, although there are no large scale plantations in Ghana. Wawa has been heavily exploited in Ghana for many decades. Available records show that in 1959 Ghana exported 650,000 m^3 of logs and 30,000 m^3 of sawn timber (Bosu and Krampah, 2005). In 1992, wawa accounted for 64% of Ghana's log exports (Richards, 1995). As a result, a ban on the exports of wawa in log form was implemented in 1993, which subsequently shifted the exports to secondary and tertiary wood products. In 2009, about 36,733 m^3 of wawa was exported in the form of kiln-dried lumber, earning about €10 million (Ghana Forestry Commission, 2010) constituting 52% of all kiln-dried lumber exports and by far the number one exported timber species from Ghana in lumber form that year. In addition, smaller quantities of wawa were exported as sliced veneer, plywood, mouldings and kindling. Exports of wawa wood products in 2010 increased to over 54,000m^3 mostly in kiln-dried lumber and moulding with about 133,000m^2 of sliced veneer. These exports from wawa earned over €17 million (Ghana Forestry Commission, 2011).

4.7.2. General Description

Triplochiton scleroxylon (Schum) belongs to the family *Sterculiaceae*. It is a large deciduous forest tree commonly attaining 45 m in height and 1.5 m in diameter. The boles of mature trees are often heavily buttressed but usually free from branches (Orwa et al., 2009). The timber is whitish to pale straw with no difference between heartwood and sapwood, while the texture is medium to coarse with grain typically interlocked that gives a striped figure. The wood has an unpleasant smell when green but usually does not persist after drying (USDA, 2010). The heartwood is whitish to pale yellow, indistinctly demarcated from the sapwood, which is up to 15 cm thick (USDA, 2010). Leaves are 10-20 cm long and broad, palmate with 5-7 lobes, cordate and 5-7 nerved at base, lobes broadly ovate, triangular or oblong, rounded or obtusely acuminate at the apex; glabrous; stalk 3-10 cm long (Orwa et al., 2009).

4.7.3. Distribution

Wawa is widely distributed in the West and Central African forest zones from Guinea east to the Central African Republic, and south to Gabon and the Democratic Republic of Congo. It is commonly planted in its natural area of distribution (e.g., in Côte d'Ivoire, Ghana and Nigeria), and occasionally elsewhere, such as in the Solomon Islands (Bosu and Krampah, 2005).Within its natural limits, wawa is found mainly in forests at low and medium altitudes in the monsoon equatorial forest belt. Throughout its natural range, there is always a marked dry period between December and April. Wawa is referred to as a pioneer species, and it has been suggested that shifting cultivation in West Africa has influenced the natural distribution (Orwa et al., 2009). Hall and Bada (1979) note that where extensive forest disturbance results from human activity, wawa may invade areas where it was formerly rare or absent, becoming closely associated with a group of species quite different from that with which it grew originally.

4.7.4. Site Requirements

Wawa occurs up to 900 m altitude in regions with an annual rainfall of up to 3000 mm, but is most abundant at 200–400 m altitude and in areas with an annual rainfall of 1100–1800 mm and 2 rainy seasons (Bosu and Krampah, 2005). It prefers more fertile, well-drained, ferruginous soils with light or medium texture and acid to neutral pH (Bosu and Krampah, 2005). It does not tolerate water logging, and in general avoids swamps. It is a light-demanding pioneer species, and is unlikely that the species ever occurs naturally at elevations as high as 1000 m (Hall and Bada, 1979).

4.7.5. Propagation

Due to its fast growth under plantation conditions, wawa is one of the indigenous species most favoured for artificial regeneration in the forest zone of Ghana (Hall and Bada, 1979). However, problems of availability and storage of seed have in the past frustrated efforts to initiate large-scale planting schemes (Hall and Bada, 1979). Wawa is known for its irregular flowering and erratic fruiting (Mackenzie, 1959; Jones, 1974). The seeds are viable for only 2–3 weeks at room temperature, although they may be stored for up to 18 months if dried sufficiently (Howland et al., 1977). Seeds from mature fruits lose their viability after 5 years of subsequent storage at 0-5 °C (Orwa et al., 2009). Vegetative propagation techniques have been developed for the species to overcome difficulties of seed supply, and to encourage reforestation efforts (Nketiah et al., 1998). A wawa seed orchard has been established at the FORIG and the testing for the production of high quality seedlings via mist vegetative propagation techniques is in progress (FAO, 2002a).

Most fruits collected from the ground have been attacked by insects. The fruits can be collected from the trees when still green just before maturation. Seeds start to germinate 1–2 weeks after sowing, but the germination rate is often low (Bosu and Krampah, 2005). Germination rate and speed increase when the seeds are pre-treated by moistening between layers of damp cotton wool (Bosu and Krampah, 2005).The erratic seed production is a major drawback for the establishment of plantations, but methods of relatively cheap vegetative propagation are possible and will offer great opportunities when superior germplasm becomes available (Bosu and Krampah, 2005).

4.7.6. Management

In tests, wood material from plantation-grown trees was found to not be inferior to that from trees harvested in natural forests (Bosu and Krampah, 2005). The high growth rates, allowing comparatively short cutting cycles, the generally good form of the boles, and the possibility of planting in mixtures with other timber species make wawa even more promising. The species has been used both for plantation establishment and in enrichment planting in Ghana and Nigeria (FAO, 2002a). From 1967 to 1995 about 3000 ha was planted in Côte d'Ivoire, where they are grown on a cutting cycle of less than 40 years. In Nigeria, wawa is planted in agroforestry systems with cocoa (Bosu and Krampah, 2005).

4.7.7. Growth and Yield

Under natural conditions seedlings may reach 15 m tall and 15 cm in stem diameter after 4 years. Mean annual diameter increment in the forest averages 1 cm. In a 19-year-old plantation in Ghana with 600 stems/ha, the trees were on average 21.8 m tall with a bole diameter of 27 cm (Bosu and Krampah, 2005). In Ghana the total standing volume in natural forests was estimated at 39.30 m^3/ha in inventories in 2001, and an exploitable volume of 16.50 m^3/ha (Bosu and Krampah, 2005). The final harvest in plantations in Côte d'Ivoire yields 200–250 m^3/ha of timber, of which 170–200 m^3 from the bole, with an annual volume increment of 8–13 m^3/ha (Bosu and Krampah, 2005). In the moist evergreen forest zone of Ghana, the MAI was estimated to be 20m^3/ha/yr while it was about 12m^3/ha/yr in the moist semi-deciduous zone (FAO, 2002a).

4.7.8. Uses

The most important value of wawa in Ghana is for the production of lumber for both domestic use and for exports. In addition, veneers, plywood and mouldings are manufactured from the timber of wawa. The wood is also used for fibreboard and particleboard, block board, boat and ship building, boxes and crates, cabinet making, plywood, furniture components, marquetry, mouldings, bedroom suites, building materials, casks, chests, cutting surfaces, excelsior, furniture, interior construction, radio, stereo and TV cabinets (Orwa et al., 2009). The leaves are prepared as a cooked vegetable or sauce in traditional cuisine in Côte d'Ivoire and Benin, while the bark is used to cover the roof and walls of huts in southern Ghana (Bosu and Krampah, 2005).

4.7.9. Pests and Diseases

In Ghana, the moth *Anaphe venata* causes extensive defoliation, and *Trachyostus ghanaensis* (the wawa borer) weakens the wood by causing tunnels in standing trees (Orwa et al., 2009). Heartwood is not resistant to attack by termites and other insects and it is susceptible to attack by pinhole borers, long-horn beetles and sap-stain fungi (Orwa et al., 2009). Sapwood is prone to powder-post beetle attack. A die-back fungus, *Botryodiplodia theobromae*, which reduces most mechanical properties, is sometimes present in this species.

The roots are very sensitive to fungal rot. In Nigeria, the cricket *Gymnogryllus lucens*, the grasshopper *Zonocerus variegatus* and the psyllid *Diclidophlebia* sp. can cause serious damage to seedlings, and the wood borers *Eulophonotus obesus* and *Trachyostus ghanaensis* also cause damage to adult trees (Bosu and Krampah, 2005).

Table 1. Site and climatic characteristics of the major vegetation zones in Ghana

Vegetation zone	Rainfall (mm)	Mean annual temperature (°C)	Soils
Wet Evergreen (WE)	1750-2200	26-29	Highly leached and acidic. pH 4-5.5. They are low in cation exchange capacity, available phosphorus, nitrogen and organic matter
Moist Evergreen (ME)	1500-1750		
Moist semi-deciduous (MSD)	1250-1700		
Dry semi-deciduous (DSD)	1250-1500		
Guinea savannah (GS)	960-1200	26-32	Well-drained, friable, porous loams; sandy or silty loams. Slightly acidic with pH 4-6
Sudan savannah (SS)	960-1000		

Source: FAO (2002a); Nanang and Asante (2000).

Table 2. Summary of site requirements for the six selected plantation species

Species	Optimum site requirements				Uses	Suitable Ecozone
	Rainfall (mm)	Elevation (m)	Temp (°C)	Soil		
Teak	Optimum 1200-1600 Tolerates: 500-5000	0-1300 m	13-40	Well-drained and fertile soils Neutral to acidic pH of 6.5-7.5	-Timber production - Electric and telephone transmission poles -Fuelwood	All vegetation zones. However, productivity will be lower in the GS and SS due to reduced rainfall
Neem	Optimum: 450 – 750 Tolerates: 450-2000	50-1500	0 – 44	Well drained soils, neutral to alkaline pH<6	-Building construction -Fuelwood -Windbreaks -Pest control -Soil improvement –Medicinal uses	All vegetation zones, but better suited to the GS
Gmelina	Optimum: 750 – 4000 Tolerates: 6-7 months of dry season	0-1200	21-28	Deep, well-drained soils, pH of 5-8	-Timber production –Fuelwood –Pulp	All vegetation zones
Cedrela	Optimum: 1200-2400 Tolerates: 1000 – 3000	0-1900	22-26	Fertile, free-draining soils that are weakly acidic	-Timber – Construction -Shade -Windbreaks –Ornamental tree	All vegetation zones, but less suited to the GSZ due to rainfall limitations
Ceiba	Optimum: 1500 Tolerates: 750-3000	0 – 900	18 – 38	Deep permeable volcanic sandy to loamy soils that are free from water logging. Slightly acidic to neutral	-ornamental – timber – plywood, -veneer – fodder – fibre -medicines	All vegetation zones, but less suited to the evergreen and moist-semi-deciduous zones
Wawa	Optimum: 1000 -3000 Tolerates: 4 -6 months of dry season	0 – 900	24 – 27	Fertile, well-drained, ferruginous soils with light or medium texture and acid to neutral pH	-furniture components – timber – plywood -boxes and crates – particle and fibre board – artificial limbs	All vegetation zones, but less suited to the savannah zones due to limited rainfall

4.8. Matching Species to Vegetation Zones

Based on the critical silvics of the six tree species, it is possible to make some general recommendations regarding the suitability of these species for use in plantation forestry across the different vegetation zones in Ghana.

Table 1 summarises the site and climatic characteristics of the major vegetations zones in Ghana. Based on these characteristics, Table 2 then summarises the site and climatic requirements of the six plantation species discussed in the previous sections. All six species can be grown in all parts of Ghana, although as noted in Table 2, the species should be grown in areas where their productivity can be optimised, especially if large-scale commercial plantations are envisaged. For example, gmelina has the potential to grow well in northern Ghana based on its site requirements, although the species has not been widely promoted. Cedrela on the other hand will not do very well in the Sudan savannah zone, due to the limited rainfall of about 1000 mm/year.

4.9. Conclusions

Successful plantation establishment and management depend on several factors, most of which are decided by the manager. However, information on the critical silvics is also important for matching species with their most productive site. The critical silvics information can also be used to produce maps of the best ecological zones for the target species. The information on the six species in this chapter contributes to that effort.

Chapter 5

NURSERY, PLANTING AND TENDING OPERATIONS

This chapter provides an overview of plantation forestry practices that include seed collection and storage, germination and the types of seedling stocks raised in nurseries. It also describes the methods of site preparation, planting and tending operations (weed control, pruning, and thinning). Stand density management is one important tool used to ensure that the optimum potential of forest sites is achieved. The different types of stand density indexes are described and a case study that develops a stand density management for teak is presented.

5.1. SEED COLLECTION

5.1.1. Methods of Seed Collection

Most plantations are regenerated from seed, and hence it is important that seeds that are used for plantation establishment are chosen carefully. Ideally, seeds should be collected from parent trees of superior phenotypes that show vigorous growth and high yields, physical attributes suited to the intended purpose of the plantation, adaptable to the site to be planted, and resistant to local pests and diseases (Nyland, 2007). Understanding the reproductive biology and ecology of the tree species is central to proper seed collection, handling, germination and subsequent nursery practices (Evans and Turnbull, 2004). The two methods used to collect seeds are: from the ground and from standing trees. The factors that determine the seed collection method include the location of the seed source, the size of the trees, and the type of seed or fruits and their maturity (Evans and Turnbull, 2004).

Species which produce large heavy fruits or seeds that usually fall to the ground can be collected by spreading polythene sheets. This method is used frequently for collecting the seeds of teak, gmelina, and wawa (Evans and Turnbull, 2004). Seed collection must be timed according to the characteristics of the species and can be combined with other forestry operations in order to ensure efficiency and minimise costs of the operations. Collecting seeds from standing trees can either be accomplished by using long-handled tools, manual shaking of the trees, or climbing the trees to collect the seeds (Evans and Turnbull, 2004). This method has been used for collecting the seeds of various species of pines and eucalyptus.

Before commencing seed collection from standing trees, it is critical to verify that the seed has reached maturity.

To maximise seed viability and genetic quality, seed should be collected from a stand of trees rather than isolated individuals; single trees tend to self-pollinate and produce seed that is less viable and genetically inferior (Rudolf, 1974). The importance of using the best seeds and the need to produce seed in large quantities to satisfy plantation development programmes suggest that seed selection for plantations be carried out by those with the requisite expertise. The traditional practice in Ghana has been for forestry technical officers and forest guards to collect seeds from parent trees for propagation in tree nurseries. This practice of not confining seed collection to superior phenotypes could result in poor quality seeds being collected, problems with seed germination and low quality trees in the plantations.

5.1.2. Seed Storage

The timing of seed collection and effective storage methods are critical components of achieving planting objectives. Some seeds lose their viability soon after falling off the tree and hence seed collection should rely on a good knowledge and understanding of the ecology of the tree species. For tree species that lose their viability quickly, the seeds should be collected before the fruits fall off the trees. Neem seeds, for example, are short lived and do not retain their viability for long periods (Troup, 1921), with a two week upper limit. The seeds begin to germinate as soon as they fall from the trees (Evans, 1992). Some species such as wawa (*Triplochiton scleroxylon*) fruit irregularly and their seeds may need to be collected and stored in only seed years.

Seed storage can prolong the viability of seeds up to ten times longer under controlled than in field conditions (Nyland, 1996). Unfortunately, the only controlled conditions for storing tree seed in Ghana at the FORIG can only contain 20kg of seed (FAO, 2002a). In addition, storage facilities need to incorporate the ability to monitor the seeds to ensure that they do not deteriorate over time. In order to meet demand for seeds at a national level, the FORIG has established seed orchards of teak, wawa, cedrela and gmelina. Two private companies, Dupaul Wood Treatment and Subri Industrial Plantation Ltd. (SIPL) also have seed orchards of teak and gmelina, respectively (FAO, 2002a).

The FORIG has a Tree Improvement and Seed Technology Laboratory, but this is ill-equipped and there are no control environments for the proper testing of seeds. FAO (2002a) reports that the estimated average annual production of seeds from FORIG is about 1000kg of teak, 20kg of cedrela, and 20-30kg of indigenous species. It is obvious that this level of seed production is woefully inadequate to meet any large-scale plantation development needs in Ghana. To successfully implement a national plantation programme, it is imperative that more research be undertaken to develop improved seeds, and increase storage and testing facilities. In addition, tree species that will do well in many parts of Ghana are able to produce the desired wood products, and are resistant to local pests and diseases should be developed.

5.1.3. Quantity of Seed Required for Plantations

Large-scale national level plantation development programmes will require large quantities of genetically superior seeds to be successful. Given the long process involved in a plantation establishment, it means that adequate seeds have to be procured at least one year ahead of the season in which planting is expected to occur. Forestry authorities therefore need to be aware of the seed requirements and whether they can be obtained locally, and if not, make arrangements to import them.

Evans and Turnbull (2004) provide a formula for computing the weight of seeds required for plantation programmes. This formula is given in Equation (1) as:

Weight of seed needed in kg =

$$W = \frac{H \ x \ N_s}{S \ x \ PS} \tag{1}$$

where:

H = the annual planting target in ha

N_s = intended number of seedlings to be planted per ha

S = estimated survival rate as a decimal

PS = estimated number of plantable seedlings per kg of seed

While data on H, N_s and S are usually available at the local level, the number of plantable seedlings (PS) is more difficult to obtain, and can be estimated from Equation (2) as:

$$PS = N \ x \ G \ x \ P \ x \ (1 - M) \tag{2}$$

where:

N = number of seeds per kg

G = germination percentage as a decimal

P = purity percentage, as a decimal (the proportion of seed which is not other matter)

M = estimate of seedling mortality during nursery, as a decimal

The planting target under the NFPDP in Ghana was set at 20,000 ha per year. Given that most of the planting is teak, we can estimate the quantity of teak seeds required per year to plant that area. The following assumptions are made: the number of seedlings per kg (N) is an average of 1500 (Evans and Turnbull, 2004 provide a range of 1000-1900 seeds/kg for teak). Given that most of the plantations are planted out at 2 x 2 m initial spacing, N_s is 2500. Survival (S) is estimated at 0.80, P is set at 0.90; M is 0.20, and G is set at 0.70. Based on these assumptions, the total quantity of seed required to support a national plantation programme to meet the annual planting target is 82,672 kg per year. The quantity of seeds actually required for the various species used in the plantation programme can be calculated using this method once the exact areas to be planted to each species are known.

5.2. Seedling Production

Producing the growing stock for forest plantations involves several stages including preparing the nursery site, sowing the seeds, tending, lifting, transplanting, and packaging for storage or transport. The exact requirements will depend on the stock type being produced (container, bare-root or stumps), but in each case special care must be taken at each stage to produce good quality stock that has high chances of survival and growth rates after outplanting. Nursery practices should be geared towards producing the best quality and quantity of seedlings to meet demand.

5.2.1. Pre-Sowing Treatment

The seeds of some tree species germinate slowly or incompletely because they are completely or partially dormant (Evans and Turnbull, 2004). These types of seeds may require pre-sowing treatment to overcome such dormancy problems and improve germination results. One of the causes of seed germination failure is that the seed may need to be "after-ripened" before they can germinate. *Stratification* is the process of mixing the seed in a moist medium (such as sand or saw-dust) and placed in a storage container and stored in a controlled temperature to help "ripen" the seed and break dormancy. Other methods of breaking dormancy, is by soaking or treatments with laser beams. For example, in an experiment in Tamale, pre-sowing treatment significantly improved germination performance of dawadawa (*Parkia biglobosa*) seeds. Seeds soaked in cold water for 12 hours had the highest germination (91.3%), compared with untreated seeds, which had 80.0% germination rate (Korankye-Gyamera, 1997). For neem seeds, no pre-treatment is usually required, although pre-soaking the seeds in water for 24 hours and removing the endocarp or cutting the seed coat at the round end with a sharp knife also increases the germination capacity. Seeds that have hard shells may have to be scarified by soaking in water, acid or by mechanically means before stratification. *Scarification* is an artificial way to prepare seeds with hard coats for germination. Other reasons for treatment are to protect seeds from destruction by fungi, insects and rodents or preserve their viability.

A key consideration in nursery practices is to decide what type of seedlings to produce. When seedlings are raised in the nursery, they are usually of four main types: container stock, bare-root stock, striplings and stumps, and vegetative propagation. Sometimes unrooted cuttings are also produced.

5.2.2. Bare-Root Stock

The bare-root method is the oldest seedling production method and was used before the advent of motorised vehicles mainly to reduce the cost of transporting seedlings. It has been the traditional way of raising planting stock in the Sahel-Sudan zone of Africa though in dry years survival is often poor (Evans and Turnbull, 2004). It is usual when producing bare-root stock, to first broadcast the seeds on a germination bed before lifting them out unto transplant beds about a week or more after germination. The seedlings are allowed to grow to the desired size and age on the germination bed, where they are then lifted up (bare of any soil on the roots) to the field for transplanting. Under the RAP in the 1990s, nursing of seedlings usually begun in the dry season (December/January) for outplanting in the next rainy season (May-September). The most popular method of planting neem in northern Ghana is as bare-root stock (Nanang, 1996). Other species that are suited for planting as bare-root stock are cedrela and mahogany species such as *Swientenia macrophylla*. The main advantages of bare-root stock are (after Nyland, 1996): lower costs of production and transportation; easy adaptability for size and condition; and easy handling during field operations. Usually, bare-root stock takes longer to grow compared to container stock.

5.2.3. Container Stock

The usual practice in Ghana when producing container stock is to first broadcast the seeds on a germination bed and lift them into polythene bags about a week after germination (Nanang, 1996). The seedlings grow to the desired size in the polythene or biodegradable container before being transported to the planting site. The most common container type is polythene tubes of various sizes. If polythene tubes are used these are removed just before the seedling is put into the soil during transplanting. However, the use of biodegradable containers (such as paper tubes) would allow the seedling and container to be put into the planting hole, thereby speeding up the planting operation. In the RAP nursery in Tamale, container seedlings were produced using one litre polythene bags filled with 3/6 soil, 2/6 rice chaff and 1/6 composted cow dung. When raising container seedlings, care has to be taken to avoid poor drainage and over-watering as this often results in die-back of the seedlings. This was a major problem observed by the author in some nurseries in northern Ghana.

Figure 1. Tree nursery near Bontanga in the Northern Region. Foreground: container seedlings of neem. Background: bare-root seedlings of teak. In the picture are Mr. James Amaligo (left), Prof. Robert Day (second from left) and some nursery labourers.

Container grown seedlings have the following advantages: the containers protect the roots from injury and desiccation until planted, the rooting medium around the seedling provides a microsite that favours root-soil contact, easy to plant without twisting roots, and more adaptable to mechanised handling (Nyland, 1996). These advantages increase the survival rates of container seedlings over those of bare-root stock. For example, in an experiment with dawadawa in Tamale, survival of seedlings after three months of outplanting was found to be highest in container seedlings (83.0%) than bare-root (40.0%). Only 20% of stumped seedlings survived after the same period of time (Korankye-Gyamera, 1997). Despite these advantages, container stock need more logistical support to produce and transport and hence is more costly than bare-root stock.

5.2.4. Striplings and Stump Stocks

Most seedlings of tree species in the family *Miliaceae*, such as cedrela and Khaya species are usually stripped off most, if not all, of their leaves before planting. The main objective of using striplings is to reduce the amount of transpiration stress following out planting of the seedlings. This practice is used in places where animal damage is common, there is weed competition and stem borers are a problem (Evans and Turnbull, 2004). This type of seedling stock is recommended for transplanting in semi-arid conditions where the seedlings are likely to be subject to desiccating weather conditions as pertains in the savannah vegetation zones of Ghana.

Figure 2. Teak seedlings in a nursery in Tamale.

The term stump is used to describe a type of cutting applied to a seedling, which is usually 15-25 cm long with about 80% root and 20% shoot with the shoot striped of leaves and the lateral roots trimmed off (Evans and Turnbull, 2004). This method is particularly suited for tropical species that are capable of producing new roots and shoots when in contact with moist soil. Stump planting is especially suitable for taproot dominated species and is frequently used when establishing teak, gmelina and a number of other important tropical genera (e.g. *Afzelia, Cassia, Chlorophora, Khaya, Lovoa, Pterooarpus, Terminalia, Triploohiton* and many Leguminosae) (Chapman and Allan, 1978). Plantation grown teak is established using stump plants rather than direct sowing of teak seeds, which has been found to give less satisfactory results (Borota, 1991). Neem seedlings can also be planted as stumps by trimming the tap and lateral roots, pruning the stem to leave 20-30 cm of the stem above the root collar, and any side branches and leaves are cut off as close as possible to the stem, without damage to the stem. This leaves a sufficient number of buds for sprouting. Leaving a longer stem may enable the wind to whip and loosen the plants. In India as little as 2.5 cm of the stem is left with apparently good success (FAO, 1995).

5.2.5. Vegetative Propagation

Vegetative propagation is an asexual reproduction in which a vegetative part of the original plant is used in the reproduction process, rather than the seed. This is slightly different from the three methods of raising planting stock discussed above as it does not necessarily use seedlings in a nursery. The three main types of vegetative propagation used in forestry are grafting, air-layering and cuttings. FAO (1993) describes the three main types as follows. Propagation by cuttings is the most convenient and cheapest method and usually preferred when possible. Air-layering is a variation of propagation by stem cuttings in which root formation is initiated before the plant part is separated from the mother tree. In grafting, the shoot (scion) of the desired tree is joined with a root (stock or root stock) of different genetic origin. The use of unrooted cuttings is currently the subject of research in Ghana by FORIG, which is researching into the vegetative propagation of *Triplochiton scleroxylon, Milicia excelsa, Khaya* spp., *Ceiba pentandra, Terminalia superba, Baphia* spp. and *Ficus* spp (FAO, 2002a).

5.3. Site Preparation

Before planting, it is essential that the site be prepared to a high standard. Site preparation refers to preparing the land before regeneration (planting trees in the case of plantations). It is any measure that makes the physical environment suitable for germination, and later survival and growth of seedlings (Nyland, 1996). Site preparation may be needed to reduce or redistribute slash, reduce competition from existing vegetation, prepare the soil, and facilitate and reduce the cost of regeneration and future management operations (Nyland, 2007; Day, 1996). Benefits such as ease of planting, increased tree survival, better weed control and improved growth rates may far outweigh the time and effort involved in site preparation. The following factors affect the type of site preparation that may be required: the amount, size and species distribution of existing vegetation; the method of harvesting existing forests; and the nature of the stand desired in the next rotation (Day, 1996).

In situations where no active measures are taken to prepare the site (known as passive site preparation) foresters rely on conditions created by previous forest management practices such as logging to achieve sufficient regeneration. In many cases however, active site preparation is required when existing conditions favour less-desirable species, and the vegetation or soil conditions might limit the regeneration of the desired species (Nyland, 2007). Despite the advantages of site preparation, it should be noted that site degradation and loss of nutrients and organic matter could result from the practice (Nyland, 1996). Loss of nutrients varies with the nature of site preparation and management practices and can have a dramatic effect on the growth of the next stand. Nutrients lost by removal or leaching can be replaced by fertilisation, but best results are usually obtained when fertilisation is combined with other practices such as weed control or site preparation (Tiarks et al., 1998). There are generally three methods of site preparation used by foresters: mechanical, chemical applications, and the use of prescribed burning.

5.3.1. Mechanical Site Preparation

Mechanical site preparation practices affect the soil surface layers, either to modify the seedbed conditions or alter the physical attributes of the rooting zone (Nyland, 1996). This generally includes scarification which loosens the upper soil or breaks up the organic layer; removes undecomposed litter and humus to expose mineral soil; or mixes surface organic materials with mineral layers beneath them (Ford-Robertson, 1971). There is the potential to increase soil compaction with mechanical equipment, erosion, or siltation to nearby waters (Williams, 1989). Furthermore, removing and burning slash and other organic materials prior to planting may cause some nutrients to leach, and reduce the amount available for tree growth. With the exception of large-scale industrial plantations, small-holder and community plantations may not be able to afford the cost of mechanical site preparation in Ghana.

5.3.2. Chemical Treatment

Chemical site preparation involves the use of chemical applications, mainly herbicides, to kill competing vegetation. Herbicide use in forestry is limited to not more than a couple of applications per rotation. Herbicide use is regulated by health and environmental authorities, and hence only authorised chemicals can be applied. Herbicides are used to kill dense and unwanted vegetation, control harmful insects and animals, alter the habitats for insects and animals, prevent the development of understory vegetation, inhibit germination of seeds for weeds, and supplement available nutrients by fertilisation (Lowery and Gjerstad, 1991). The general technique for applying herbicides is by aerial spraying, but injection into bark of woody plants (as was done in the application of the tropical shelterwood system in Ghana), or incorporating it into the soil have also been used. The main advantages of herbicides is that they control a broad range of weeds and prevent sprouting from stumps and root systems, while not interfering with the productive capacity of soils, and are generally cost effective (Day, 1996).

5.3.3. Prescribed Burning

Prescribed burning in forestry refers to any burning that is carried out intentionally for a specified purpose. Prescribed burning is used to prepare seed beds for planting, reduce fuel loads, control natural succession, improve grazing, manage wildlife, and improve sanitation on a site. In plantation establishment, bare-root stock or container stock are usually used and unimpeded access to the site is often required. Furthermore, enough of the surface organic matter has to be removed to enable the planters to reach the mineral soil with their tools. In most cases, burning can achieve this cheaper than mechanical treatment (Day, 1996). Moreover, in Ghana, fire has been used for generations to prepare land for agricultural purposes, and hence its use is already widespread and could easily be applied to plantation site preparation. The main risk associated with using prescribed fire is the safety of workers and the possibility that the fire might get out of control and burn areas that were not intended.

5.3.4. Examples of Site Preparation in Ghana

In Ghana, site preparation can take several forms depending on the scale of the plantation and the nature of the land to be prepared. For small-scale farmers, site preparation would involve clearing the land with simple farm implements such as hoes, cutlasses and ploughing manually or with animal traction or the use of fire. Industrial plantation investors will more likely rely on mechanical or chemical methods to prepare the sites. For example, sites of the plantations belonging to Bonsu Vonberg Farms Ltd. were site prepared by stumping and ploughing. It was observed that the plantations on such treated sites performed better (by as much as one and a half times) than those that did not receive similar treatment at the beginning of the project (FAO, 2002a). A new technique of site preparation in Ghana is the Subri conversion technique whereby degraded natural forests are cleared, and lines are cut through the debris and planted up without any burning. This conserves the humus layer and the extra debris on the forest floor (Nwoboshi, 1994). In Northern Ghana, plantations are commonly established on either farmlands or abandoned lands which are out of cultivation mostly as part of the cycle of shifting cultivation. In either case, lands are prepared in March/April prior to the rains. Land preparation is often done by the farmer with his family or the community in 'communal labour' programmes in the case of community plantations.

5.4. OUTPLANTING

The planting exercise should be planned to give seedlings the best chance of survival. Seedling quality directly affects survival of the trees. Experience has shown that seedlings and cuttings of poor quality would not survive and grow well even under the best site conditions. Also, the initial spacing and stocking of the plantation is determined at this stage, taking into consideration the management regime envisaged for the plantation enterprise. The tree planting exercise should be well thought out ahead of time. Planting should be carried out after a rain when the soil is moist. It is advisable to soak seedlings in their pots prior to planting out. Several factors to be considered include species selection, stock type, spacing, arrangement of trees, site preparation, seedling handling, and other operational considerations. The techniques for planting differ between container and bare-root stocks. Bare-root seedlings are planted by digging a small (20-25 cm deep) planting pit or open up a slit with a hoe or spade and insert seedlings up to the root collar level with the ground, using one hand (Evans and Turnbull, 2004). Soil is carefully filled back with the other hand and pressed with the foot around the seedling (Evans and Turnbull, 2004). For container seedlings, the planting pit should be slightly bigger than the container. Planting is carried out by removing the container and positioning the seedling in the pit with the root collar level with ground; the soil is filled back and firmed with the foot around the seedling (Evans and Turnbull, 2004).

5.4.1. Stock Type

Foresters have a choice between direct seeding, transplanting seedlings, and vegetative reproduction when regenerating forests. The method chosen depends on which one gives the best results for the chosen species and site. The stock type also has implications for when the trees can be planted out and the cost of the operation. For example, direct seeding does not require nursing the seedlings in a nursery, and hence can be carried out more readily and cheaply compared to raised seedlings. Direct seeding means artificially spreading seed over an area where a landowner wishes to regenerate a new forest stand artificially (Ford-Robertson, 1971). The most important factors affecting the success of direct seeding are the rate of seed applied, the degree of site preparation, and the weather. The seeding methods include: broadcast sowing by hand or aerial sowing; spot sowing with or without covering; drilling by hand or with small mechanical seeders (Day, 1996). Direct seeding is unsuitable in droughty conditions or on seedbeds that dry up quickly, on erodible surfaces where seed can easily be washed off, on soils that become inundated during the rainy season, and on sites where seed-eating organisms can easily consume the seed (Nyland, 2007). Direct seeding by spot sowing is used by farmers in Ghana to plant most fruit trees on farms and home gardens. However, for large areas, this method would be labour intensive and costly, except where the main objective is to create jobs.

5.4.2. Planting Season

Outplanting of seedlings must take into consideration the season of the year and the rainfall pattern in the location. In northern Ghana, there is a single rainfall season, followed by a long dry season, when there is no rain and dry desiccating conditions that are exacerbated by rampant bushfires. In southern Ghana, there are two rainfall seasons, a major and a minor one. The temperatures are moderated by the vegetation cover and the impact of the dry winds is less severe on the soil. The season of planting would therefore have important ramifications for survival.

Outplanting of seedlings in the savannah zones is done exclusively during the rainy season [from May to end of August]. The beginning of the planting season each year however depends on when the rains start and therefore varies from year to year. In most years, planting starts before the end of May. Sometimes planting is carried on into September but this practice is not recommended since the newly transplanted seedlings do not receive enough rainwater to establish themselves well to survive the long, dry season starting from October to March. Late planting (after September) should be avoided whenever possible. There is more flexibility in when to plant in southern Ghana, but the considerations are identical.

5.4.3. Initial Spacing

One of the critical decisions in plantation establishment is the initial spacing at which the trees will be planted. The ability to determine initial spacing and hence the stocking levels of the plantations is a major advantage of plantations over natural forests. Initial spacing is a function of the chosen species' growth rate and form, its habitat and shade tolerance, site fertility, planned time to first treatment, the planned silviculture, intended products from the plantation, and costs considerations. Other considerations include the need to plant trees in a way that facilitates subsequent operations and enables general access to different parts of the plantation. Initial spacing also affects the costs of seedlings, area to be planted and labour and machinery costs of outplanting seedlings. Costs for plants and labour tend to increase with decreasing planting distances, but on the other hand, costs of weeding tend to increase with wider spacing (Chapman and Allan, 1978). Closer-spaced plantations produce biomass in less time but wider-spaced plantations can produce larger trees in less time. Therefore, in general, if the intended purpose of the plantation is to produce fuelwood within a short time, closer spacing should be used, while wider spacing should be used if bigger diameter trees are required for sawlogs or electric transmission poles.

The growth rate of the species planted should influence the initial spacing. In general, slower growing species tend to be planted at closer spacing than faster growing species, and for this reason spacing in the tropics tend to be greater than in temperate regions (Chapman and Allan, 1978). Furthermore, species that branch heavily need to be planted closely to promote the formation of a well-defined leading stem. Most tropical species are self-pruning (e.g., *Terminalia superba*) and can therefore be planted more widely apart than temperate trees (Chapman and Allan, 1978). If mechanised management operations such as weeding are intended, then the spacing at which the trees are planted should take this into consideration. A distance of 2.8 m between rows is considered a minimum spacing where weeding is mechanised (Chapman and Allan, 1978). Wider spacing is used on poorer, less fertile, and arid soils to allow more room for trees to develop roots to access nutrients.

Table 1. Plantation spacing and stocking levels in English and metric systems

English system		Metric system	
Feet	Trees per acre	Metres	Trees per hectare
2 x 2	10890	0.3 x 0.3	111,111
3 x 3	4840	0.5 x 0.5	20,000
4 x 4	2723	1 x 1	10,000
5 x 5	1742	2 x 2	2,500
6 x 6	1210	3 x 3	1,111
7 x 7	888	4 x 4	625
8 x 8	680	5 x 5	400

The initial spacing determines the stand density (number of trees per ha) of the plantation. Closer initial spacing promotes slower diameter growth, compared to wider spacing. Table 1 gives examples of various spacing in English and metric systems and their corresponding stocking levels in acres and hectares. In the metric system, stocking levels of trees/ha are determined as 10,000/ (spacing in m)^2, while in the English system, trees/acre is calculated as 43,560/(spacing in ft)^2.

For many decades, the initial spacing for plantation grown trees in Ghana has been 2 x 2 m apart, giving 2500 trees/ha. Most species are planted out as pure stands of exotic species such as teak, neem, or lueceana. Some farmers grow food crops such as maize and millet until canopy closure. Shade tolerant food crops [e.g., pepper] are also grown under older trees on farmlands. Since almost all plantations are established in close proximity to villages, intensive silviculture is not only feasible, but is beneficial to the communities. For the purposes of producing fuelwood and poles, the initial spacing of 2 x 2 m is considered wide and can be reduced to 1.75 x 1.75 m to give an initial planting density of 3265 stems/ha (Nanang, 1996). A higher density than this will lead to increased plantation costs and also likely produce inferior poles and rafters since there will be a proliferation of many small stems and branches rather than increases in diameter of individual trees. The adoption of closer initial spacing and thinning regimes will circumvent the current problem of pole and rafter wood removal by villagers which interferes with the development of fully stocked plantations (Nanang, 1996).

5.5. Weed Control

5.5.1. The Concept of Competition in Forestry

Successful plantation establishment requires that planted trees are protected from undesirable competing vegetation. Competing vegetation is defined as unwanted or undesirable vegetation which suppresses or inhibits the growth and survival of desirable tree crops (Coates and Haeussler, 1986). One objective of silviculture, both from biological and economic viewpoints is to restrict the composition of forest stands to those species that are best adapted to a particular site (Smith et al., 1997). Notably, sites that are most favourable to tree growth tend to be also favourable for competing vegetation, especially in the early successional stages (Day, 1996). This vegetation often competes with tree crops for growth resources such as water, light, and nutrients and hence directly interferes with growth of the desired species. Undesirable vegetation can also negatively interfere with crop trees by directly competing with seedlings or outplants, causing physical injury to seedlings or outplants, smothering seedlings or outplants, producing or exuding toxins that limit the growth of seedlings or outplants, increasing the fire potential of the site, and providing a favourable habitat for biological pests which have the potential to damage or kill seedlings or trees (Day, 1996; Bell, 1991; Ross and Walsted, 1986; Sutton, 1985).

On some sites, the tree crop will eventually grow through the weeds, dominate the site and become established; and on such sites the main function of weeding is to increase crop uniformity and speed up the process of establishment, while on other sites, the type or density of the weed growth is such that in the early stage of a plantation it will suppress and kill some or all of the planted trees, and in such areas the main purpose of weeding is to reduce mortality and maintain an adequate stocking of trees to establishment (Chapman and Allan, 1978). Weed control provides the most benefits when it is done before competing vegetation has had an opportunity to impede tree growth, i.e., before crop trees display visible signs of suppression, as the response of trees to weed control is rather slow once they have already been suppressed. Weed control is not always necessary. Competing vegetation does not always have a level of impact on crop tree survival or growth that warrants an investment in weed control.

There are three opportunities for vegetation control: during site preparation; release or cleaning operations after the trees have established; and during pre-commercial thinning. Competing vegetation have characteristics such as the ability to produce abundantly and at frequent intervals, shade tolerance, grow faster than tree crops, and have the ability to survive various kinds of natural disturbances (Nyland, 1996). Competition can occur between plants of the same species or of different species, and also occur above or below ground. Intraspecific competition is the negative interaction that occurs between plants of the same species, while interspecific competition occurs between plants of different species (Radosevich and Osteryoung, 1987).

The timing and frequency of weeding depends on the site preparation that was carried out as well as the tree species, initial and subsequent spacing, the nature of the competing vegetation, and climate. An effective site preparation controls the existing competing vegetation and ensures that the trees have a good start ahead of weeds. Therefore, weed control could be delayed until later. Without site preparation, weeding may need to be undertaken immediately to avoid suppression of the trees by weeds. Sites that have previously been forested tend to be invaded by pioneer species and have more woody regrowth than sites that did not previously contain a forest. Also, some tree species are more competitive, and tolerate competition better than others. Moreover, some species achieve crown closure faster than others, and are able to suppress weeds much sooner. For example, for successful eucalyptus establishment, clean weeding is required for good early growth (Evans and Turnbull, 2004). The timing and frequency of weeding also depends on the amount of rainfall in the area. Weeds, just like other plants, require adequate rainfall, and hence in years of high rainfall, more frequent weeding may be required compared to drier years.

In northern Ghana, most small-scale farmers only weed their plantations if they are intercropped with food crops. However, weeding is used more as a fire management tool. Most farmers and communities make fire belts around their plantations to exclude fire from the trees using hoes and cutlasses to clear brush in the form of belts one to two metres wide. This is often done at the end of the rainy season (October). Despite these preventive measures, some plantations especially those established on unfarmed lands are still burnt by annual bushfires during the dry season.

5.5.2. Weed Competition in Ghana's Forested Ecosystems

Weed invasion into forest and savannah ecosystems presents a serious threat to successful plantation establishment. However, information on the diversity, modes of introduction, spread and impacts of invasive weeds in Ghana is limited (Anning and Yeboah-Gyan, 2006). Anning and Yeboah-Gyan (2006) examined the diversity and distribution of invasive weed species in the humid forest environment of Ashanti Region, Ghana and found a total of 43 species belonging to 19 families, 41 genera and six life/growth forms in cultivated, degraded, aquatic, ruderal and forested ecosystems. The dominant invasive weeds were *Chromolaena odorata* (L.) (12.71%), *Centrosema pubescens* Benth. (10.42%) and *Rottboellia cochinchinensis* (Lour.) Clayton (6.39%).

Generally after forest clearance, tropical forests are dominated by pioneer species which are light demanding with fast growth rates. For example, in Ghana, forest landscapes are invaded by *Chromolaena odorata* Linn (Acheampong), a highly competitive weed species noted as the primary cause for poor natural regeneration (Honu and Dang, 2000). *Chromolaena odorata* forms a dense canopy, which induces severe competition for below ground growth resources such as water and nutrients. The grass cover also suppresses and prevents tree seedlings from growing through the canopy (Honu and Dang, 2000; 2002). Competition from *C. odorata* may account for 64% mortality in tropical tree species (Honu and Dang, 2000). Thus, seedling mortality is high when shade-intolerant pioneers regenerate under *C. odorata* canopy (Richards, 1996).

In many sites across Ghana, grass competition is also intense from such species as *Imperata cylindrica* (speargrass) which invades after forest and woodland clearing and must be controlled in order to re-establish tree species. Speargrass was once ranked as the world's seventh worst weed (Holm et al., 1977). The persistent and aggressive rhizomes of speargrass are the main mechanism of spread, and, coupled with their resilience, speargrass is very difficult to control (Bolfrey-Arku et al., 2006). Managing competition from competitive vegetation may be required to increase natural regeneration success or before and after establishing forest plantations. For example, removal of 50% of *C. odorata* cover reduced competitive effects, increased seedling height growth and number of leaves (photosynthetic area will enhance CO_2 capture) three fold in released relative to control plots in Ghana (Honu and Dang, 2000).

5.5.3. Methods of Weed Control

Even after adequate site preparation, undesirable vegetation will still likely compete with desirable crop trees, and there may be a need to control this vegetation in order to optimise the productivity of the planted trees. Several kinds of grasses and shrubs that are abundant in both plantations and natural forest ecosystems will tend to pose threats to trees. In general, four methods of weed control are used in forestry: manual, mechanical, chemical and biological controls. Any of these methods could be applied as spot treatments around individual seedlings or in the case of chemical treatments, can be broadcast across the entire stand of trees (Nyland, 2007).

***A. Manual weed control*:** This refers to cutting back or removing weeds using manual implements such as hoes and cutlasses. This is the most common weed control method, and is identical to weed control for agricultural crops in Ghana. It is suitable for small-holder plantations, and agroforestry settings. Labour is usually supplied by the farmer or his household and relatives. Occasionally, hired labour may be used such as in the modified taungya system where the government provides incentives and pays for labour to undertake weeding of planted trees. In most cases, cutlasses are used to cut off the competing vegetation, while hoeing, which is a more effective method, is used to remove the weeds and also helps aerate the soil. In addition to increased soil aeration, hoeing is especially important in the savannah areas that have a long dry season, where it can increase rainfall percolation and reduce evaporation from the soil. Manual weeding is cheap, but more labour intensive and ineffective where the weeds are cut off, as the weeds may regrow following the weeding.

B. Mechanical control: Mechanical weed control involves the use of machines to remove weeds and cultivate the ground between trees by harrowing, rotovating and shallow ploughing (Evans and Turnbull, 2004). Since machines are used, plantations must have been planted at a spacing that makes this feasible. Machines are more effective than manual weeding, but are also more expensive and beyond the reach of the small-scale farmer. At the BVFL plantations in Somanya in the Central Region of Ghana, mechanical weeding is carried out during the first two years before the dry season using a tractor and slasher attached to it. However, this type of weeding does not usually affect the weed growth at the base of the trees, which are the crucial ones with respect to competition and hence such weeds are generally removed by manually. Mechanical weed control with large machinery carries risks associated with harm to wildlife, potential soil compaction, increased erosion and also excessive burning of fossil fuels.

***C. Chemical control*:** Chemical weed control involves the use of herbicides to reduce, eliminate or suppress competition. Chemicals can be applied through aerial spraying, or through on-ground treatments involving vehicle-mounted equipment, backpack sprayers, or other hand applications tools. Weed control by chemical methods is more effective than other weed control techniques because the chemicals kill the weeds (Evans and Turnbull, 2004). The use of chemicals to control weeds is widespread in agriculture in Ghana, while its use in forestry is still limited. For example, the use of *glyphosate* to control water hyacinth in areas with heavy infestations is accepted and approved in Ghana (Madsen and Streibig, 2003). Glyphosate is used around the world in forestry in countries such as Canada. It is a non-selective, systemic herbicide, that translocates (or moves throughout) plants very effectively once it penetrates the waxy cuticle of plant leaves or stems. As such, it is particularly useful for control of weedy plant species that re-sprout from roots, rhizomes or cut stumps and it exhibits a high degree of effectiveness on most of the key competitive species in Canadian forest regeneration sites (Thompson, 2009). Herbicide use includes the risk that weeds could become resistant to the active ingredients used. It is good practice to use spot application and apply herbicides to the as minimum areas as are required to achieve weed control.

The risks of herbicide use are generally associated with the potential for direct or indirect effects on wildlife species or to humans that may be inadvertently exposed to herbicide residues. However, such risks are significantly mitigated by the extensive scientific research that enhances and defines biological effects thresholds for herbicides (Thompson, 2009). Furthermore, in Canada, operational practices have put into place to reduce the probability that actual exposures will exceed such thresholds (e.g., buffer zones, signage, use of minimum effective rates, advanced application technologies to optimise targeting and reduce drift potential, etc.) (Thompson, 2009). Economic evaluations have shown that the returns to small-scale farmers could be considerably increased by the use of *glyphosate* for weed control in agriculture in Ghana (Darkwa et al. 2001), although there is no similar evaluation for forestry use yet.

D. Biological control: Biological control involves the use of biological agents such as the natural enemies of the weeds or cover crops to control weeds. In Ghana using biological agents to control weeds in plantations of cedrela is an option that may help manage competitive effects of the species (Timbilla and Braimah, 2000), but this approach must be implemented with caution. The oil palm and para rubber industries in Ghana and elsewhere have used and continue to use the creeper plant *Peuraria phaseoloides* for weed control and soil improvement. But forestry in Ghana has not applied this technique (FAO, 2002a). In Ivory Coast, the performance of the leaf mulches of *Leucaena leucocephala, Gliricidia sepium* and *Flemingia macrophylla* in weed control has been tested by Budelman (1988). Of the three mulch materials only that of *F. macrophylla* showed promise in retarding weed development. The effective lifespan of a mulch layer of 3 tonnes was found to be between 12 and 13 weeks, while that of treatments 6 and 9 tonnes had effective life-spans of over 14 weeks. The value of mulching in weed control is limited to the control of weed species that multiply by seed since regrowth originating from roots or stumps from former vegetation are unlikely to be checked by a mulch layer (Budelman, 1988).

5.6. PRUNING

5.6.1. Definition and Importance

Pruning is a silvicultural activity that removes branches in order to improve tree form or wood quality, health and vitality of a forest. It is used almost exclusively for conifer plantations. As trees grow, branches become incorporated into the wood of the tree trunk and form knots. When trees are widely spaced, the lower branches stay alive longer, thereby producing larger diameters, but also increasing the number of knots, the rotation age and the time to produce clearwood (knot-free timber) through natural pruning (Nyland, 1996). Branches may have to be removed artificially if there is a desire to obtain clear wood sooner than it would otherwise occur under natural conditions. Clearwood is valuable for producing veneers, which are obtained by cutting thin layers of timber from logs. Veneers improve the finished appearance and value of lower quality timber products.

Pruned timber will result in a higher proportion of clean, knot-free timber that is easier to plane and can be put to high-value end-uses. Knots cause structural and visual defects that can lower the value of the timber; a correctly pruned tree can potentially be worth 15%-50% more than the value of an un-pruned tree (Rodney-Bowman, 2007). The timing of pruning is important to the overall management of the stand; premature pruning wastes time and effort, while delayed pruning results in large loses of valuable clearwood (CQFA, 2009). Pruning also serves to reduce the chances of ground fires reaching the crowns, and to facilitate access to the stand. It is a costly operation, which should be perceived as an investment to improve the quality of the final product. It is therefore justifiable only when the extra revenues involved out-balance the costs (Centeno, 2009)

The frequency, season and standard of pruning vary between species, growth rates and the length of clearwood to be produced (CQFA, 2009). In general, removing live branches of trees also removes part of the photosynthetic capacity of the plant, and therefore, severe pruning will decrease the ability of the trees to produce carbohydrates and decrease growth (B. C. Ministry of Forests, 1995). However, removing live branches in a pruning would not affect the radial increment unless more than 50% of the crown or the live crown ratio is reduced to less than 40% (Daniel et al., 1979; Smith et al., 1997). Pruning only the lowest branches on a tree will have little effect on tree growth since these branches produce few carbohydrates (B. C. Ministry of Forests, 1995). With regards to timing of pruning, dead branches can be removed at anytime, but it is advisable to remove live branches during dormant periods, such as during the dry season in order to reduce bark damage resulting from peeling.

If the spacing is appropriate, hardwood species normally tend to self-prune (Nyland, 1996). However, for small-scale plantation growers in the tropics, biomass from pruning could serve as fuelwood to households. In agroforestry practices, lower branches of trees are pruned to grow high quality, high value timber, provide fuelwood, reduce shade and promote pasture growth, improve access and visibility, reduce fire hazards and increase tree stability in windy areas (Evans, 1992). Pruned material can be used as mulch in an agroforestry system, although the biomass on the forest floor could increase fuel loads and hence the risk of fires. From an ecological point of view, pruning removes lower branches that serve as perching and nesting sites for some birds and contributes to structural diversity of the forest (Hunter, 1990).

Pruning is currently not a major forestry operation in Ghana. The plantations established by the Forestry Department were neither pruned nor thinned until recently when market was established for teak poles and cedrela logs (FAO, 2002a). Centeno (2009) recommends pruning in teak plantations to remove branches up to a desired height near the time of canopy closure. For cost effectiveness, pruning should be done selectively, coordinated with the intended method of thinning, to clear 2 to 3 metres of stem at a time. Centeno (2009) notes that teak has the propensity to produce adventitious branches and epicormic shoots next to the scars caused by pruning; therefore a balance is necessary between the need to produce knot free timber, the stem length pruned at a time, and the need to prevent a slowing down of growth due to an excessive reduction of the crown.

5.6.2. Methods of Pruning

There are two basic types of pruning: form pruning - the selective removal of branches to produce a single straight trunk and aids future pruning by controlling tree form; and clearwood pruning - the removal of lower branches in such a way as to grow the maximum amount of knot-free wood (CQFA, 2009). Pruning is further divided into low and high pruning, depending on the height at which pruning takes place. Low pruning is the removal of branches up to 2m up the stem of the tree, or just after canopy closure in plantations (Evans, 1992). It is usually carried out by hand or a chainsaw to provide easy access to the plantations, reduce the fire hazards, facilitate felling and markings for thinning and produce knot-free timber at the base of the tree (Evans, 1992). Not all tree species would require low pruning. For example, in teak plantations, low pruning is not required except the removal of shoots during weeding or cleaning (Evans, 1992).

Pruning can be carried out manually, using saws, axes, cutlasses, shears, chainsaws, pole loppers, etc., or by mechanical methods using powered hydraulic shears and mobile mechanical pruners. For taller trees, these simple implements can be mounted on long poles to reach branches from the ground. Pruning is labour intensive, and even when machines are used, it is slow. For example, analyses of the time required for mechanical high pruning various species of eucalyptus plantations in Australia show that about 15-22 trees per hour were pruned with experienced contractors (Rhodey-Bowman, 2007). This suggests that a one-hectare plantation with 2500 stems would need between 14-21 days per ha, working at 8 hrs per day. The time requirements would even be higher with manual pruning.

Pruning should first remove structurally weak and dead branches. Those branches that appear dead or have leaves that appear unhealthy and distressed should also be removed. Pruning cuts should be made just outside the branch collar and nearly, but not completely, flush to the trunk, so as to provide viable growing branch bark that would encourage early closure of the wound (Rodney-Bowman, 2007). Generally, pruning for form should start in the second or third year, depending on the species and other site characteristics. Forked stems should be reduced to only one dominant trunk in order to minimise future pruning requirements and encourage healthy stems. Due to the potential for disease infestation following pruning, permanent branches should be shortened by pruning them back to a lateral branch or bud where an immediate growth response will be initiated (Rodney-Bowman, 2007).

There is no consensus in the literature regarding the profitability of pruning as a silvicultural exercise. Because pruning is labour intensive, it is an expensive operation. For it to be profitable, the benefits must outweigh the costs of doing so. The benefits includes the increase in value of the pruned trees relative to unpruned stands and the value of the pruned biomass if used for fodder, fuelwood, or other non-commercial benefits. The main factors affecting the costs of pruning are the number of branches to be removed per tree, the number of trees to be pruned, and the height at which pruning is done (low versus high pruning). The economic basis of pruning is to produce a higher proportion of clearwood and hence increase the value of the log. However, in addition to a knot-free log, other factors affect the value of the log as well, such as the size and shape of the log. The value of clear lumber produced from a pruning treatment can be three to five times greater than comparable knotty boards (Forest Practices Code, 1995). However, when viewed at the stand level, the net present value (NPV), measured in dollars per hectare, of a pruned stand were found to be lower for lower site qualities in British Columbia, Canada. There is an optimum stand density to obtain maximum NPV for a properly pruned stand; which is a balance between sacrificing the size of the clear wood shell and the overall stand volume density (B. C. Ministry of Forests, 1995). It is reasonable to conclude that pruning will only be profitable under limited circumstances. Given the analyses above, it is advisable to use planting densities that will completely eliminate the need for pruning.

5.7. Stand Density Management

Conceptually, stand density management is the process of controlling the level of growing stock through initial spacing or subsequent thinning to achieve specific management objectives (Newton, 1997; Newton *et al.*, 2005). Manipulation of stand density with thinning affects stand structure, yield, tree size and rotation length (Kumar et al., 1995). Determination of appropriate levels of growing stock at the stand level is a complex process involving biological, technological and economic factors specific to a particular management situation (Castedo-Dorado, 2009). The process requires selection of upper and lower limits of growing stock while taking into account the fact that the upper limit is chosen to ensure site occupancy, to maintain adequate stand-level yields and to produce trees of a desired form, whereas the lower limit is chosen to promote individual-tree growth rates (Dean and Baldwin, 1996).

Once the initial spacing decision has been made, the only remaining tool to control the density of forest stands is through thinning. Thinning is the removal of some stems in an immature forest stand in order to give the residual trees improved conditions for growth and increase wood quality, improve plantation health, or remove product prior to mortality due to crowding or natural events. The main purpose of thinning therefore is to allow fewer stems to utilise the full potential of the growing site and ensure that volume increment is put on residual stems of high quality. Thinning also ensures the utilisation of all the merchantable material produced by the stand during the rotation.

Thinning is not new, it has been practiced in forests since the first century; the first recorded thinning regimes date to the 14th Century and have been progressively developed since that time (Day and Nanang, 1997). The development of the principles of modern thinning since the 1920s by Craib (1939), MarMoller (1954), Assmann (1970) and many others, now provide a basis for manipulating the growth and development of natural and managed forest stands to favour almost all objects of management. Three very important principles for optimum timber production are (after Day and Nanang, 1997):

1) determining the optimal density range for the species to be grown throughout the proposed rotation;
2) thinning natural regeneration of the species to, or outplanting it at, spacings that minimise intra-specific competition and maximise growth to time of first thinning. The time of first thinning is then scheduled to occur when the species reaches the maximum permissible density; and
3) implementing subsequent thinnings to maintain the species in an optimum band of density that maximises growth and stem quality through the remainder of the rotation.

5.7.1. Objectives of Thinning

Thinning is expected to result in the following bio-economic benefits (Day and Nanang 1997):

(i) increases yield per ha by harvesting trees that would otherwise die because of intense competition;
(ii) favours the trees with the best growth potential and higher economic return;
(iii) shortens the time needed to reach a specified diameter or log volume;
(iv) extends the time to reach peak current annual increment;
(v) improves the quality and value of the timber by concentrating growth on the most valuable trees;
(vi) maintains high tree vigour and permits the removal of trees susceptible to insects and disease;
(vii) generates income early in the rotation, which can defray investments in silviculture and often results in higher economic returns;
(viii) reduces mortality almost to zero;

(ix) addresses age-class imbalances and shortening technical rotations, thereby alleviating impending wood-supply problems;
(x) creates employment opportunities thereby contributing to community stability;
(xi) minimises the inoculum potential of facultative parasites, which often cause disease in over-dense natural stands
(xii) strengthens the bole and branches of residual trees, making them resistant to breakage and wind throw.

Despite the many advantages of thinning, there are some subtle disadvantages associated with the practice. First, intervention may break the canopy and render the stand more vulnerable to windthrow, or harvesting may cause physical damage to and initiate biological deterioration of residual stands (Price, 1989). If stands are over- thinned, it reduces the amount of photosynthesis and also forces the thinned branches to elongate, which creates substantially weakened branches that may break easily during storms. Furthermore, if the thinnings have no economic value, thinning occurs at a net cost to the forest manager, especially if it occurs in unroaded areas of the forest. Finally, since thinning generally leaves large trees, volume increment due to thinning may do little to enhance the value of the volume to which it is added, since large trees may have already reached the plateau of the price-size relationship (Price, 1989). Due to these disadvantages, no general case can be made for or against the practice, since thinning may be viable and justified for a particular circumstance and may not be in other cases.

5.7.2. Types of Thinning

Foresters traditionally distinguish between five types of thinning (Smith, 1986, Ford-Robertson, 1971): thinning from below, thinning from above, selection (or improvement thinning), mechanical or systematic thinning and free thinning. Three very important factors in thinning are the timing of the practice, the intensity (how much volume to remove), and the frequency.

Low thinning (German thinning) or thinning from below is the removal of smaller, weaker, and deformed trees whose crowns are in the lower portion of the stand canopy to ensure the right planting density is maintained and to allow resources to be concentrated on the better trees. This type of thinning leaves trees that are larger and more vigorous, with the most developed crowns in the stand (Nyland, 1996). This is the oldest thinning technique, and is not widely used in commercial forestry because the removal of these lower trees provides little extra growing space for the dominant trees in the stand. Thinning from below is identical to pre-commercial thinning, where the thinnings would not have a commercial value and is done at a net cost to the landowner. There is evidence that hardwood species on poorly drained soils do not respond to low thinning (Tepper and Bamforf, 1959). Unlike in natural forests where trees often regenerate at high densities outside the control of the forest manager, in plantations, the initial spacing can be chosen so as to completely eliminate the need for low thinning – and in fact, this should be done. Unless the low thinnings have value as fuel, fodder, or mulch in an agroforestry system, it would make little economic sense to plant trees at high densities and costs, only to thin them out within a few years at an additional cost to the landowner.

Crown thinning is also known as the "French" method or "thinning from above", and is targeted at reducing crowding within the main canopy and involves the removal of intermediate and upper level trees (co-dominant and dominant trees). By removing the overstory trees, light and air can penetrate to the lower and middle crown classes. The increased light and air circulation within the crowns often reduce the incidence of diseases and pests. Tree selection is based on health and growth potential, with healthy and higher potential trees left in the stand. Weak or diseased trees in the lower canopy are also removed at this stage for convenience. Crown thinning improves the diameter growth of upper-crown residual trees (Nyland, 1996).

Selection, or improvement thinning, removes trees based on size, quality, and their position in the canopy. Usually, dominant trees are felled in order to release trees in the lower canopy. This is not a widely used practice and is generally used in plantations with some shade-tolerant species. In agroforestry or community plantations, this method of thinning can be applied by pre-determining a size limit for trees. Once trees reach that size, they are harvested. This practice was found to be common in Northern Ghana on neem plantations planted on individual or community plantations, where trees that reached pole sizes were harvested, leaving only the smaller trees (Nanang, 1996). Evans (1992) also reports that this method is applied on eucalyptus plantations in Ethiopia once the trees have reached 12 cm dbh (Evans, 1992).

Mechanical (systematic or line thinning) is a method in which trees are thinned based on an objective and a systematic procedure, where individual tree quality is not considered (Evans and Turnbull, 2004). Trees may be removed within a fixed spacing interval, or by strips within fixed distances between them (Ford-Robertson, 1971; Smith et al., 1997). In natural forests, mechanical thinning is applied as spacing thinning, while in plantations, it is applied as row thinning (Nyland, 2007). In plantation applications, the landowner decides on the interval for the rows that will remain, and then harvest all trees between those rows. For example, the landowner could decide to remove every second or third row in the plantation. The thinning intensity desired will inform how many rows are removed. This practice fits well with mechanised operations such as weeding, thinning, and logging.

A final method of thinning is free thinning, which releases selected crop trees, without regard to the position of the trees in the crown. It may require a combination of thinning methods to achieve this. Once the crop trees have been selected based on a set criteria, thinning is undertaken to favour those crop trees, leaving the remainder of the stand unthinned (Smith et al., 1997). Irrespective of the criteria used for selecting crop trees, landowners usually mark out those trees that interfere with the designated crop tree and cut them out. This practice allows the landowner to release only a minimum number of choice trees per unit area without investing in the removal of undesirable trees as well (Nyland, 2007).

5.7.3. Effects of Thinning on Stand Development and Volume Production

All recent major textbooks of silviculture (e.g., Nyland, 2007; Smith et al., 1997) agree that merchantable timber volume can be optimised by periodic reductions in stand density that maximise the diameter and volume growth of residual trees without sacrificing gross volume production. If plantations are planted at close densities, they would be subject to severe intra-specific competition and wasteful mortality, and hence thinned stands often produce more gross growth in volume than unthinned stands. Thinning is complex because of the many variables that determine the success and outcome of the practice – frequency, intensity, type, method and timing. Therefore, the growth and yield of individual stands and their ability to respond to thinning depends primarily on the species, the productivity of the site, the stand's age, density and historical development, and the nature and intensity of the thinning. The main benefits of thinning is to increase the size of individual trees, and redistribute the growth potential of the stand to optimum advantage, which ultimately enhances the value of the final crop, while lowering its harvest and milling costs (Johnstone, 1997).

Thinning provides more space for the growth of the residual trees. It opens the canopy for improved photosynthesis and the soil for improved absorption of water and nutrients. In the years after thinning, residual trees form more and larger buds and extend new foliage and branches filling the canopy, and roots filling the soil space (Nyland, 2007; Smith et al., 1997). The current annual increment per hectare declines immediately after thinning and then increases as the residuals extend. The loss in increment occurring immediately after thinning is soon recovered as the residuals extend and fill the growing space.

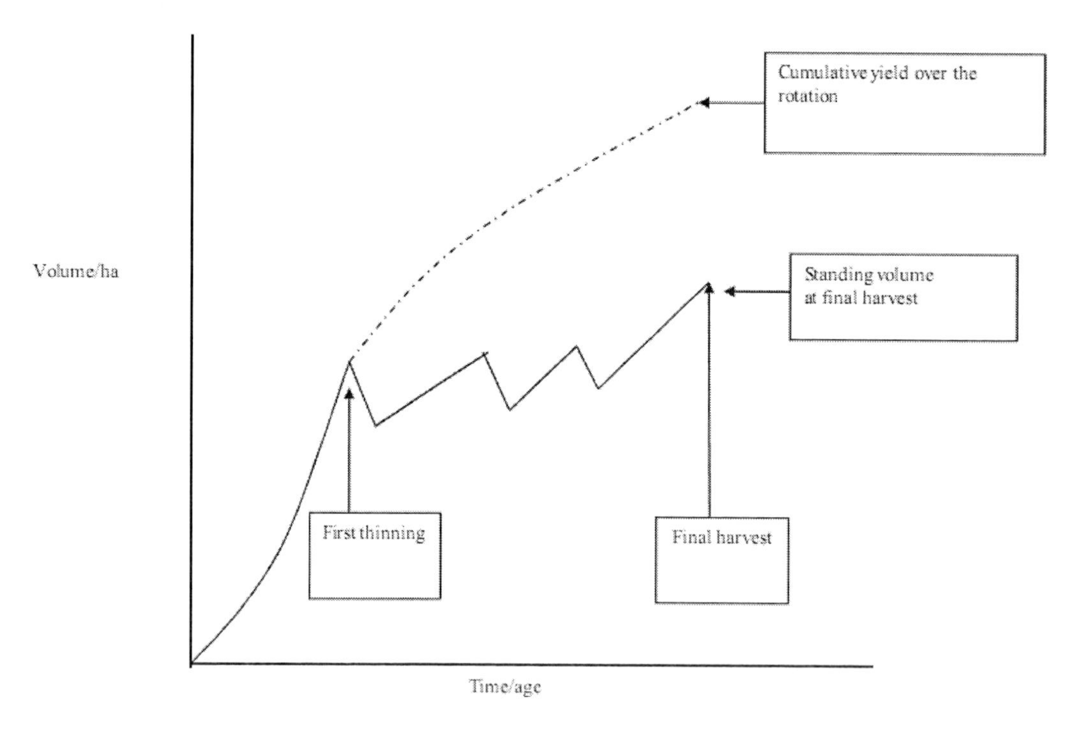

Figure 3. Hypothetical relationship between volume thinned, net productivity and gross productivity of a forest stand following thinning.

Over the life of the rotation, landowners can recover more usable volume by periodically thinning to control stand density and by selling the excess and potentially mortality tress (Nyland, 1996). In Figure 3, the upper line depicts a yield curve and shows the gross volume produced during the life of the stand. This volume includes the volume currently in the stand, plus the amount removed through thinnings. Three thinning regimes are shown in Figure 3 by the bottom curve. It shows that thinning removes some of the standing volume, and then the stand is allowed to regrow before the next thinning. The gap between these two curves show the amount of wood recovered in thinnings. If the stand had not been thinned, this volume would have been lost to wasteful mortality. It is important to point out that although the final harvest volume is less when the stand is thinned than unthinned, the cumulative volume is actually higher for the thinned stands. In other words, the combined volume of thinnings and final harvest is greater than final harvest in the unthinned stand.

5.7.4. Determining Optimal Stand Density

When trees are young, they need less space to grow than when they are older. The amount and timing of thinning older trees depends on growth rates and initial spacing. Crowding of the tree crowns gives a rough indication of when thinning is necessary. Deciding the optimum stand density for managing any tree species remains one of the most controversial issues in forest management. The question of how many trees per ha constitutes the ideal number for any tree species is continuously debated among foresters. This debate has led to the development of indices to assess the densities of forest stands and predict when stands should be thinned.

Measures of Stand Density

Several measures of stand density have been used in forestry. These include: a) trees per ha, b) volume, c) basal area, d) crown competition, e) stand density index, and f) relative stand density index. Density measures are used to estimate economic timber production and evaluate or estimate total biomass, design management regimes and evaluate other non-timber values such as carbon sequestration, wildlife habitats, etc. (Davis et al., 2001).

With regards to stand density and relative stand density indices, Kira et al. (1953) and Shinozaki and Kira (1956) studied the effect of density (number per area) on the mean dry weight of plants in competing populations. These Japanese population biologists have shown that there is a linear relationship between the reciprocal of mean plant dry weight (1/w) and the density of a wide range of plant species. This relationship is referred to as the "Reciprocal Yield Law" (Harper, 1977). In other words:

$$\frac{1}{w} = Ad + B \qquad (1)$$

where w = mean plant dry weight

d = plant density (number per unit area)

A and B = species dependent constants

Kira et al. (1953) and Shinozaki and Kira (1956) further showed that when plant populations are grown at high densities, mortality causes self-thinning in accordance with the reciprocal yield law. From this law, it has been demonstrated that when the logarithm of the mean dry weight of the survivors in a competing plant population are plotted over the logarithm of their density, there is a linear relationship with a slope of $-^{3}/_{2}$ or -1.5 (Harper,

1977, Yoda et al., 1963). This relationship is defined as the $-^3/_2$ Power law. In forestry, the Power law often defines the self-thinning line, which has proven valuable in understanding the dynamics and development of forest stands and their projection through time (Solomon and Zhang, 2002). The slope of the self-thinning line represents the largest number of trees of a given size that can occupy a hectare at any one time.

Several density indices have been used in forestry to describe measures of stand density: the stand density index (Reineke, 1933), the self-thinning rule (Yoda et al., 1963), the relative density index (Drew and Flewelling, 1977; 1979) and the relative spacing index (RS) (Wilson, 1946). The advantage of these density indices is that they are independent of site quality and stand age (Long, 1985; McCarter and Long, 1986). The self-thinning line has been commonly expressed by the following mathematical equations:

$$\ln V = k - 1.5 \ln N \tag{2}$$

(Drew and Flewelling, 1977)

$$\ln W = k - 0.5 \ln N \tag{3}$$

(Zeide, 1987)

$$\ln D_q = k - 0.625 \ln N \tag{4}$$

(Jack and Long, 1996)

$$\ln N = k - 1.605 \ln D_q \tag{5}$$

(Reineke, 1933)

where:

V = mean tree volume

W= total tree volume or biomass

N=number of trees per unit area

D_q = quadratic mean diameter

k= species specific constants (note: the value of k is not the same across equations)

In general, tree species that are very intolerant will have their Reineke's stand density index (RSDI) in the 1000 to 2000 range, while very tolerant species will range in RSDI of between 8000 and 10,000 (Day, 1996).

The relative spacing index (expressed as Spacing Factor %) is defined as the mean distance between trees in a uniform canopy crop expressed as a percentage of the stand's top height (Day, 1996). Mathematically, this relationship is expressed as:

$$N = k\, Top Ht^{-2.0} \tag{6}$$

or

$$\log N = k - [2.0 \log(Top\,Ht)] \quad (7)$$

where k is calculated as:

$$k = \log N + [2.0\,(Top\,Ht)] \quad (8)$$

In Wilson's SF% relationship, the number of trees per ha is plotted over the logarithm of the top height of the stand in metres on a Wilson's graph, with the slope of the SF% lines being -2.0. Tree species that are very intolerant will have SF% values of between 30 and 20%, while very tolerant species will have SF% from 15-10% (Day, 1996).

Stand Density Management Diagrams

Stand density indices are the basis for developing stand density management diagrams (SDMDs). These have greatly facilitated stand density management decision making (Castedo-Dorado, 2009). SDMDs are average stand-level models that graphically illustrate the relationships among yield, density and density-dependant mortality at all stages of stand development (Newton and Weetman, 1994). Their utility has been largely limited to evaluating density management outcomes in terms of mean tree size and stand-level volumetric yields (Newton, 1997). In addition to volumetric yield, maximisation of product value is becoming an important management objective, which requires estimation of the underlying diameter distribution, given the inherent relationship between monetary value and tree size (Castedo-Dorado, 2009). Newton et al. (2005) have made recent innovations to include structural yield prediction in the development of SDMDs and of traditional stocking guides which enable forest managers to estimate the number of trees in each diameter class at any point during stand development (Castedo-Dorado, 2009).

SDMDs are constructed by characterising the growing stock with indices that relate the average tree size (e.g., mean weight, volume, height or diameter) to density (e.g., number of trees per ha). While stand density management diagrams have been developed for some Japanese and North American tree species (e.g., Drew and Flewelling, 1979; McCarter and Long, 1986; Newton and Weetman, 1994), there are no such diagrams for tropical species, with the exception of teak in India (Kumar et al., 1995).

5.7.5. Profitability of Thinning Operations

There is evidence in the forestry literature regarding the ability to increase gross timber production through optimal stand density management (e.g., Frank 1973, Van Cleve and Zasada, 1976). From an economic perspective, however, there is still controversy over the economic profitability of intensive forest management practices such as thinning. For example, Anderson (1992) argues that the net benefits of silvicultural investments are at best marginal. The economic consideration in applying enhanced silviculture is very critical because it determines whether firms or landowners involved in forest management will invest in thinning activities. Forest managers, researchers, industry, and governments are interested in knowing whether thinning makes economic sense, and if so under what conditions.

The profitability of thinning operations is seldom evaluated in isolation from the overall profitability of the plantation enterprise. That is, instead of assessing whether the revenues received from thinning operations are greater than the costs, it is often the case that the analysis is done at the plantation enterprise scale, to determine whether the overall profitability of the plantation is higher with and without the thinning operation. If the thinning operation is profitable in isolation, it increases the chances that the whole plantation enterprise would be profitable at the rotation age. Conversely, if the thinning results in a net cost to the landowner, the chances that the operation would be judged profitable when evaluated at the final harvest are reduced. The methods for undertaking this kind of analysis are presented in the chapter on economics of plantations (Chapter 7).

In cases where thinning has been shown to be profitable (e.g., Day and Nanang, 1997; Ghebremichael et al., 2005), the increase in profitability of the thinned over the unthinned stands was mainly due to the increase in total merchantable volume (thinnings plus final harvest) of the thinned stands. Secondly, the bigger trees (greater volume/tree) at the end of the rotation resulted in reduced final harvesting costs. In addition, intermediate revenue from thinnings offset the cost of regeneration early in the life of the plantations and therefore reduced the interest on capital. The profitability of thinning therefore depends on a combination of site productivity, final timber price, whether the thinnings have a market value, timing of the thinning operation, the rotation length, and site stability. Thinning increases revenues but incurs a risk in terms of wind throw on unstable sites (Phillips, 2004).

It has been argued (e.g., Braathe, 1957) that even in cases where the thinnings do not give a direct profit, it may pay to carry them out because the first thinning in a stand may be more than repaid by a better development of the stand in the immediate future. This increase in value results from reduced harvesting and transportation costs, and increased value due to the bigger trees at final felling. The financial viability of thinning is also determined in part by the distribution of fixed and variable costs of the operations. Thinning involves relatively large fixed costs, such as movement of equipment to treatment sites. In this regard, there are economies of scale in the application of thinning treatments.

5.7.6. Case Study: Stand Density Management for Teak

Experience from India and Vietnam

This section begins with examples of stand density management from Asia, where they have had more experience managing teak than we have in Ghana. It has been found that the initial planting spacing and the timing of thinning strongly affects the pattern of growth and yield of teak plantations in India (Krishnapillay, 2000). Wider planting spacing and/or early heavy thinning may result in heavy side branching and epicormic shoots resulting in reduced volume yield. On the other hand, closer spacing and /or late light thinning may result in a decline in growth rates (Krishnapillay, 2000). Kumar et al. (1995) provide guidance on the maximum and minimum levels of the RSDI for teak in India as 1200 and 720, respectively. The lower limit for full site occupancy to occur is 420 trees/ha, while the lower limit for onset of competition is 300 trees /ha. An initial planting density of between 1,200 and 1, 600 trees per hectare is suggested for teak in India (i.e., an initial spacing of about 8.3 m x 8.3 and 6.25

x 6.25 m, respectively). These suggested spacing for teak are far wider than those used in Northern Ghana during the RAP of 2 x 2m (2500 trees/ha).

Krishnapillay (2000) describes the thinning regimes in India for teak to consist of three thinnings. The timing of the first thinning is often determined by the height of the trees and is commonly carried out when the trees reach 9.0 to 9.5 m. The second thinning may be carried out when the trees reach 17 to 18 m (Krishnapillay, 2000).The mean basal area is often allowed to reach 20 to 22 m^2/ha after the second thinning. A third thinning is then carried out to reduce the mean basal area to between 13 and 15 m^2 per hectare. A final stocking of about 300 trees per ha would be the ideal (Krishnapilla, 2000). In Vietnam, the thinning schedule for teak includes three thinnings at plantation ages 6, 12, and 20 years, leaving 800, 400, and 200 trees/ha on good sites, and 1200, 600 and 300 trees/ha on medium sites (FAO, 1998).

Stand Density Management for Teak in Ghana

Stand density management for teak in Ghana was investigated using data from Nunifu (1997). The data for these analyses were collected through a field study in the Guinea Savannah zone of Ghana. Sample plantations were selected from four Forest Districts; Tamale, Savelugu, Yendi and Damongo, all located in the Guinea Savannah vegetation zone. Data were collected from a total of 100 temporary sample plots from 25 plantations, ranging in ages from 3 to 40 years. Plantations were stratified into one- year age classes and sampled with an effort to equal allocation of three sample plantations to each age group. For each age group, efforts were made to include the full range of site conditions (from the poorest to the best). A pre-sample inspection was done to assess the conditions of each sampled plantation. Plantations that were found to be badly understocked due to mortality or selective harvesting were replaced.

For each plantation selected, four circular plots each of radius 7 m were sampled at random. For each plot, all teak trees were measured for diameter at breast height (dbh) and total height. Three trees were selected at random from each plot as sub-sample trees. These were felled for detailed measurements of total height and diameters at stump level, breast height and half the height above breast height. In all, 12 trees were felled and measured for each plantation. In addition, plantation records from the erstwhile Rural Forestry Division were examined. Growth and yield survey data from about 42 plantations in the high forest zone as well as records of planting and stand tending practices were obtained from the Forest Services Division Planning Branch in Kumasi. The ages of these plantations ranged from 13 to 26 years and were mostly from the Offinso Forest District of the Ashanti Region. The data were mainly stand level summaries including top height, average DBH, basal area per ha, stem density per ha, age and gross volume.

Based on this information, the stand characteristics as well as the RSDI and the SF% were calculated and presented in Table 2. The calculations were done using the following equations:

$$RSDI = TPH(D_q / 25)^{1.605} \tag{9}$$

$$D_q = \sqrt{(40000/\pi)(G/TPH)} \tag{10}$$

$$SF\% = \frac{SI}{TH} x100 \qquad (11)$$

$$SI = \sqrt{\frac{10000}{TPH}} \qquad (12)$$

where: D_q = the quadratic mean diameter in centimetres
G = stand basal area in m^2/ha
TPH = the number of trees per ha
TH = top height in m
SI = spacing interval (average distance between trees)

From Table 2, the maximum RSDI for teak in the savannah vegetation zone was about 1200, whilst that in the high forest zone was 373. The SF% was much larger in the HFZ than the savannah due mainly to the lower densities of trees in the HFZ.

The RSDI and /or the SF% can be used to develop stand density management regimes for teak. A generalisation is that about 60% of a species' maximum RSDI approximates the lower limit of the zone of imminent competition mortality (Drew and Flewelling, 1977). This means that below 60% of maximum RSDI, a stand experiences little or no density-related suppression, while above 60%, self-thinning would be expected to occur. For the savannah zone, the lower limit for growing teak is a RSDI of 720 (60% of 1200) and the upper limit is a RSDI of 1200 in order to avoid competition that leads to self-thinning. This information can now be used to set upper and lower limits of RSDI for thinning. Stands should be thinned as soon as their RSDI reaches 1200, and reduced to a RSDI of 720, and allowed to grow back up until 1200 again, and thinned, until the rotation is reached. It is also usual to assume that 35% of the maximum RSDI represents the lower limit for full site occupancy (i.e., the density that captures the most growth potential of the site). For teak, this would be about RSDI of 420 in the savannah zone. If however, the emphasis of management is on individual tree growth at the expense of some potential yield, then it might involve limiting the upper density to about 25% of the maximum RSDI (Kumar et al., 1995). In the case of teak, this will be an RSDI of 300.

A Thinning Schedule for Teak in the Savannah Zone of Ghana

Below, I describe a simple schedule and a SDMD for thinning teak plantations in the savannah zone based only on the RSDI and quadratic mean diameters calculated in Table 2. It was not possible to develop a similar diagram for teak in southern Ghana because the stand density data shows that these stands are already understocked. However, the same techniques can be used if stand level data for fully stocked plantations in southern Ghana were available. Similar schedules can be developed using the SF% or any of the other stand indices previously described. In practice, more complicated SDMDs are developed using 3 or more of these indices at the same time (e.g., Kumar et al., 1995; Castedo-Dorado et al., 2009; Newton et al., 2005 and Newton and Weetman, 1994). The example below is simplified for illustrative purposes.

Table 2. Summary of stand parameters of teak plantations in the savannah and high forest zones

Stand age (years)	Average DBH (cm)	Top Height (m)	Density (trees/ha)	Basal area (m^2/ha)	Gross volume (m^3/ha)	RSDI	Spacing factor %
Savannah Zone							
3	2.76	4.10	1510	0.99	4.13	47	62.77
4	3.25	4.82	1613	1.50	5.49	67	51.66
6	5.95	7.36	2100	6.87	29.49	239	29.65
7	7.67	9.45	1765	8.34	44.33	270	25.19
8	7.69	9.62	1705	7.94	43.12	258	25.17
9	7.82	9.33	1591	8.22	38.09	261	26.87
17	10.78	10.29	1488	14.21	73.74	400	25.19
26	18.21	16.82	1413	37.89	339.42	870	15.82
31	21.06	17.19	1396	50.14	472.28	1086	15.57
38	23.58	24.53	1299	58.83	776.72	1218	11.31
40	19.53	18.37	1429	43.83	220.33	980	14.40
High Forest zone							
13	21.00	16.80	291	10.38	84.00	225	34.89
14	27.00	19.90	265	13.03	114.00	265	30.87
15	22.50	17.45	311	11.71	91.50	251	32.50
16	27.33	21.03	276	15.07	136.67	301	28.62
17	29.33	21.22	243	15.57	142.17	301	30.23
18	28.00	20.65	246	14.08	126.75	278	30.88
19	30.00	19.40	177	11.91	100.00	228	38.74
20	29.40	20.82	264	16.23	148.00	316	29.56
21	30.50	23.10	243	16.79	164.50	320	27.77
22	31.00	24.40	271	19.83	204.50	373	24.90

The thinning schedule is developed for teak using information on the upper and lower limits on the RSDI and the diameter of the tree at the rotation age. In this example, two scenarios are evaluated. First, the maximum RSDI of 1200 is used as the upper limit for managing teak. However, other authors have used lower percentages of the maximum RSDI as upper and lower limits. For example, Kumar et al. (1995) used 35 and 20% of maximum RSDI as upper limit and lower limits, respectively in developing a SDMD for teak in India. Dean and Baldwin (1993) used 45 and 30% of maximum RSDI as the upper and lower limits, respectively for a SDMD for loblolly pine plantations in the United States. The limits are usually determined based on experience with the species in the location where it is grown.

To illustrate how this is done with upper limit of maximum RSDI =1200 and lower limit of SDI=720, we assume that the plantation is grown for sawlogs and the desired diameter (D_q) at the rotation is 40 cm. It is further assumed that no thinning will take place until the trees reach a minimum diameter of 10 cm. From Figure 4, the point at which the 40 cm D_q line intersects the 100% RSDI line (1200) represents the point at which final harvesting should take place (rotation). In order to find the sequence of thinnings to get to the final point, we stair-step backwards between the upper and lower limits until D_q falls below 10 cm. The thinning is represented by the horizontal portion of the stair-step and approximates the number of trees removed in the operation. The vertical portion of the stair-step represents the post-thinning growth phase of the operation.

To determine the timing of thinning and the final harvest, we use the site index curves and a relationship between D_q and top height of the plantations. For this data, top height can be predicted from D_q using:

$$TH = 7.3159\ln(D_q) + 0.3487 \qquad R^2 = 0.73 \qquad (13)$$

We know that all thinning occurs when the RSDI reaches the maximum of 1200 and these occur at points G, E, C and A (final harvest) in Figure 4. At these points, the corresponding D_q's are 6, 18, 29, and 40 cm. Using these D_q's, we predict the corresponding top heights of the plantations using Equation 13 as 14, 22, 25 and 27m, respectively. The plantation ages to thin are then the ages on the site index curves that correspond to these top heights in Figure 5 (using a site index of 20).

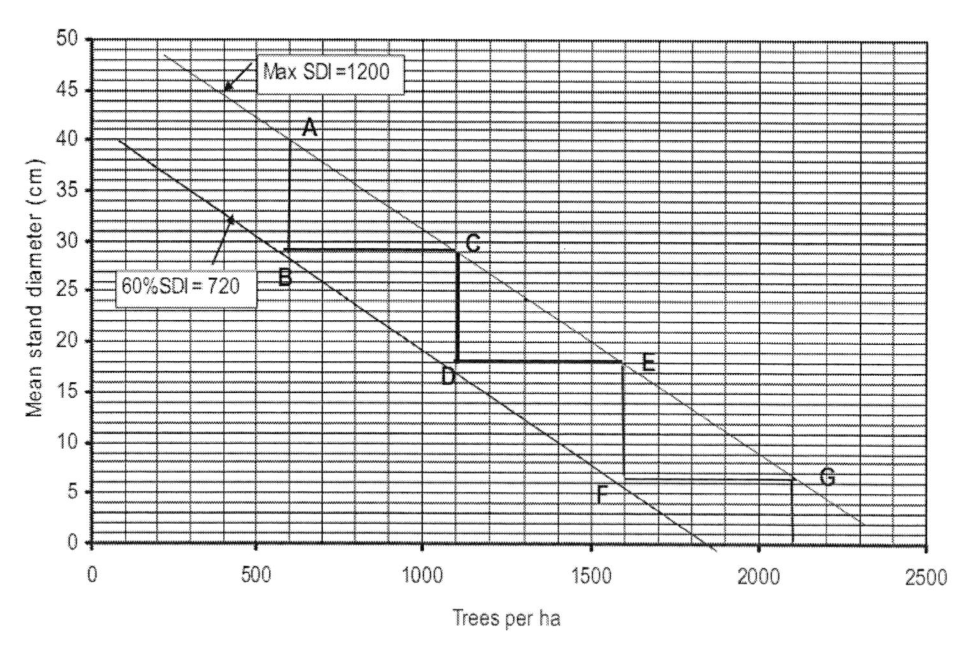

Figure 4. Stand density management diagram for teak in the savannah zone of Ghana.

From Figure 5, the first thinning should occur at about 4 years after planting, followed by a second thinning at about 15 years and a third thinning at about 34 years. The final harvest should occur in year 50. We can determine that the initial stocking will be about 2100 trees/ha (at an initial spacing of about 2.2 x 2.2 m). The number of trees to be removed at each thinning is determined in Figure 4 and are 400, 420 and 450 in the first, second and third thinning, respectively. Assuming no natural mortality, about 600 trees will remain for the final harvest. The information is summarised in Table 3.

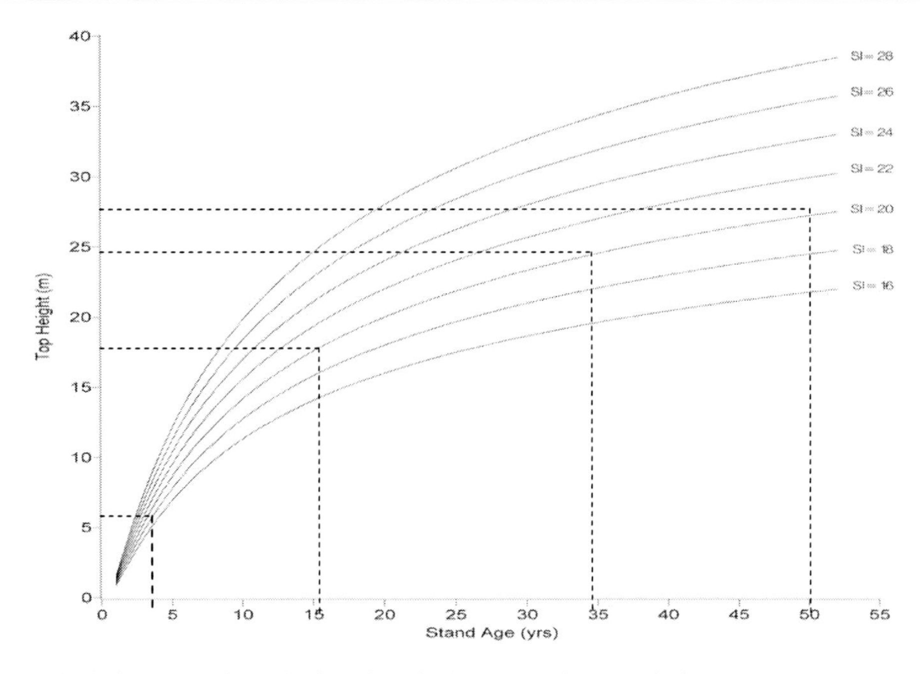

Figure 5. Site index curves for teak plantations in the Savannah Zone of Ghana.

Table 3. A comparison of two density management regimes for plantation teak grown for sawlogs in the savannah zone of Ghana

Operation	Age (year)	Top height (m)	TPH Before	TPH After	No. of trees removed /ha
Upper limit of maximum SDI =1200; lower limit of 60%SDI=720					
Planting	1	-	2100	2100	-
First thinning	4	14	2100	1600	500
Second thinning	15	22	1600	1100	500
Third thinning	34	25	1100	600	500
Final Harvest (rotation)	50	27	600	0	600
Upper limit of 60% SDI =720; lower limit of 30%SDI=360					
Planting	1	-	1600	1600	-
First thinning	4	12	1600	1100	500
Second thinning	15	21	1100	600	500
Third thinning	34	25	600	100	500
Final Harvest (rotation)	50	27	100	0	100

Figure 6 presents a regime for stand density management for teak using 60% of the maximum SDI as the upper limit and 30% as the lower limit. Under these circumstances, self-thinning would be avoided. In this scenario, the minimum diameter below which thinning would not take place is 5 cm. The plantation ages at which thinning take place were determined using the same procedure as the case where the upper limit was the maximum RSDI (but not shown in Figure 5). A comparison of the two density management regimes shows that when the upper limit is 60% of maximum SDI, stocking levels reduce accordingly, and more growth is put on individual trees at the expense of overall volume yield. This is an obvious balance that forest managers need to keep in mind when setting density management regimes.

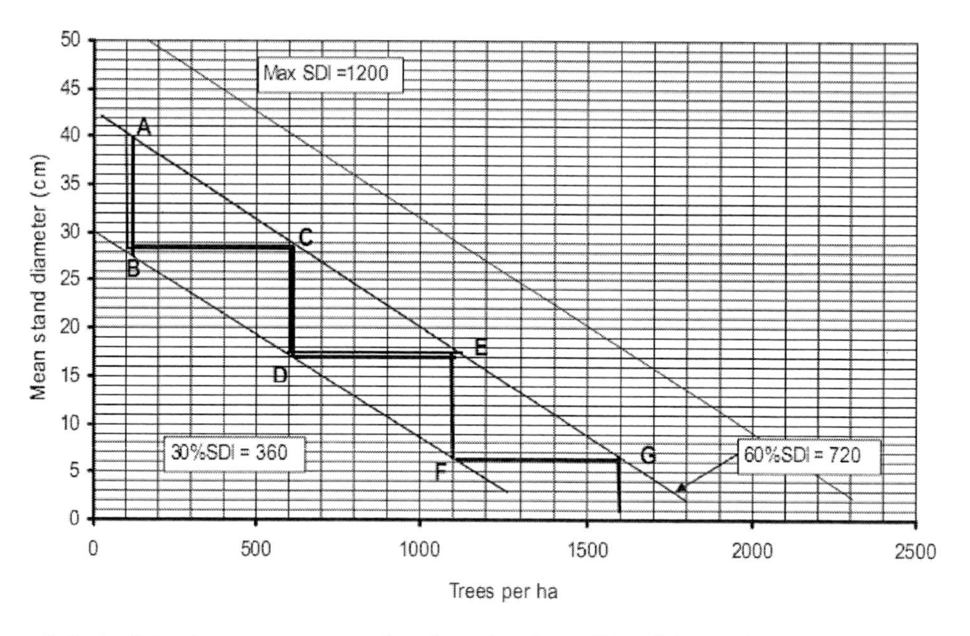

Figure 6. A stand density management regime for teak using a 60% of the maximum SDI as the upper density limit.

5.9. Conclusions

This chapter described the practices in plantation silviculture and management, starting from seed collection, through nursery practices to stand density management. Understanding and implementing these practices are essential to optimising the growing potential of forestry sites and improve the productivity and the economic profitability of the plantations. The concepts outlined here are of general application, and therefore the specific practices will depend on the tree species under consideration. Hence some of the knowledge of critical silvics discussed in Chapter 4 will be crucial to determining the management regime that should be pursued.

Chapter 6

PLANTATION GROWTH DYNAMICS AND PRODUCTIVITY

Understanding and predicting the behaviour of forest stand dynamics is one of the most important tools of forest management. Unfortunately, these tools are limited for majority of the tree species grown in plantations in Ghana today. The development of such tools and techniques has been slow, relative to the rate at which the natural forests decline and the need for such tools increase. In this Chapter, I provide the basic concepts of stand growth dynamics and productivity of even-aged forest stands. The purpose is to describe the concepts used in forest stand dynamics, and provide a theoretical basis for modelling forest stand growth and yield. To illustrate the application of these concepts, two case studies of modelling the growth and yield of neem plantations and diameter distributions are presented in the last two sections of this chapter.

6.1. FOREST STAND DYNAMICS AND GROWTH

6.1.1. Forest Stand Dynamics

Forest stand dynamics are the changes in forest structure, function and composition through time, including stand behaviour before and after disturbances (Oliver and Larson, 1996). Knowledge of forest stand dynamics is applied in many areas of forest management, including plantations for timber production (e.g., Cannell and Last, 1976), semi-natural woodlands where conservation objectives are a priority (e.g., Smith et al., 1997) and multi-functional woodlands managed for both timber and biodiversity values (e.g., Kerr, 1999). The usefulness of understanding forest dynamics is to be able to make reliable predictions about how forests will change over time and in the face of natural and anthropogenic factors.

Most descriptions of stand development characterise it as a progression through stages toward an older forest, possibly an old forest, in the absence of disturbance (O'Hara, 2004). Disturbances, from human or other causes, can move stand development backward or forward in the process, depending on their type, severity, and timing (O'Hara, 2004). Oliver and Larson (1996) describe the four generally recognised phases of forest stand dynamics. The first is the stand initiation phase, which corresponds to the phase of recruitment of stems to the stand. In plantations, planting ensures that the stand achieves full stocking of the site in one or more planting operations. In this phase, the trees are still young and growing; hence full site occupancy will not have occurred. The second stage is the stem exclusion phase, where the stand develops a closed canopy and the deep shade in the understory prevents further recruitment of trees to the stand while competition, site factors and genetic differences lead to differentiation of crown dimensions and stem diameters (Wilson and Leslie, 2008). The stand then continues on to an understory re-initiation stage, where herbs, shrubs and advanced regeneration of trees appear and survive as a result of gradual thinning of crowns or occasional gaps that allow increased levels of solar radiation to reach the forest floor (Oliver and Larson, 1996). Finally, the stand reaches an old growth stage where the overstory trees die in an irregular pattern, either from natural causes or disturbance, creating space for recruitment into the canopy of trees from lower strata (Oliver and Larson, 1996). The transition from one stage to the next generally holds true for plantations, if retained for long enough (Kerr, 1999), and those natural forests that regenerate after a large-scale, stand-replacing disturbance such as a forest fire (Kimmins, 2003).

However, each stage of the sequence may be disrupted as a result of managed interventions or natural disturbance events (Johnson and Miyanishi, 2007; Kimmins, 2003). Because plantations are usually made up of one tree species, even-aged and most often of uniform spacing, the trees are often clear-felled before reaching the understory re-initiation stage in order to maximise the financial return on investment costs (Savill et al., 1997; Matthews, 1989; Evans and Turnbull, 2004).

Figure 1 shows the growth patterns of neem plantations in Northern Ghana in terms of basal area and gross volume. The data were based on temporary sample plot data from plantations of neem planted at 2 x 2 m spacing. This represents the natural progression of a stand from planting until age 9 years, without thinning. The volume available for harvesting in year 9 is approximately 100m^3/ha. The diameter and basal area growth follows identical patterns to that of volume. The figure provides a baseline for how the stand would evolve without any disturbance, and could be the basis for predicting the impacts of management decisions on the development of identical stands in the savannah zone.

6.1.2. The Concept of Growth

Growth is the increase in a particular stand characteristic over a period of time. This includes increases in height, diameter, volume, crown or any other forest attribute of interest. In simple terms, if a forest stand was measured in 2001 and it contained 100m^3/ha of wood and a subsequent measurement in 2006 showed a volume of 150m^3/ha, then the growth can be said to be 50 m^3/ha over the five years or on average, 10m^3/ha/year (assuming no mortality).

Individual trees grow by means of photosynthesis through adding woody biomass to all parts of the tree- stem, branches, roots, etc. Most of these growth components are difficult if not impossible to measure on the field (Davis et al., 2001). Growth is affected by how many trees die between the measurement periods for which growth is being assessed, i.e., mortality. In single species plantations, mortality is easier to estimate than in natural forests with high species diversity. For example, if a plantation is planted at an initial density of 2500 trees/ha, and after five years there are 2000 trees, then the mortality is 500 trees over the 5-year period.

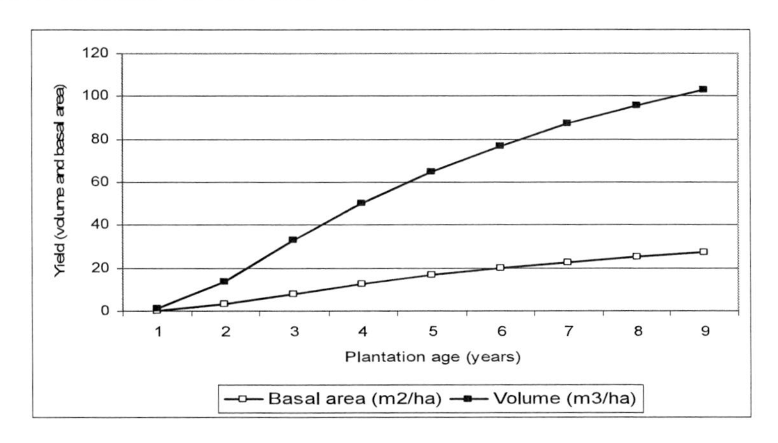

Figure 1. Growth patterns of neem plantations in Northern Ghana.

Depending on the purpose of measurement, growth can be categorised into total growth, potentially usable growth, and growth that is actually removed and used (Davis et al., 2001). Total growth refers to all biomass produced by the trees, including roots, stump, bole, branches, leaves, etc. Potentially usable growth refers to woody components that could be used by a manufacturer, given current technology, while the amount of the growth that is actually removed and utilised constitutes the third kind of growth (Davis et al., 2001).

6.1.3. Measuring Forest Growth

Growth measurement is complicated by mortality. Because the increase in volume, for example, between any two measurement periods has to consider that some trees might have died in that time, while within that interval some trees might have grown into the measurable size as well (Davis et al., 2001). Following Beers (1962) the following framework is defined for estimating volume growth within forest stands:

V_1= volume of living trees at the beginning of the measurement period
V_2= volume of living trees at the end of the measurement period
M = volume of mortality over the measurement period
C = volume removed during the measurement period
I = volume of in-growth over the period

Depending on the type of growth required, the following five components of growth are defined (after Davis et al., 2001).

1. Gross increment including in-growth = $V_2 + M + C - V_1$
2. Gross increment of initial volume = $V_2 + M + C - I - V_1$
3. Net increment including in-growth = $V_2 + C - V_1$
4. Net increment of initial volume = $V_2 + C - I - V_1$
5. Net change in growing stock= $V_2 - V_1$

In order to understand and predict growth, stand growth models are used. A stand growth model represents an abstraction of the natural dynamics of a forest stand, and depicts growth, mortality and other changes in stand composition and structure (Ministry of Forests and Range, 2009).

6.2. Stand Volume Estimation

According to Spurr (1952), few subjects in the entire field of forestry have received as much attention as the estimation of tree volume. The large number of approaches to the problem of volume estimation may be taken as an indication that no one approach has received more than partial recognition. To assess the volume of a complex and highly variable geometric solid like a tree in terms of very few measurements and by simple algebraic techniques is by its very nature difficult if not impossible (Spurr, 1952).

There are three general approaches to the estimation of stand volume in use. In practically all inventory and management work, volume tables are used. To use a volume table for volume estimation, samples of trees are measured, but only from one to three measurements are taken on each tree (Spurr, 1952). If only one measurement is taken, it is usually the diameter at breast height (dbh)[1]; if two, height is also measured; if three measurements; a measure of form is also included. From these few variables, volume is estimated by reference to volume tables[2]. Volume tables give the mean regression of volume on dbh, height and form (in form-class tables only) for a series of carefully measured sample trees.

The second general approach is called the mean tree approach. The underlying theory of this method is that, if the tree of mean volume can be isolated, then the volume obtained by careful measurement of this tree can be multiplied by the number of trees in the stand to give the volume of the stand (Spurr, 1952). One method of determining the tree of mean volume is to use the mean sample tree, which is based upon the assumption that the tree of mean basal area is also the tree of mean volume (Crow, 1971). Although fairly good results have been obtained by this method, the fallacy of the basic assumption has long been recognised (Spurr, 1952). Crow (1971) agreed that the mean tree method can be used for expedient estimates of tree biomass in plantations or uniform, natural stands or even-aged species. Especially promising are techniques based on individuals at one standard deviation from a mean stand dimension.

Another variation of the mean tree approach is to first estimate the volume of each sub-sample tree (mean tree) using standard formulae such as the Smalian's formula, which is popular because of its simplicity. Stand volume per ha can then be estimated from these volumes in three ways:

a)A volume equation can be developed using the individual sub-sample tree dbh, height and calculated volume:

$$V = \alpha + \beta D + \gamma D^2 + \lambda D^2 H \quad (1)$$

[1] The diameter at breast height is the diameter of the tree measured at 1.3 m above the ground.

[2] It is assumed that volume tables for the species in that geographical area are already in existence otherwise this method cannot be applied.

where D is the dbh, H the total tree height, V is the calculated tree volume, α, β, γ and λ are regression coefficients. Equation (1) can be estimated by ordinary weighted least squares (OWLS) (Cunia, 1964). The fitted volume equation is then used to predict the total volume of each tree in each sample plot. The individual tree volumes within each plot are then aggregated to estimate average stand level volume (in m^3/ha).

b) When two-stage simple random sampling is used to obtain data from plantations, the stand volume can be estimated using the two-stage simple random sampling formula (Cochran, 1977).

$$V = \frac{N}{n}\sum_{i=1}^{n}\frac{M_i}{m_i}\sum_{j=1}^{m_i}V_{ij} \qquad (2)$$

where V is the total volume per ha, n the number of sample plots per plantation, N is the number of plots that could potentially be sampled in 1 ha obtained as $1/a$, where a is the individual plot area expressed in ha, M_i the number of trees per sample plot, m_i the number of sub-sample trees per plot, V_{ij} the volume of the jth tree in the ith plot.

c) If however, a two stage sampling with probability proportional to basal area at the second stage (PPG) of sampling is used then the formula for computing stand volume would be:

$$V = \frac{N}{n}\left(\sum_{i=1}^{n}\left(\sum_{j=1}^{M_i}BA_{ij}\frac{1}{m_i}\sum_{j=1}^{j=m_i}\frac{V_{ij}}{BA_{ij}}\right)\right) \qquad (3)$$

where, V_{ij}, M_i, m_i, Y, N and n are as defined before and BA_{ij} is the basal area of the jth tree from the ith sample plot.

6.3. Yield Estimation

6.3.1. Definition and Data Acquisition

Yield refers to the total amount of material available for harvest at a given time (Avery and Burkhart, 1994). Mathematically, growth can be calculated as the first derivative of the yield equation, while yield is calculated as the integral of the annual increments [sum of growth] (Clutter, 1963). A yield table is usually a table showing the volume/ha at different ages for even-aged stands of trees growing on forest land of different productive capacities (Chapman and Meyer, 1949). It is one of the oldest approaches to yield estimation, and modern yield tables often include additional information such as number of trees/ha, basal area/ha, stand height, diameter, and current and mean annual increments. A normal yield table shows the yields capable of being produced on forest sites when the "normal" capacity of site is fully utilised by even-aged stands of forest-grown trees. Empirical [or variable-density] yield tables usually refer to yield tables that apply to "average" rather than full, or normal stocking (Avery and Burkhart, 1994).

The best method of obtaining data for yield table construction is to measure permanent plots located in the stands of interest at intervals of 5 or 10 years over the entire period of growth of the stands (Chapman and Meyer, 1949). The actual increases in the various yield variables and the number of trees that die from year to year and the volume of the surviving stand at any age are then a matter of record. However, this entire process would take a long time to complete, and the stand might be destroyed at any time by fire, disease or other agencies (Chapman and Meyer, 1949). Yield tables constructed by this method are referred to as real growth series yield tables (Turnbull, 1963).

The second type of yield tables, known as "abstract" growth series yield tables, are constructed by measuring a number of permanent sample plots (PSPs) located in stands of different ages and to re-measure these plots at 5- or 10-year intervals (Chapman and Meyer, 1949; Turnbull, 1963). By the overlapping of the ages chosen, the trend of the development for one or two decades for plots of all ages is obtained after the second or third measurement. It is however essential that lands for PSPs be in stable ownership, preferably in public forest, so that the owner's whim may not interfere with the experiment (Chapman and Meyer, 1949).

Chapman and Meyer (1949) contend that the two methods described above are too time-consuming and other methods are needed if yield tables are to be prepared for immediate use. To fill this need, a third standard method which has been extensively used is based on the principle of comparison of plots of different ages. The averages of stands for the same site but differing in age are combined into a curve assumed to show the trend of growth. If many plots of different ages on different sites are measured, each plot will show the yield obtained in a similarly stocked stand at the given age and for the site in which it is located.

6.3.2. Types of Growth and Yield Models

Growth and yield models are generally classified into three types: whole stand models, diameter class models, and individual tree models. Whole stand models usually include stand variables in the model such as age or basal area per ha, while diameter class models include average tree within each diameter class (Davis et al., 2001). Individual tree models use each individual tree in a sample or stand to estimate yield (Davis et al., 2001). Whole-stand models can further be divided into density-free and variable-density models. The basic Schumacher-type model given by $Q= f(A,S,D)$ is an example of a variable-density yield model, where Q is some measure of yield (height, dbh, basal area, volume, and fresh weight), A is stand age in years, *S* is some function of site index, D is some function of stand density. If however, stand density is fixed, then $Q= f(A,S,)$ would be a density-free whole stand model.

Diameter class models often simulate the growth in each diameter class by calculating the characteristics, volume, and growth of the average tree in each class and multiplying this average tree by the inventoried number of trees in each diameter class (Davis et al., 2001). To obtain the volume of the stand, the computed volumes in each diameter class are summed over all diameter classes in the stand. Davis et al. (2001) draw a distinction between diameter *distribution* models and diameter class models. The former is considered a whole stand model because the number of trees in each diameter class is wholly a function of the stand variables and all the growth functions are for stand variables. Diameter *class* models on the other hand have the number of trees in each class empirically determined and independently model the diameter classes subject to some aggregate stand influences (Davis et al., 2001).

The individual tree models are more complex to model than the whole stand or diameter class models. Most individual tree models calculate a crown competition index for each tree as a measure of how well the tree can compete for light and nutrients and growing space relative to other trees in the stand. This index is used to project whether the tree lives or dies, and if it survives, it predicts is growth characteristics (Davis et al., 2001). The total yield of the stand is then the aggregate of the individual tree characteristics.

6.3.3. Volume Tables

One of the most important yield variables of interest to foresters is the amount of woody material available in a plantation, i.e., volume. For species that have volume tables, volume is often read off the tables for a given site class using measurements of dbh and /or height of the plantation in question. Credit for the first modern volume table is generally given to Heinrich Cotta, who published one for beech in 1804 and in 1817, developed a set of standard volume tables. Three types of volume tables are recognised: local, standard and form-class volume tables (Husch et al., 1982). Local volume tables give tree volume in terms of dbh only. The term *local* is used because such tables are generally restricted to the local area for which the height-diameter relationship hidden in the table is relevant (Husch et al., 1982). Local volume tables are usually derived from standard volume tables though they can be prepared from raw data - that is from volume and diameter measurements for a sample of trees. Standard volume tables give volume in terms of dbh and merchantable or total height. Tables of this type may be prepared for individual species, or groups of species, and specific localities (Husch et al., 1982). Form-class volume tables give volumes in terms of dbh, merchantable or total height, and some measure of form, such as Girard form class or absolute form quotient (Husch et al., 1982).

6.3.4. Volume Table Construction

The problem of constructing volume tables is a statistical one. The construction of volume tables involves directly relating volume to height and diameter by means of graphs, alignment charts or equations (Spurr, 1952). The possibility exist however, of relating tree diameter and height to an indirect measure of volume such as a form factor or taper, and then constructing the volume table as a separate step from the form-factor table or taper curves (Spurr, 1952). Historically, three methods that have been used for volume table construction: graphic, alignment –chart methods and mathematical modelling based on least-squares techniques.

Graphic Techniques

Of the three general approaches to volume table construction, the graphic method is the oldest and requires less mathematical skill than the least-squares or alignment chart techniques (Spurr, 1952). Basically, the harmonised-curve method involves the sorting of data into groups according to diameter and height classes, and the preparation of a series of curves which give volume for any combination of height and diameter. Another advantage of the graphic technique is that the curves are fitted to the actual data rather than being forced to conform to any set pattern (Spurr, 1952). Spurr (1952) however, notes that the method is not only subjective, but considerable experience is necessary especially for a small sample, and a large number of tree measurements are needed to provide good trends for each of the diameter and height classes.

Alignment-Chart Techniques

Alignment charts provide an efficient means of expressing an equation (Spurr, 1952). They were introduced as tools to correct for curvillinearity in multiple regression equations by Bruce and Reineke (1931). (Spurr, 1952) indicates that this method has been generally accepted as a method of constructing volume tables. It requires little mensurational training and generally gives satisfactory results. Spurr (1952) discussed several disadvantages of this method. Firstly, prepared base charts are needed which are not always available. Furthermore, the charts cannot be read too accurately and are quite subject to error because of dimensional changes in the paper.

Least-Squares Techniques

The graphic and alignment-chart techniques have been generally discarded in favour of mathematical functions and models. Almost all volume tables are now constructed using standard regression techniques based on the method of least squares. These techniques have gained wide acceptance for the construction of volume tables (Unnikrihnan and Singh, 1984). This is because the method is free from the subjective bias of fitting curves by hand (Spurr, 1952). A large number of different equations have been proposed for volume table construction and considerable difficulty may arise in attempting to decide which equation is the most appropriate for a particular set of data (Furnival, 1961). Considerable difference of opinion also persists regarding not only the function to be used but the proper criterion of comparison (Spurr, 1952).

Classical least squares estimation is based on the assumptions of independently and normally distributed errors and the property of homoscedasticity, i.e., the variance of the dependent variable is constant for all values of the independent variables. The fourth assumption is that the sample is a simple random sample (Cunia, 1964). In practice, however, most, if not all of these assumptions are not satisfied. Cunia (1964) found that tree volume for a given dbh is not normally distributed and highly skewed: large trees tend to deviate more on the average from the regression surface than do small trees (i.e., heteroscedasticity). It is also usual in forest inventory to replace the simple random sampling by stratified, cluster or systematic sampling (Bruce and Schumacher, 1950; Spurr, 1952; Cochran, 1953).

Theoretically, weights should be employed that are inversely proportional to the variance of the residuals in order to achieve a homogenous variance. But in practice, it may be difficult to determine the most appropriate way to weight a particular function (Furnival, 1961). Another option is the use of logarithmic transformations of the equations. When logarithmic transformation is used to ensure that the assumptions of ordinary least squares are satisfied, it is often necessary to express estimated values of the variable of interest in original measured [i.e., untransformed] units (Baskerville, 1972). However, the conversion of the logarithmic estimates back to measured units produces a bias that must be corrected. This results from the fact that if the distribution of the residuals about the log-transformed model is normal, then the distribution of the untransformed residuals is skewed (Furnival, 1961; Baskerville, 1972). It is argued that the transformation from the logarithmic form back to measured units by simply determining the antilogarithm has, by failing to account for skewness of the distribution in arithmetic units, yields the median rather than the mean value of the estimates (Finney, 1941; Brownlee, 1967; Baskerville, 1972). This produces systematic underestimates of the dependent variable. Therefore a correction must be made for this inherent bias which is proportional to the amount of variation associated with the regression (Schlaegel, 1981).

The following correction for the skewness that results from this retransformation was proposed by Baskerville (1972).

$$\hat{Y} = \exp(\mu + \sigma^2/2) \quad (4)$$

$$\sigma_A^2 = \exp(2\mu + 2\sigma^2/2) - \exp(2\mu + \sigma^2/2) \quad (5)$$

where:

$\hat{Y}$ = estimated mean in measured units

σ_A^2 = estimated variance in measured units

μ = estimated mean in logarithmic units

σ^2 = sample variance of the logarithmic equation

Beauchamp and Olson (1973) extended Baskerville's work and concluded that unless the variance of the residuals is quite large the correction provided by Baskerville (1972) gives close approximations to the unbiased value.

6.3.5. Model Selection Criteria

When modelling volume of forest stands, several options exist in terms of choice of functional forms and explanatory variables to be included in the regression equations. A natural question that arises is how to choose the most appropriate model among the potential models. Some of the criteria that have been suggested for selecting appropriate biomass and volume models are reviewed below. It is difficult however to find a single criterion or statistic to demonstrate that one model is definitely better than another (Schlaegel, 1981).

Coefficient of Determination

Probably the most commonly used model selection criterion is the coefficient of determination, R^2 value. This statistic indicates the proportion of the total sum of squares of the dependent variable explained by the regression. One major disadvantage of this statistic is that it can be used to compare two or more models only if the units of the dependent variable are the same between models. Secondly, inclusion of additional independent variables never decreases the R^2 value, even though they may not be statistically significant (Schlaegel, 1981). Almost all statistical packages today allow for the user to provide the R^2 as part of the output of the model estimation process.

Standard Error

According to Schlaegel (1981), the use of the standard error of the estimate ($s_{y.x}$) as a selection criterion for models is only second to the coefficient of determination. The standard error is calculated from the residual sum of squares of the regression of any equation by:

$$s_{y.x} = \sqrt{\frac{\text{Residual sum of squares}}{(n-p)}} \tag{6}$$

where:

n = the number of observations

p = the number of coefficients estimated in the model

(n - p) = degrees of freedom of the residual term

This is a measure of the variation in the observed dependent variable values not accounted for by the linear relationship with the independent variable(s) (Husch, 1963). The standard error is a function of the number of coefficients estimated from the model since the denominator is dependent on sample size and number of regression coefficients in the model. Transforming the dependent variable changes the magnitude of the standard error for the same equation making it inappropriate for comparing equations in different units or with different dependent variables. As Schlaegel (1981) noted, the standard error is difficult to interpret without additional information such as the distribution of the data, the mean and range of the dependent variable, etc.

Fit Index

An index of fit (Fit Index) comparable to the R^2 value can be obtained for every equation. In the case of untransformed linear regressions, the Fit Index is equal to the coefficient of determination (Schlaegel, 1981). To calculate the FI, the predicted values are transformed back to the original units and corrected for bias if needed. The total corrected sum of squares is given by:

$$TSS = \sum(Y_i - \overline{Y})^2 \tag{7}$$

and residual sum of squares by:

$$RSS = \sum(Y_i - \hat{Y}_i)^2 \tag{8}$$

where

Y_i = value of the *ith* observation in actual units

$\overline{Y}$ = arithmetic mean of Y, in actual units

$\hat{Y}_i$ = the *ith* predicted value of Y_i converted to actual units

The fit index is:

$$FI = 1 - \frac{RSS}{TSS} \qquad (9)$$

Standard Error of Estimate in Actual Units

Using the residual sum of squares of the regression in the actual units of measure, a standard error of the estimate in actual units, s_e may be calculated as follows:

$$s_e = \sqrt{\frac{\Sigma(Y_i - \hat{Y}_i)^2}{(n-p)}} \qquad (10)$$

where Y_i and $\hat{Y}_i$ are observed and predicted values of Y in actual units respectively, n is the number of observations, and p is the number of coefficients in the model (Schlaegel, 1981).

Coefficient of Variation

A useful statistic for making quick comparisons between models is the coefficient of variation (CV) expressed in actual units as per cent:

$$CV = \frac{s_e}{\bar{Y}} x100 \qquad (11)$$

The CV is an index of the variation among means of the predicted Y's after accounting for the variation due to the measured variables. This statistic should be much smaller than the coefficient of variation of the sample tree mean.

The Furnival Index

The usual index of fit, the root mean square residual (standard error), and the coefficient of determination can only be used to compare equations that have the same dependent variable, but are not suitable when transformations of the dependent variables are involved (Furnival, 1961; Crow, 1971).

Furnival (1961), proposed an index for comparing equations used in constructing volume tables with different dependent variables, based on the maximum likelihood principle. The index, known as the Furnival Index (I) is computed in three stages. First, the standard error of the residuals is obtained by fitting all equations to the data. Next, the geometric means of the derivatives of the several dependent variables with respect to volume are computed with the aid of logarithms. Finally, the standard error is multiplied by the inverse of the appropriate geometric mean (Furnival, 1961). The Index is given by:

$$I = [f'(V)^{-1}]s \qquad (12)$$

where I is the Furnival Index, $f'(V)^{-1}$ is the reciprocal of the derivative of the transformation applied to the dependent variable [volume] with respect to volume and s is the standard error of the fitted regression. For the common transformations, the corresponding geometric means are presented in Table 1 (Furnival, 1961; Alder, 1980).

Table 1. Examples of geometric means of common transformations used in volume table construction

Transformation	Geometric mean
Log V	$anti\log = \frac{2.30\sum \log V}{n}$
lnV	$anti\log = \frac{\sum \log V}{n}$
V/D^2	$anti\log = \frac{\sum \log D^2}{n}$
V/D^2H	$anti\log = \frac{\sum \log D^2 H}{n}$

where *D* is the dbh, *H* is height and n is the number of observations. The equation with the smallest index is selected as the one that gives the best fit to the data. In the case of an untransformed equation (i.e., where the dependent variable is volume), the Furnival Index reduces to the usual estimate of the standard error of the regression.

6.4. Site Quality

6.4.1. The Concepts of Site, Site Quality and Site Productivity

According to Skovsgaard and Vanclay (2007) the term *site* refers to a geographic location that is considered homogeneous in terms of its physical and biological environment; and in forestry, site is usually defined by the location's potential to sustain tree growth, often with a view to site-specific silviculture. Sites are therefore classified into *site types* according to their similarity regarding climate, topography, soils and vegetation. *Site quality* refers to the combination of physical and biological factors characterising a particular geographic location or site, and may involve a descriptive classification (Skovsgaard and Vanclay, 2007). The properties that determine site quality are generally inherent to the site, but may be influenced by management. *Site productivity* in forestry is usually evaluated in terms of how much wood biomass (or any other desired material) the site can produce. However, the history of the stand such as when the stand originated, management regimes, pests and diseases, weather etc., all affect the productivity of a site (Davis et al., 2001).

6.4.2. Methods of Site Quality Assessment

Clutter *et al.* (1983) describe both direct and indirect methods of site quality evaluation. Estimation of site quality based on historical yield records, stand volume data and stand height data constitute the direct methods. The indirect methods include the use of overstory interspecies relationship, lesser vegetation characteristics and from topographic, climatic and edaphic factors using a) site index, b) non-tree-vegetation, and c) basic environmental and land attributes (Jones, 1969). These indirect methods result from the difficulty of measuring the productivity of a site growing at its maximum potential through volume production. Non-tree vegetation typing as an indicator of site quality has been used extensively in Canada and Scandinavia. This method is based on the knowledge that in certain ecosystems, certain plants or plant communities are associated with certain forest types and to some extent, with site quality (Davis et al., 2001). The method of using physical environmental factors as measures of site quality evolved from the difficulties of correlating understory vegetation with site quality (Coilie, 1938). The most widely used land attribute is to link soil type with site productivity, as soil has a large and often controlling influence on tree growth, especially as it relates to moisture and nutrients (Davis et al., 2001).

In the remaining sections, I will focus on the use of site index as a measure of site quality, because, of all the indirect methods of site quality that have been investigated, the use of tree height growth appears to be the most practical, consistent and useful indicator of site quality (Davis et al., 2001). Consequently, site quality estimation from stand height data is the most popular method and usually involves the use of site index curves. Any set of site index curves is simply a family of site development patterns with qualitative symbols or numbers associated with the curves for referencing purposes (Clutter et al., 1983). Site index curves are derived either graphically or by statistical curve-fitting procedures (Alder, 1980; Clutter et al., 1983). Data for the development of site index curves are obtained from three main sources: temporary sample plots (TSPs), permanent sample plots (PSPs) and stem analysis. TSPs provide the most inexpensive and quickest data, but the use of such data involves the assumption that the full range of site indices is well represented in all age classes within the sample (Alder, 1980; Clutter *et al.*, 1983; Avery and Burkhart, 1994).

Although site index is still the most widely used index of site productivity, it is important to point out the inherent assumptions in this method and concerns that have been raised by previous researchers. The main assumption in using site index is that height growth correlates well with stand volume growth (Skovsgaard and Vanclay, 2007). Some empirical studies by Assmann (1955, 1959) have shown that this assumption does not always hold, and this has led to the development of the concept of *yield level*, which is the stand volume growth per unit of height growth. To apply this concept, a stand density index based on the combination of stem number and quadratic mean diameter is used to provide an indication of the yield level, which may be used to adjust height-age – based estimates of site productivity (Skovsgaard and Vanclay, 2007).

6.4.3. Developing Site Index Curves

The statistical methods used for site index curves are of three general types: the guide curve, the difference equation and the parameter prediction methods (Clutter et al., 1983). The most common equation forms used to fit site index curves are the Schumacher (1939) and Chapman-Richards equations (Richards, 1959; Chapman, 1961).

To develop site index curves using the proportional curves method described by Alder (1980), a single equation is fitted to the plot level top height[3]/age data, using the logarithmic transformation of Schumacher's (1939) equation given as:

$$\ln(H_{top}) = \ln(H_{top0}) + \beta A^{-k} \quad (13)$$

where H_{top} is the mean top height, H_{top0} is the maximum top height the species could reach on the site (the asymptotic top height), A is the age of the stand, β is a regression coefficient and k is a constant. Using nonlinear regressions, we can estimate the value for k iteratively. Nanang and Nunifu (1999) have found that the value of k for teak and neem plantations in Northern Ghana was approximately 0.5. The site index (SI) for each plantation is determined by the algebraic re-arrangement of Equation (13) as follows:

1. A reference age of for the species is fixed. In this example, a base age of 20 years is used.
2. SI of each stand is then defined as the top height of the stand at the base age (20).
3. By substituting H_{top} = SI and A = 20 into Equation (13), eliminating $\ln(H_{top0})$ and re-arranging terms, the expression for SI is obtained as follows:

$$\ln(SI) = \ln(H_{top}) + \beta\left[\left(\frac{1}{A}\right)^{0.5} - \left(\frac{1}{20}\right)^{0.5}\right] \quad (14)$$

Equation (14) can then be used to calculate the site index of each stand.

6.4.4. Choosing a Functional Form to Fit Site Index Curves

Foresters have been interested in the specification and estimation of site index curves for a long time now. The proportional or guide curve method of site index curve development described above has become the most popular method for even-aged single-species plantations. The reason has been the lack of re-measured permanent plot databases and the inability of most species to show annual or seasonal growth rings. To address urgent needs for these curves for management, the common practice has been to collect data from temporary plots from stands of different ages to cover, at least, a major part of the rotation of the species. The basic requirement is to assume that the distribution of plots with respect to site quality is identical across all ages (Alder 1980). Failure of this assumption (when certain site/age combinations are unavailable) will bias the fitted guide curve away from the true average height growth trend (Walters et al., 1989). Other factors that can lead to biases in the site index curves when temporary sample plot data are used and alternative explanations of these biasing effects are discussed by Beck and Trousdell (1973), Smith (1984) and Walters et al. (1989). However, biases that arise due to the use of an inappropriate functional form were only addressed in a comprehensive way by Nanang and Nunifu (1999). The main conclusion is that the mathematical properties of the models provide some guidance for the

[3] The average height of the largest 100 trees per hectare.

selection of the most appropriate functional form using both the data-related criteria and the characteristics of the models themselves (see Nanang and Nunifu, 1999 for details).

6.5. Case Study 1: Modelling Growth and Yield of Neem Plantations

In this section, an application of stand dynamics modelling techniques is presented. The data used in the analyses were collected from neem plantations planted in the savannah zones of Ghana.

6.5.1. Sampling Procedure

The data were collected from 120 temporary sample plots selected from 30 plantations within the study area using a stratified two-stage sampling design. Table 1 shows the distribution of sample plantations, plots and the number of trees which were measured.

Table 2. Distribution of sample plantations, plots and trees measured by town/village

Town/Village	No. of Plantations Observed	No. of sample plots measured	Total No. of trees measured
Tamale	8	32	580
Nyankpala	6	24	563
Kumbungu	5	20	500
Kumbungyili	3	12	300
Katariga	2	8	185
Vitin	1	4	100
Nyashie	1	4	100
Tarikpaa	1	4	100
Jangyili	1	4	100
Dalogyili	1	4	100
Choggo	1	4	65
Total	30	120	2, 693

All plantations of neem in the Tamale Forest District were stratified by age into 5 groups (1-, 2-, 3-, 4-, and 5-year age groups). Five plantations were selected randomly for measurement in each stratum. In addition, 3 and 2 plantations from 6- and 9-year age groups respectively were also selected randomly for measurement. In all plantations selected for study, the following procedures were carried out:

a) four square plots with a 10 m x 10 m sides (1/100 ha) were set up in each stand. To avoid the effect of errors in locating plot boundaries, plot boundaries were laid exactly half-way between the lines of trees. Since the usual initial spacing for neem and other plantation species in the area is 2 m x 2 m, in the absence of mortality, each sample plot contained 25 trees. It was realised after measuring a few plantations that selective felling had been done in some of the plantations above 3 years. It was therefore decided that instead of setting up random plots within the plantations, selective sampling should be done in areas as fully stocked as possible. It was therefore not possible to estimate mortality as this was confounded with human effects.

b) all trees in each plot were measured for diameter at 50 cm above ground and diameter at 130 cm (diameter at breast height [dbh]) with either digital callipers or a diameter tape (for the larger trees) to the nearest mm, and for total height with height poles to the nearest cm. For trees that were less than or equal to 130 cm in height, only diameter at 50 cm and height were measured. Each stem of a tree forking below breast height was measured at 130 cm and recorded separately and the height of only the tallest stem was measured. A single diameter corresponding to the diameter of the tree of the same basal area as the total basal area of all the forked stems was then calculated using the formula:

$$D_m = 2x\sqrt{\sum_{i=1}^{n} D_i^2} \tag{15}$$

where

D_m is the mean diameter, D_1, D_2,.. D_n are the diameters at breast height of the first, second etc. to the *n*th forked stem.

c)the basal areas of all trees measured on each plot were calculated using the diameters at 50 cm. One tree from each plot with dimensions as close to the tree of mean basal area (at 50 cm) and height as possible was selected and felled close to the ground with a cutlass. Just after felling the sampled tree, the following variables were recorded:

i). diameter at 50 cm above ground level,
ii). diameters at 0, 25, 50 and 75% of the total height,
iii). diameter at breast height,
iv). total height,
v). fresh weight of the whole tree,
vi). fresh weight of a 10 to 30 cm stem section in the centre of the stem for determination of dry to fresh weight ratio.

Figure 2. Dr. Thompson Nunifu (left) and Mr. James Amaligo taking measurements of a 5-year-old neem plantation.

The stem sections were air dried and their air-dry weights taken.

6.5.2. Data Analysis

Individual Tree Volume Computation

Smalian's formula was used in conjunction with the dbh, height and diameters at 0, 25, 50 and 75% of total height to compute the volumes of the 120 sample trees. As the sample trees were measured at equal intervals along the stems, the formula was reduced to:

$$V = \frac{H}{8}(B_1 + 2B_2 + 2B_3 + 2B_4) \qquad (16)$$

Where:

V = total stem volume

H = total height

B_1, B_2, B_3, and B_4 = basal areas at 0, 25, 50 and 75% of total height

Individual Tree Volume Equations

The individual tree volumes were used to develop local and standard volume table equations using regression analysis. Local volume equations estimate volume using only diameter, whilst standard volume equations use both diameter and height as explanatory variables. Prior to the analysis, 12 of the sample trees were discarded because they were shorter than 130 cm and therefore had no diameters at breast height. The remaining 108 trees were used for the development of the volume tables. Fifteen of the common regression models used in volume estimation (Unnikrihnan and Singh, 1984) were fitted to the computed volumes in order to determine the most appropriate model. Six of these models have single independent variables and the remaining nine have two independent variables. The independent variables used were either dbh or height or combinations of both. The criterion for selecting the best regression models was the Furnival Index (Furnival, 1961).

Plot Volume Computation

The best standard volume equation among the 15 models compared on the basis of the Furnival Index was:

$$\ln V = -1.689 + 1.165 \ln D + 1.124 \ln H \qquad (17)$$

where V is the total tree volume in dm^3, D is diameter at breast height and H is total height. This equation was used to estimate the volumes of each tree on each measured plot. Plot volumes were then calculated as the sum of the volumes of all individual trees on that plot.

The computed chi-square value between the actual sample tree volumes and the volumes predicted by Equation (17) was 28.78, which was not statistically significant at the 0.05 probability level [$\chi^2_{(0.95,\ 104)}$ = 124.34]. The average error in using Equation (17) to predict volumes was - 6.07%. This means that on average, Equation 17 underestimated individual tree volumes by about 6%.

Allocation of Stands to Site Classes

All plantations were designated Site Class I, II or III in relation to their mean dominant height/plantation age. There are two main ways to measure the dominant height of forest trees: top height and predominant height. Top height is defined as the average height of a specified number per unit area of the trees with the largest diameters at breast height. Predominant height, on the other hand, is the average height of a specified number of trees per unit area of the tallest trees in the stand (West, 2003). Practically, it is easier to determine top height than predominant height, because it is easier to identify the largest diameter trees in a stand than the tallest ones (West, 2003). Therefore, in this study, top height was used.

The method used was the minimum - maximum procedure described by Alder (1980). In each age class, the minimum, mean and maximum top heights were calculated. Three separate linear regressions were then fitted to the minimum, mean and maximum sets of observations using the logarithmic transformation of Schumacher's (1939) equation:

$$H_0 = H_{max} \exp(\beta A^{-k}) \tag{18}$$

where H_o is the mean top height, H_{max} represents the maximum height the species could reach on the site, A is the age of the stand, β and k are coefficients. Taking natural logarithms of both sides gives:

$$\ln H_0 = \ln H_{max} + \beta A^{-k} \tag{19}$$

For each site class, this equation was fitted using regression techniques. Pecked lines were then drawn half-way between the site classes to show the boundaries between them. In this study, k was estimated to be 1 using the procedure suggested by Alder (1980). This procedure requires that the residual sum of squares for Equation (19) for various trial values of k be computed. The value of k at which the minimum sum of squares is observed provides the best estimate of k. The regressions are then recalculated using this value of k to give the corresponding best estimates of the α and β parameters. If $\alpha = \ln H_{max}$, the equation becomes

$$\ln H_0 = \alpha + \beta A^{-1} \tag{20}$$

6.5.3. Yield Table Construction

The basic form of the Schumacher (1939) equation was used for the development of yield tables for Site Classes I, II and III:

$$\ln Q = \alpha + \beta_1 A^{-k} + \beta_2 S + \beta_3 D_s \tag{21}$$

where Q is some measure of yield (height, dbh, basal area, volume, or fresh weight), A is stand age in years, S is some function of site index, D_s is some function of stand density and α, β_1, β_2, and β_3 are regression coefficients.

Site and stand density were excluded as explanatory variables from Equation (21) since individual equations were developed for each site class and a uniform stand density of 2500 stems/ha was used in all site classes. The reciprocal of stand age had a high correlation with *lnQ*. For these reasons Equation (21) was reduced to:

$$\ln Q = \alpha + \beta_1 A^{-1} \qquad (22)$$

According to Alder (1980), the β_1 parameter is always negative and α is usually between 2 and 7.

6.5.4. Results

Diameter and Height Statistics

The diameter and height statistics of the sample plantations are summarised in Table 3.

Table 3. Diameter and height statistics of measured neem plantations

Age	Dbh (cm)				Height (m)		
(years)	Mean	st. error	Range		Mean	st. error	Range
1	0.82	0.02	0.16 - 2.65		1.42	0.02	0.53 - 3.11
2	2.26	0.03	0.35 - 4.60		2.14	0.03	0.60 - 5.21
3	4.22	0.05	1.14 - 7.05		4.67	0.04	2.11 - 7.76
4	5.43	0.08	2.00 - 8.72		4.74	0.04	3.11 - 9.34
5	6.07	0.09	2.50 - 11.61		5.41	0.05	3.63 - 10.61
6	8.18	0.14	4.71 - 12.26		5.85	0.10	3.71 - 11.73
9	11.40	0.27	9.05 - 16.81		9.21	0.17	7.64 - 12.57

The annual diameter increment ranged from 0.64 to 2.11 cm and averaged 1.41 cm. The standard error of the mean dbh ranged from 0.02 to 0.27 cm and increased with age. The mean annual height increment was 0.93 m and ranged from 0.07 to 2.53 m. The standard error of the mean height increased with age from 0.02 m in the first year to 0.17 m nine years after planting [Table 3]. The mean diameter at 4 years was 5.43 cm. This is slightly higher than the mean diameter at the same age of 5.14 cm reported for neem in north-eastern Nigeria by Verinumbe (1991). In this study, the mean heights at 2 and 5 years were 2.14 and 5.41 m respectively. These were lower than mean heights reported for the same ages as 3.6 and 7.5 m respectively by Streets (1962). It is difficult to compare these statistics with those of Streets (1962) because it is not clear which part of northern Ghana nor the type of neem plantations from which the measurements were taken.

A simple linear relationship between height and dbh was developed to predict height for any given diameter. This equation is given as:

$$H = 1.167 + 0.675D \qquad (23)$$

$R^2 = 0.97$

where H is the total height and D is the diameter at breast height.

Stem Form Factors

The ideal tree trunk from a timber production point of view is a cylinder. The comparison of tree bole form with cylinders may be expressed as form factors. Form factors are usually measured as the ratio of an upper-stem diameter with the dbh. Higher form factors indicate lower rates of stem taper and vice versa. In this study, both cylindrical and conical form factors (when stem is compared to a cone) were computed for each sample tree and the means for each year calculated. The summary is presented in Table 4.

Table 4. Summary statistics of form factors

Age (years)	Cylindrical Form Factors				Conical Form Factors		
	Mean	SE	Range		Mean	SE	Range
1	0.25	0.01	0.20 - 0.33		0.76	0.05	0.59 - 0.99
2	0.29	0.01	0.26 - 0.32		0.87	0.03	0.78 - 0.96
3	0.31	0.01	0.21 - 0.48		0.94	0.02	0.69 - 1.44
4	0.30	0.01	0.25 - 0.34		0.90	0.02	0.74 - 1.04
5	0.28	0.01	0.24 - 0.32		0.93	0.03	0.73 - 0.95
6	0.38	0.04	0.28 - 0.42		1.15	0.11	0.84 - 1.35
9	0.41	0.05	0.35 - 0.48		1.25	0.15	0.99 - 1.42

The mean cylindrical form factors varied from 0.25 in the first year of growth to 0.41 in the ninth year. The conical form factors ranged from 0.76 to 1.25 in the ninth year. The form factors tend to increase with age though they are uniform within the age groups as shown by their standard errors [Table 4]. The current initial spacing of 2 m x 2 m for neem appears to be wider than optimum because the form factors show that the trees are more conical than cylindrical. Closer initial spacings will ensure straighter and more cylindrical boles which are both desirable qualities for poles and rafters.

Relationship Between Stem Dry Weight And Fresh Weight

From this study, a simple linear relationship was established between dry weight and fresh weight of the stems of the sample trees. A linear regression model that can be used to predict stem dry weight from fresh weight is given by:

$$W_D = 0.716W_F \qquad (24)$$

$R^2 = 0.96$

where W_D is the dry weight of the stem and W_F is its fresh weight.

Local and Standard Volume Table Equations

The Furnival Index was used as the criterion to select the best equations from 15 equations compared for volume table construction. The following two equations were judged the most appropriate for the construction of local and standard volume tables, respectively:

$$\ln V = -0.780 + 1.711 \ \ln D \tag{25}$$

[n = 108, R^2 = 0.96, Correction Factor = 1.038]

$$\ln V = -1.689 + 1.165 \ \ln D + 1.124 \ln H \tag{26}$$

[n = 108, R^2 = 0.97, Correction Factor = 1.027]

where:

V = overbark stem volume in dm^3,
D = dbh in cm, and
H = total height in m.

In order to correct for log-normal bias in Equations 25 and 26, correction factors were calculated as $e^{s^2/2}$ where s^2 is the variance of the respective logarithmic equations (Baskerville, 1972) using Equations 4 and 5.

Table 5. Local volume table for neem in northern Ghana

Diameter Class (cm)	Volume (dm^3)
1	0.48
2	1.56
3	3.12
4	5.10
5	7.47
6	10.21
7	13.29
8	16.70
9	20.42
10	24.46
11	28.79
12	33.41

Equation (25) was selected as the best local volume table equation to provide quick volume estimates in places where height measurement is difficult or not possible to undertake. Separate local volume table equations were developed for each site class and compared with Equation (25) using the test of coincidence. The test showed that there was no significant difference in coefficients between the overall equation [Equation (25)] and the separate equations. Equation (25) underestimates volumes compared to the standard volume table (Equation (26)) but is satisfactory for field volume estimations. It is however recommended that the standard volume equation be used whenever precise volume estimates are required.

Equation (26) was selected for the standard volume table (Table 6) because it had the lowest Furnival Index. This equation had the following characteristics: dbh and height explained 97% of the variation in stem volume. Test of coincidence showed that there was a significant difference between Equation (26) and separate equations for the different site classes. Though these differences were statistically significant, they were so small that there is no practical justification for the use of separate equations for each site class.

Table 6. Standard volume table for neem in Northern Ghana [Volume In Cubic Decimetres]

Diameter Class (cm)	Height (m)									
	1	2	3	4	5	6	7	8	9	10
1	0.19	0.42	0.65	0.90						
2	0.43	0.93	1.46	2.02	2.59	3.19				
3		1.49	2.35	3.24	4.17	5.11	6.08			
4			3.28	4.53	5.82	7.15	8.50	9.88		
5			4.25	5.88	7.55	9.27	11.02	12.81		
6				7.27	9.34	11.46	13.63	15.84		
7					11.18	13.72	16.31	18.96		
8					13.06	16.03	19.06	22.15		
9						18.39	21.72	25.40		
10						20.79	24.72	28.77	32.79	36.90
11							27.62	32.10	36.64	41.24
12							30.57	35.52	40.55	45.64

Without the necessary correction for log-normal bias, Equations (25) and (26) underestimate volumes by only 3.8 and 2.7% respectively and therefore the correction factors can be neglected when precise estimates are not required. To correct for bias, the correction factor is multiplied by the geometric mean to obtain the unbiased arithmetic mean. These correction factors were applied in the construction of both volume tables in Table 5 and 6.

The addition of form factors to the independent variables in both Equations (25) and (26) showed that change in form had an insignificant effect on volume estimates.

Top Height/Age by Site Class Curves

Equation (17) was used to develop the top height/age curves in Figure 3. This figure shows means and ranges of top height in Site Classes I, II, and III. The top height over age by site class curves shown in Figure 3 are to be used to classify stands into site classes for yield estimation. To determine the site class of a particular stand, it is necessary to determine the top height and age of the plantation. For example, a plantation that is 7 years old with a top height of 11m will be classified as Site Class I, while it will be classified as Site Class III if the same plantation had a top height of 8m, using Figure 3. It might be difficult to determine plantation age on the field since there are no clearly visible annual growth rings. The ages of plantations can usually be obtained from records at the Forestry Office in Tamale or by the Forestry Technical Officer in charge of the area where the plantation is located. Individual farmers and community leaders will also be able to help determine the ages of their plantations.

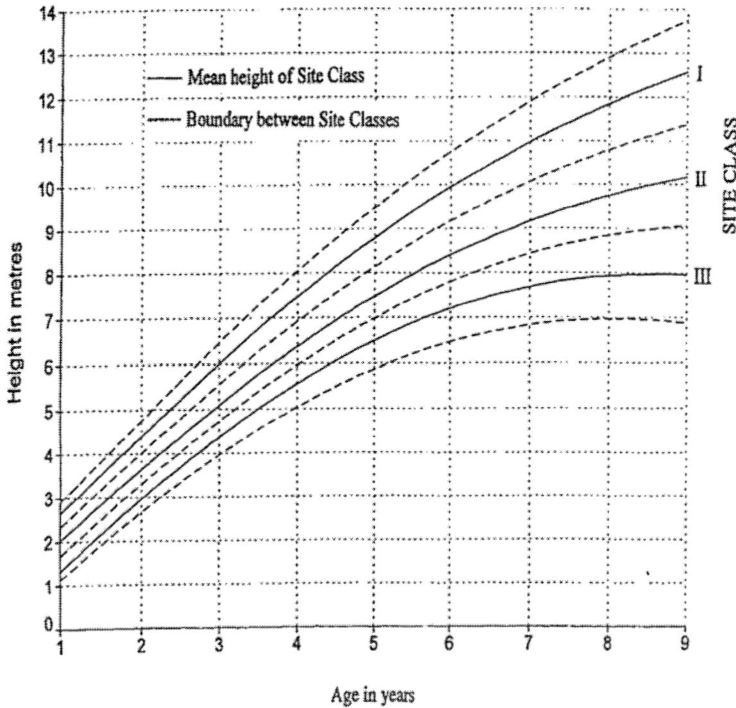

Figure 3. Top height/age curves site class curves for neem in Northern Ghana.

The accuracy of the method used for the site index curves depends critically on the assumption that all sites have an equal likelihood of being represented in each age class (Alder, 1980; Clutter et al., 1983). Since the plantations were randomly selected with no conscious effort to include any particular site type, the above assumption is likely satisfied for the 1-5 year age groups. However, it cannot be guaranteed for the 6- and 9-year age groups where only 3 and 2 plantations were measured respectively. For example, if older-age plantations are found on poorer sites and the younger ones on good sites, it can lead to bias in the site index curves. Several factors operate to prevent equal representation of all sites in the various age classes. Good sites produce bigger and taller trees more rapidly than poorer sites and therefore both selective felling and harvests take place earlier on good sites. Secondly, most of these lands on which neem is planted have been removed from agricultural production and since the poorer lands are removed first, older age plantations are more likely to be found on poorer sites. For these reasons, the site index curves produced by this method should be regarded as provisional. Permanent sample plot data will be needed to validate these curves.

Yield Tables

The coefficients of Equation (20) given in Table 7 can be used to develop empirical yield tables for neem in Northern Ghana. The R^2 values for the various yield parameters were high, ranging from 0.78 to 0.94. This means that between 78 - 94% of the variation in yield was explained by variation in age.

Table 7. The coefficients of Equation (20) used to develop the empirical yield tables for neem in Northern Ghana

Variable	Site Class I [Yield Class 12]			Site Class II [Yield Class 8]			Site Class III [Yield Class 4]		
	α	β_1	R^2	α	β_1	R^2	α	β_1	R^2
Diameter	2.75	-2.78	0.94	2.56	-3.17	0.92	2.34	-3.18	0.89
Height	2.50	-1.89	0.93	2.37	-2.14	0.87	1.99	-2.10	0.94
Basal Area	3.90	-5.45	0.89	3.45	-6.33	0.84	3.07	-6.23	0.82
Volume	5.20	-5.17	0.90	4.97	-6.53	0.78	4.80	-10.52	0.98
Fresh Weight	5.29	-4.83	0.92	4.84	-5.26	0.94	4.74	-7.42	0.91

Note: When these equations are used to estimate basal area, volume and fresh weight, the results are on a per ha basis.

The yield table is divided into Site Classes I, II and III; these are equivalent to Yield Classes 12, 8 and 4 m^3/ha respectively at optimum rotation. Both yield classes and site classes are used to refer to the productive capacity of a forest. Yield classes are measured in m^3/ha/year and are divided into steps of two m^3/ha. For example, Yield Class 12 has a mean annual timber increment of 12 m^3 per hectare. The biologically optimum rotation length for each site class is given by the estimated volume β_1 coefficients in Table 7. These rotation lengths are 5, 7 and 11 years for Site Classes I, II and III respectively. Because the MAI at optimum rotation for Site Class III occurs outside the range of the data (at 11 years), it was estimated as 4.23 m^3 /ha from Equation (20) using the coefficients in Table 7. Yields can be estimated by use of Equation (20) [coefficients in Table 7] for each stand attribute. The first step in estimating yields is to determine the site class to which the stand belongs. To do this, it is necessary to determine the mean height of the dominant and co-dominant trees and the stand age. Once the mean height and age are determined, Figure 3 is used to estimate the site class to which the stand belongs. The yield equation or table applicable to that site class is finally used to estimate the required yield variable.

The Yield Table probably underestimates the growth and yield potential of neem plantations in Northern Ghana as some of the plantations measured had been selectively harvested. Selective harvesting usually removes the largest trees for poles and rafters. Secondly, most of these plantations are established on either abandoned or infertile agricultural lands. The mean diameters and heights given for the plantations on Site Classes I and II indicate that trees on these good and medium sites can be used as small rafters at age three. Site Classes I and II can probably be thinned in the third year, Site Class III in the fourth. The time for thinning suggested here coincides with that being practised by the farmers.

In using either the equations or tables, caution should be exercised in predicting yields beyond the range of the data used in their derivation. This is important because the Schumacher model used here assumes asymptotic convergence and hence predictions of yield beyond age 9 years show that the yields flatten out after about 10 years of plantation age. It should also be appreciated that the yield equations represents averages to be expected for all stands in a given site and age class. Finally it should be noted that these yield table/equations are applicable to pure neem stands with an initial spacing of 2 x 2 m. Neem plantations with different initial spacing or those intercropped with other tree species or food crops will be expected to differ in their growth and yield characteristics. For these reasons, it is possible for observed yields for a given stand on the field to differ from those predicted by this equation.

6.6. Case Study 2: Comparing Statistical Distributions for Modelling Stand Structure of Neem Plantations

6.6.1. Introduction

Diameter and height are the two most important distributions for describing the horizontal and vertical structural characteristics of even-aged single species stands, respectively (Chen and Rose, 1978). However, there has been more extensive discussion of diameter distributions in the literature than height distributions. The most common continuous univariate distribution functions that have been used to describe diameter distributions are the Weibull distribution (Weibull, 1951; Bailey and Dell, 1973), gamma distribution (Nelson, 1964), lognormal distribution (Bliss and Reinker, 1964), beta distribution (Clutter and Bennett, 1965; Zöhrer, 1969) and Johnson S_B distribution (Hafley and Schreuder, 1977). Height distributions have been studied and reported by Chen and Rose (1978), Hafley and Schreuder (1977) and Kassier and Bredenkamp (1994), among others.

Forest managers and communities may also be interested in the distribution in size of both diameter and height. This information can be generated using bivariate distributions

(Schreuder and Hafley, 1977). Bivariate distributions are also useful for determining regression relationships between diameter and height. In the literature, Hafley and Schreuder (1976) compared several bivariate distributions potentially useful for describing joint frequency distributions of tree diameters and heights. Their results showed that the Johnson S_{BB} distribution was more flexible for describing joint frequencies of diameter and height compared with the other bivariate distributions they examined.

In this case study, the temporary sample plot data described in Case Study 1 were used to develop theoretical diameter and height distribution models for neem plantations in Northern Ghana. Height distributions are modelled using direct and indirect approaches. The direct method involves directly fitting height observations to the normal, lognormal, Johnson S_B and gamma distributions. The indirect method uses a relationship between diameter and height to indirectly derive a height distribution from a diameter distribution. Because diameters are more frequently measured (due to their ease of measurement) than heights, the ability to satisfactorily predict a height distribution from a diameter distribution will be of importance to local foresters. Finally, three bivariate distributions; the normal (SNN), lognormal (SLL), and Johnson SBB (Johnson, 1949b) are compared, with the aim of determining which one best describes the joint diameter and height frequency distributions of these plantations.

6.6.2. Methods

Direct Method of Fitting Univariate Diameter and Height Distributions

The classical approaches to diameter and height modelling are based on the assumption that at any point in time, the underlying diameter or height distribution of a stand under study can be adequately characterised by a probability distribution function from which the number of trees between specified diameters or height can be obtained (Knoebel and Burkhart, 1991). The direct method, therefore, involved the use of observed measurements of either diameter or height to estimate frequencies directly from the four theoretical distributions under study.

The probability density functions (pdf) of the univariate gamma and S_B distributions are described below. The normal and lognormal distributions are simple to fit and have been widely discussed in the statistics literature, [e. g., Johnson *et al.*, (1994) and Bury (1975)] and so the estimation procedures for these two distributions will not be repeated here. The estimated parameters for each distribution were used with the respective cumulative distribution functions to recover the frequencies by diameter or height class. In the following brief discussion of the distributions, x refers to either diameter at 1.3m or total height, depending on the distribution of interest.

Gamma Distribution

The pdf of the two-parameter gamma model with σ and λ as scale and shape parameters respectively, is given as:

$$f(x) = \frac{1}{\sigma\Gamma(\lambda)}\left(\frac{x}{\sigma}\right)^{\lambda-1} \exp\left[-\left(\frac{x}{\sigma}\right)\right] \tag{27}$$

Bury (1975) provides maximum likelihood equations for estimating the two parameters as follows:

$$\hat{\sigma}\hat{\lambda} = \bar{x} \tag{28}$$

and

$$\ln\hat{\sigma} + \psi(\hat{\lambda}) = \ln G \tag{29}$$

where $G = \prod_{i=1}^{n} x_i^{1/n}$ is the geometric mean of x and ψ is the diagamma function.

The estimated parameters of the gamma distribution using Equations 28 and 29 for each age group are given in Table 9.

Table 8. Summary of diameter and height data used in this study

Age (yrs)	Diameter						Height				
	n	Min	Max	SD	Skewness	Kurtosis	Min	Max	SD	Skewness	Kurtosis
1	187	0.50	2.64	0.42	1.71	3.40	1.50	1.80	0.09	0.05	-1.16
2	369	0.50	4.37	0.70	1.23	1.92	1.50	3.13	0.37	0.32	-1.02
3	498	1.14	9.43	1.41	0.47	0.86	2.11	4.51	0.37	0.43	-0.89
4	400	2.00	11.00	1.21	1.12	2.16	3.11	5.25	0.21	0.01	-1.12
5	450	2.29	11.51	1.55	0.59	-0.21	2.70	6.67	0.39	0.39	-0.89
6	200	4.71	13.19	1.92	0.08	-0.69	3.71	8.17	0.45	0.68	-0.73
9	100	7.50	16.80	1.87	0.88	1.54	6.72	10.80	0.85	0.06	-1.35

Table 9. Estimated parameters of the gamma distribution

Age (yrs)	Diameter		Height	
	λ_1	σ_1	λ_2	σ_2
1	7.020	0.143	35.936	0.054
2	5.995	0.229	22.389	0.106
3	12.743	0.331	24.774	0.189
4	12.753	0.426	38.107	0.124
5	10.080	0.602	27.702	0.195
6	38.194	0.214	47.837	0.122
9	49.193	0.232	45.627	0.202

Johnson S_b Distribution

The Johnson S_B distribution (Johnson, 1949a) has four parameters, two of which are the lower limit, ξ, and the range, ε, respectively. The pdf for the Johnson S_B distribution is:

$$f(x) = \frac{\delta}{\sqrt{2\pi}}\frac{\varepsilon}{(x-\xi)(\xi+\varepsilon-x)}\exp\left[-\frac{1}{2}\left(\gamma+\delta\ln\left(\frac{(x-\xi)}{\xi+\varepsilon-x}\right)\right)^2\right] \tag{30}$$

$\xi < x < \xi + \varepsilon, \delta > 0, -\infty < \gamma < \infty, \varepsilon > 0, -\infty < \xi < \infty = 0$ elsewhere

where $\gamma + \delta \ln\left(\frac{x-\xi}{\xi+\varepsilon-x}\right) = z_x \approx N(0,1)$. The estimation procedures for the remaining two parameters are discussed later. In the fitting procedure used in this study, the smallest diameter and height were set as the lower bound for each age group. These lower bounds are represented by ξ in Table 10.

Indirect Method of Fitting Univariate Height Distributions

If a relationship exists between two variables and a distribution is known for one of the variables, then a distribution can be derived for the other variable provided that the function describing the relationship can be inverted (Chen and Rose, 1978). For a given site and stand age, an indirect height distribution can be generated from a diameter distribution based on the relationship between height and diameter. In this study, the relationship given by $D = (H/\theta)^{1/\phi}$ was used, where H is the height, D is the diameter and θ and ϕ are positive regression constants related to species, age, site and stand density. Height distributions were therefore indirectly derived from the diameter distributions using $D = (H/\theta)^{1/\phi}$ in the cumulative density functions of the four univariate distributions. This equation was used because it was the best among those developed using the three bivariate distributions given in Table 15. Chen and Rose (1978) also found this relationship to be suitable for indirectly deriving a height distribution from a diameter distribution for a 32-year red pine plantation.

Fitting the Bivariate Distributions

The bivariate distributions considered in this Case Study are those for which both marginals are normal, lognormal and S_B distributions. Johnson and Kotz (1972) provide formulae for approximating bivariate probabilities of these distributions. Hafley and Schreuder (1976) note that fitting the bivariate gamma distribution is extremely complicated; neither the parameter estimation nor the calculation of probabilities is easy. Knoebel and Burkhart (1991) also note that the bivariate gamma distribution did not fit positively skewed and symmetrical distributions well. For the first reason, the bivariate gamma distribution was not pursued despite the good performance of its univariate counterpart in describing the diameter and height data.

If z_1 and z_2 are the standard normal variates for the Johnson S_{BB} distribution where

$$z_1 = \gamma_1 + \delta_1 \ln\left(\frac{y_1}{1-y_1}\right) \tag{31}$$

and

$$z_2 = \gamma_2 + \delta_2 \ln\left(\frac{y_2}{1-y_2}\right), \tag{32}$$

then z_1 and z_2 have the joint bivariate normal distribution (pdf) with correlation coefficient ρ given as:

$$p(z_1, z_2; \rho) = \frac{1}{2\pi\sqrt{(1-\rho^2)}} \exp\left[-\frac{1}{2(1-\rho^2)}\left(z_1^2 - 2\rho z_1 z_2 + z_2^2\right)\right] \quad (33)$$

In terms of diameter and height, $y_1 = \left(\frac{D-\xi_1}{\varepsilon_1}\right)$ and $y_2 = \left(\frac{H-\xi_2}{\varepsilon_2}\right)$ where D and H represent diameter and height respectively, and ε_1 ,ε_2 and ξ_1, ξ_2 are as defined above. The two remaining parameters of the S_B distribution are estimated as: $\hat{\gamma} = -\bar{f}/s_f$ and $\hat{\delta} = 1/s_f$ where $\bar{f}$ is the mean and s_f is the standard deviation of the transformation $f = \ln\left(\frac{y}{1-y}\right)$ with y defined as above.

Johnson (1949b) and Schreuder and Hafley (1977) discuss the properties of the S_{BB} distribution. One of the properties of interest is the regression relationship between height and diameter. Johnson (1949b) derived the median regression for the S_{BB} as follows:

$$\frac{(H-\xi_2)}{\varepsilon_2} = \theta\left[\left(\frac{\xi_1 + \varepsilon_1 - D}{D-\xi_1}\right)^{\phi} + \theta\right]^{-1} \quad (34)$$

where $\theta = \exp\left(\frac{\rho\gamma_1 - \gamma_2}{\delta_2}\right)$ and $\phi = \rho\frac{\delta_1}{\delta_2}$ with $\phi > 0$

In general, the mean regression is complicated and so the median regression is often used (Johnson,1949b, Schreuder and Hafley, 1977).

In the case of S_{NN} distribution, the pdf is also given by Equation (33), with z defined as: $z_1 = \gamma_1 + \delta_1 D$, $z_2 = \gamma_2 + \delta_2 H$, $\hat{\gamma} = -\bar{x}/s_x$, $\hat{\delta} = 1/s_x$ where $\bar{x}$ is the mean of diameter or height and s_x is the standard deviation.

For the lognormal distribution (S_{LL}), the pdf is given as (Bury, 1975):

$$p(z_1, z_2; \rho) = \frac{1}{2\pi(\sqrt{1-\rho^2})DH} \exp\left[-\frac{1}{2(1-\rho^2)}\left(z_1^2 - 2\rho z_1 z_2 + z_2^2\right)\right] \quad (35)$$

where: $z_1 = \gamma_1 + \delta_1 \ln D$, $z_2 = \gamma_2 + \delta_2 \ln H$, $\hat{\gamma} = -\bar{g}/s_g$, $\hat{\delta} = 1/s_g$ and $g = \ln D$ or $g = \ln H$. The mean and standard deviation of the log-transformed diameter or height are $\bar{g}$

and s_g respectively. For the bivariate normal and lognormal distributions, the median regressions are (Johnson, 1949b):

$$H = \ln\theta + \phi D \tag{36}$$

and

$$H = \theta D^{\phi} \tag{37}$$

respectively, where H is mean or expected total height in m for a given dbh, D is the tree dbh in cm and θ and ϕ are positive regression coefficients as defined under Equation (34) and are related to species, site and density.

6.6.3. Results

Univariate Distributions

Figures 4 and 5 show the predicted and observed diameter and height frequencies from all four univariate distributions, respectively (for all age groups combined). In general, all four distributions predicted diameter and height that were close to the observed. The figures show that the Johnson S_B performed well on the lower diameter and height classes, but overestimated both diameter and height in all classes above 5 cm and 7m, respectively. Given that these plantations are young, the observed distributions are generally positively skewed (Table 8). Therefore, the lognormal distribution is the appropriate choice for the diameter and height distributions if a single distribution is needed. Since this distribution is easy to fit even on a spreadsheet, it is a definite advantage for many developing countries where fast computers and expensive programmes may often be unavailable. A drawback of the univariate lognormal distribution noted by Hafley and Schreuder (1976) is that because it is limited to describing data that is positively skewed, it generally approximates symmetrically distributed data poorly. As these plantations age beyond nine years, it is possible that the lognormal distribution will fail to satisfactorily describe the height distributions.

With the univariate gamma and lognormal distributions, it is possible to predict the number of trees per hectare in any given diameter or height class. However, the use of neem for poles and rafters require that minimum diameter and height standards be met. The use of two separate univariate distributions does not allow for the simultaneous determination of frequencies that satisfy these diameter and height requirements. This is because for a given height (diameter), diameter (height) can vary considerably, depending on stocking and other factors.

The quality of fit of the four univariate distributions for each age group was assessed using the Kolmogorov-Smirnov (KS) statistic (Sokal and Rohlf, 1981). The ranking of each distribution based on the KS statistic for the seven age groups is summarised in Table 11. Ranks are in ascending order; the distribution with the smallest KS statistic between observed and predicted frequencies is given a rank of 1. The KS criterion compares the absolute difference between the cumulative frequency of the observed and expected frequencies. The null hypothesis is rejected when the greatest absolute difference between the observed and expected cumulative frequency is greater than the critical value. At $\alpha = 0.05$, the critical value for $n > 30$ is $1.36/\sqrt{n}$.

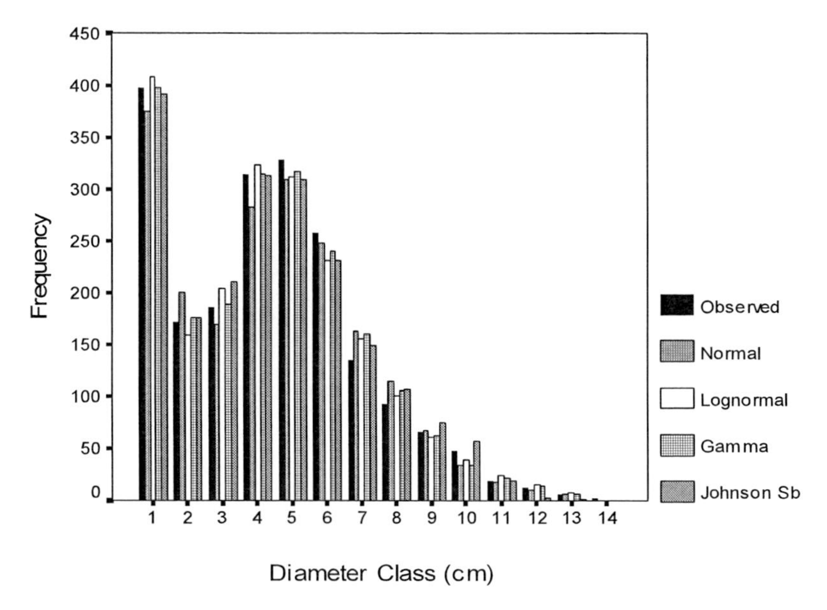

Figure 4. Comparison of observed and predicted diameter frequencies from the normal, lognormal, gamma, and Johnson S_B distributions for the neem plantations (all age groups combined).

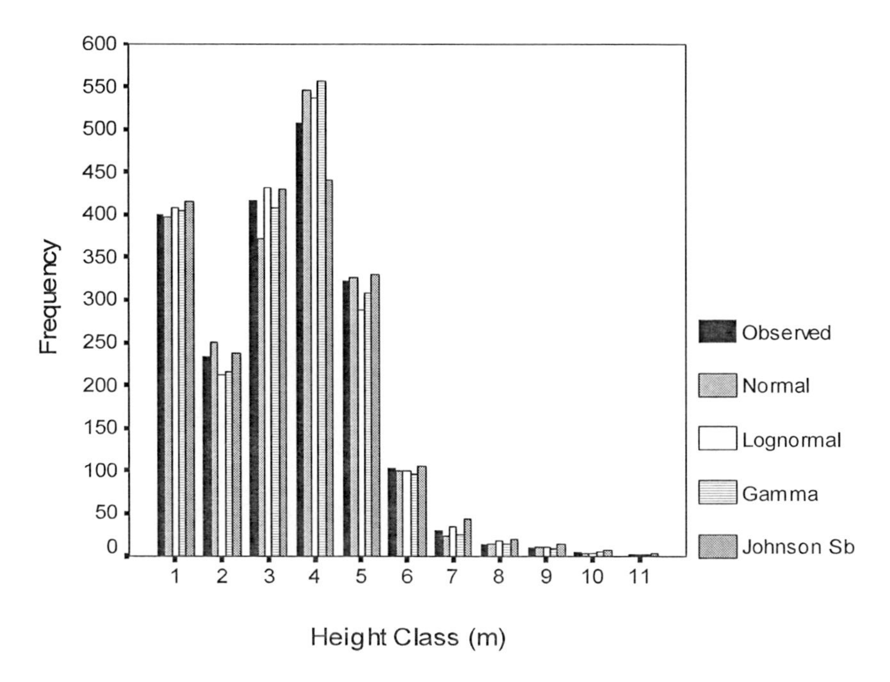

Figure 5. Comparison of observed and predicted height frequencies from the normal, lognormal, gamma, and Johnson S_B distributions for the neem plantations (all age groups combined).

The general problems with the use of the KS statistic as a goodness-of-fit test discussed in Reynolds et al., (1988) apply to this study. In particular, the fact that the KS test tends to be conservative. Almost all the univariate distributions predicted statistically insignificant differences between the observed and predicted frequencies for both diameter and height when the actual measured sample size was used to calculate the critical value (except those marked with asterisks in Table 11). However, when the data were converted to per hectare basis, most of these were significant because of the increase in sample size, which consequently lowered the critical value from 0.06094 to 0.02725. Based on the rank sums for the seven age groups, the gamma distribution was judged the best in describing the diameter data, whilst the lognormal was best for the height data (Table 11).

Table 10. Estimated parameters of the Johnson S_B distribution

	Diameter				Height			
Age (yrs)	ξ_1	ε_1	γ_1	δ_1	ξ_2	ε_2	γ_2	δ_2
1	0.500	2.150	1.129	0.780	1.500	1.620	0.921	0.747
2	0.500	3.100	1.000	0.872	1.500	2.460	0.641	0.804
3	1.140	6.870	0.311	1.276	2.110	5.650	0.257	1.282
4	2.000	9.510	0.789	1.226	3.110	4.530	0.717	1.023
5	2.290	10.90	0.804	1.086	2.700	6.450	0.475	1.319
6	4.710	6.580	-0.128	0.992	3.710	4.430	0.110	0.838
9	7.500	6.500	0.455	1.306	6.720	4.080	-0.501	0.707

Table 11. Ranking of the normal, lognormal, S_B and gamma distributions for diameter and height (rank sums in parenthesis) based on the KS statistic

	Diameter				Height			
Age (yrs)	Normal	Log normal	S_B	Gamma	Normal	Log normal	S_B	Gamma
1	1	2	3	2	2	2	1	3
2	4*	3	2	1	3	2	2	1
3	1	4	3	2	4	1	3	2
4	3	1	4	2	3	1	1	2
5	4	1	3	2	1	4	3	2
6	1	3	4	2	1	2	3	3
9	1	1	3*	2	2	1	2	2
Rank sum	15(2)#	15(2)	22(3)	13 (1)	16 (3)	13(1)	15(2)	15(2)

Note: *Significant at the 5% level. #Rank of distribution for all ages.

The observed and predicted diameter and height frequencies for the three-year age group are presented in Tables 12 and 13 respectively. This age group was chosen for presentation because it had the highest stocking among the age classes studied as a result of selective harvesting in plantations older than three years. All the four distributions performed well in predicting diameter and height in this age group, except in the tails of the distributions. In particular, the normal distribution overestimates both diameter and height in the lower tails, whilst the remaining three distributions overestimate these two variables in the upper tails.

Table 12. Observed and predicted diameter frequencies (No. of trees/ha) from the normal, lognormal, S_B and gamma distributions for the three-year age group

Diameter Class	Observed	Normal	Lognormal	S_B	Gamma
0.5 – 1.49	15	25	0	0	5
1.5 – 2.49	140	150	135	170	125
2.5 – 3.49	520	495	645	580	590
3.5 – 4.49	825	810	805	740	840
4.5 – 5.49	610	670	520	610	580
5.5 – 6.49	310	275	240	315	250
6.5 – 7.49	65	60	95	65	75
7.5 – 8.49	5	5	35	10	20
8.5 – 9.49	0	0	15	0	5
Total	2490	2490	2490	2490	2490
KS statistic		0.01606	0.04217*	0.03012*	0.01807

Note: * The critical value of the KS statistic at 5% level is 0.02725.

Table 13. Observed and predicted height frequencies (No. of trees/ha) from the normal, lognormal, S_B and gamma distributions for the three-year age group

Height Class	Observed	Normal	Lognormal	S_B	Gamma
1.5 – 2.49	5	30	5	5	5
2.5 – 3.49	280	245	245	290	230
3.5 – 4.49	830	780	905	800	880
4.5 – 5.49	865	935	850	835	910
5.5 – 6.49	405	425	360	460	375
6.5 – 7.49	100	70	100	80	80
7.5 – 8.49	5	5	20	20	10
8.5 – 9.49	0	0	5	0	0
Total	2490	2490	2490	2490	2490
KS statistic		0.024096	0.01606	0.02008	0.01807

Note: *The critical value of the KS statistic at 5% level is 0.02725.

Given that the lognormal was the best in fitting the observed height data, followed by the gamma distribution, the possibility of predicting parameters of the lognormal and gamma distributions for height from other stand attributes was considered useful. A regression method was used to predict the height parameters of the gamma and lognormal distributions from the estimated parameters of the diameter distribution and stand age. In this procedure, the parameters of both diameter and height of the gamma and lognormal distributions were estimated for all 30 plantations and linear regressions of the height parameters on stand age and diameter parameters estimated. The predicted height parameters were then used to fit a height distribution for each age class and the predicted frequencies compared with the observed using the KS statistic. The four regressions estimated to predict height parameters for the gamma (Equations 38 and 39) and lognormal (Equations 40 and41) distributions are:

$$\hat{\lambda}_H = 27.755 + 0.693\hat{\lambda}_D - 1.537A \qquad R^2 = 0.686 \qquad (38)$$

$$\hat{\sigma}_H = 0.036 + 0.167\hat{\sigma}_D + 0.013A \qquad R^2 = 0.676 \qquad (39)$$

$$\hat{\overline{H}} = 0.631 + 0.456\overline{D} + 0.047A \qquad R^2 = 0.983 \qquad (40)$$

$$\hat{s}_H = 0.148 + 0.162 s_D - 0.004A \qquad R^2 = 0.523 \qquad (41)$$

where D and H identify diameter and height parameters respectively, A is the stand age, s_H and s_D represents standard deviations of height and diameter respectively. The predicted height frequencies for the gamma distribution using parameters from Equations 38 and 39 showed that the KS statistic for the four-, five-, and nine-year age groups were not significant, whilst the remaining age groups had frequencies significantly different from the observed. In the case of the lognormal distribution, all predicted height frequencies using parameters from Equations 40 and 41 were significantly different from the observed.

Indirect Method

The results from the indirect method of fitting height distributions were a bit disappointing, as none of the four indirectly derived height distributions were satisfactory based on the KS criterion. For the normal distribution, all indirectly derived height frequencies were significantly different from the observed; whilst for the lognormal, Johnson S_B, and gamma distribution, only the two- and five- year age groups were insignificant. However, the gamma distribution was best overall with the smallest KS statistic in all age groups, though five of these statistics were significant at the 5% level. Comparisons of height frequencies from the direct and indirect methods were made using the KS statistic. The results showed that for the normal distribution, the statistic was significant for all age groups. For the lognormal, only the two- and five-year age groups were not significant, whilst for the Johnson S_B, the one- and two-year age groups were non-significant. Only the five-year age group was not significant for the gamma distribution.

6.6.4. Bivariate Distributions

The lognormal bivariate distribution was the most appropriate one to fit the diameter and height frequencies. This distribution provided predicted frequencies that were not statistically significant from the observed for six out of the seven age groups studied. The bivariate Johnson S_B was also found to provide a satisfactory fit, as four out of the seven age groups were not significant. The normal distribution was less satisfactory, with only one age group (age 2) providing a good fit. The observed and predicted frequencies for all age groups are not presented here; however, as before, the three-year age group is used as an example to present the observed and predicted bivariate frequencies for the S_{LL} distribution (Table 14).

With bivariate distributions, it is possible to estimate the number of trees/ha that meet the utilisation standards for poles and rafters. Field observations during data collection and sample measurements of poles and rafters often used in the study area show that the minimum diameters (at breast height) for poles and rafters are 5 cm and 10 cm, respectively. The minimum height requirements are 3 m for poles and 4 m for rafters. Using the information in Tables 12 and 13, the number of trees/ha that satisfy the diameter and height requirements separately can be determined. Table 14 can be used to estimate the bivariate frequencies that satisfy these requirements simultaneously for age group three. Since the lognormal distribution is a transformation of the normal distribution, bivariate probabilities of the lognormal distribution can be calculated using formulae by Johnson and Kotz (1972), or from charts and formulae provided by Owen and Wiesen (1959) or from statistical tables e.g., Pearson (1931).

Table 14. Observed and Predicted (O/P) frequencies (No. trees/ha) from the S_{LL} distribution of the neem plantations for the three-year age group

Diameter Class	Height Class (m)							
(cm)	2	3	4	5	6	7	8	Total
1	5/2	10/13						15/15
2		115/120	25/18					140/138
3		120/105	335/312	60/73	5/9			520/499
4		30/25	370/395	370/362	55/43			825/825
5		5/4	90/85	325/315	155/159	30/63	5/3	610/629
6			10/5	105/120	160/152	35/28	0/2	310/307
7				5/8	30/48	30/20		65/76
8						5/1		5/1
Total	5/2	280/267	830/815	865/878	405/411	100/112	5/5	2490

Regressions from the Three Bivariate Distributions

Median Regressions of height on diameter from the three bivariate distributions (Equations (34), (36) and (37)) were estimated and compared based on the χ^2 test and the sums of square deviations from the observed heights. Table 15 gives the estimated regression equations and the relevant statistics.

Table 15. Comparison of the three regression equations in terms of the sums of squares and χ^2 values between observed and predicted heights (all ages combined)

Bivariate Distribution	Equation	Sum of Squares	χ^2- values
Normal (S_{NN})	$H = 1.8327 + 0.5688D$	1130.60	237.82
Lognormal (S_{LL})	$H = 2.078D^{0.5252}$	993.53	194.81
Johnson (S_{BB})	$(H-1.50) = 7.732\left[\left(\frac{14.0-D}{D-0.5}\right)^{0.7245} + 0.8314\right]^{-1}$	1046.67	216.26

*Significant at 5%. Critical value for the χ^2- distribution at 5% is 2313.03

The regression from the S_{LL} distribution provided the best fit for the criteria used in the comparisons. The regressions in Table 15 were developed using the data from all age groups combined. The three regressions were used to predict heights of all observed diameters in the study. The χ^2- statistic showed that the observed heights were not significantly different from the predicted heights.

Similar regressions were estimated for each age group and subsequently developed height distributions from the predicted heights. This was done to determine whether given the diameter data alone, and the height/diameter relationship, it would be possible to generate a height distribution that is close to the observed. Based on the χ^2-test, the predicted heights were not significantly different from observed heights for all three-equation types. However, when these predicted heights were classified into height classes, the resulting distributions for all age groups were significantly different from the observed for the regression from the S_{NN} distribution. For the S_{LL} and S_{BB} regressions, three out of the seven age groups gave frequencies that were not significant. The frequencies from these regressions were also compared with predicted frequencies from the normal, lognormal and S_B distributions (direct method). That is, frequencies from the S_{NN} regression were compared with frequencies from the normal distribution for the seven age groups, the S_{LL} to the lognormal, and the S_{BB} to the S_B. Height frequencies from three out of the seven age groups from the S_{LL} and S_{BB} regressions were not significant, whilst for the S_{NN} regression, only the four-year age group was not significant.

6.7. Conclusions

The concept of stand growth dynamics is important in forestry because it allows managers to learn about the impacts of their management interventions on stand development. This chapter not only described the concepts of growth and yield and how to model them, but also provided examples of how this modelling is applied to actual plantation data from Ghana. The results of these examples have applications to the two plantations discussed. Although based on temporary sample plot data, they are still a useful start. In most other countries, the availability of growth and yield models for their major tree species is considered basic information for forest management. In Ghana, this basic information is still lacking and efforts must be made by researchers to develop these tools in order to support plantation development efforts.

PART III. PLANTATION FORESTRY ECONOMICS

Chapter 7

ECONOMICS OF PLANTATION FORESTRY

7.1. IMPORTANCE OF ECONOMIC ANALYSES IN FORESTRY

Economics is the study of how scarce resources are allocated among competing uses and how the wealth that is generated is distributed amongst users in order to further social objectives. The main reason economics exists as a field of study is because there are not enough resources for everybody to have all they want, i.e., resources are scarce. Scarcity is a result of resources having alternative uses. The application of economic principles to managing natural resources is not new, although there have been considerable shifts over the years from sustained yield management paradigms towards the concept of sustainable development. If current resources are managed sustainably, it would ensure that future generations also have access to some form of the resources to meet their own needs.

Economic considerations are important to any successful plantation development exercise, irrespective of whether the plantation is established by an individual, a community or a private firm. Forest economics applies economic principles to the management of forests. Forest economics has three main unique characteristics that distinguish it from other branches of economics: 1) a growing forest crop is both capital and an end product at the same time; 2) trees grow on long rotations; and 3) forests produce marketable and non-marketable products simultaneously. Economic analyses are applied in forestry to ensure that the economic implications of forestry practices are understood to enable forests to be grown optimally. Since resources have alternative uses (opportunity costs), it is important that if these resources are used to grow trees, the choice can be justified in economic terms. Below, I elaborate on the applications of economic principles to forest management. The list is not exhaustive and the tools used are not mutually exclusive since there may be some overlap in their application.

First, economic tools can be applied to determine *ex ante* whether undertaking particular activities associated with plantation development and investments make economic sense. Such decisions, which may include taxation, trade, investment, conservation, efficient use and optimal policy development, rely on economic principles for analysis and insights. Economic principles can be applied to choose among competing investments and between consumption and investments. For example, the decision of whether a piece of land should be used to plant trees or cultivate food crops can be made through economic analyses that examine all costs and benefits of each option. The application of benefit/cost analysis and other efficiency criteria are very important in making investment choices among forestry projects and between forestry and non-forestry investments.

Secondly, forests produce timber and non-timber products. The challenge for forest managers is to understand the nature and degree of tradeoffs that may be associated with the provision of multiple benefits. To understand these tradeoffs, we need to know the nature of the relationships between the various forest uses. Determining this relationship is a task of considerable complexity, although these relationships are known to fall into three main categories: independent, complementary, or competitive (Teeguarden, 1982). When the relationships between forest uses are independent or complementary, there are no tradeoffs. However, most forest uses are more likely competitive, especially at higher levels of use, making tradeoffs inevitable. In this case, economic tools can be used to guide efforts to assess these tradeoffs and to optimise the benefits from the production of multiple forest resources.

The free market system has been a cornerstone of economic theory for many decades. The perfect competitive market system is efficient in allocating resources only when the strict assumptions of perfect competition are met. However, markets can fail as a result of violations of some or all of these assumptions. When markets fail, inefficient outcomes that lead to over use of resources or pollution of the natural environment may result. Market failures abound in forestry due to the joint production of goods and services, some of which are not traded in markets. Under these circumstances, economic analyses can provide guidance on the best approaches for governments to intervene to correct such market failures and ensure efficient use and management of forest resources.

One interesting application of economic principles is to predict human behaviour. The assumption of utility maximisation in microeconomic theory, which applies to firms and people, provides a powerful foundation to predict human behaviour. Such theories and models can be used to predict the impact of policy measures on the forest industry as well as how people will react to the changes in policy. For example, the ability to predict how people would react to changes in legislation, taxation, subsidies, etc. is invaluable to governments in the design and implementation of forest policies.

Fifth, economic principles can be applied to choose instruments for forest policy. Once a forest policy decision has been made, there may be more than one approach to achieving the policy goal. Economics provides powerful insights into how best to structure policy measures to achieve the goals in the most effective and least costly way. In general, governments can apply command-and-control techniques or use economic instruments. Economists tend to favour the latter approach unless it can be proven that they will be ineffective. An example of economic instruments used in Ghana is the provision of incentives such as subsidies to plantation developers. On the other hand, legislation that mandates TUC holders to reforest harvested areas in their contract areas is a command-and-control measure to plantation development.

The subsequent sections discuss the basic concepts and tools needed to undertake economic analyses of plantation forestry investments.

7.2. Basics of Discounting and Compounding

7.2.1. Interest and Discount Rates

Interest is a fee for borrowing money, which is paid to the lender over and above the amount borrowed (principal). The amount of interest paid by the borrower is determined by the *interest rate*. Most people who borrow money from a bank or other financial institutions for various purposes are familiar with this concept. Investments in forest plantations are long-term in nature and require huge capital outlays in the first few years, while benefits accrue after many years. Many investors may not be able to fund these projects from their own resources hence there may be a need to take out a loan to finance the project. The loan will continue to attract an interest until it is fully paid back to the lender.

The interest rates charged by the banks or other financial institutions are made up of three main components. The first component is the real rate (or pure) interest rate. The real interest rate is the return on money borrowed that would prevail if borrowers and lenders expect repayment to be made in money having constant purchasing power (i. e., the interest rate in an inflation-free situation). This is approximately the rate of interest charged on Treasury Bills in the commercial banks. Added to this pure rate is an expected rate of inflation during the period of the loan. The third component is the risk rate, which the lender determines based on a risk assessment of the investment itself and /or the risk of the borrower. As an example, if the pure interest rate is 20%, the expected level of inflation is 10% and the risk rate is 5%, the total interest rate would be about 39%[1].

The *discount rate* is the interest rate used in determining the present value of future cash flows. This means that the discount rate is not *necessarily* the same as the interest rate. In general, the discount rate is the interest rate the Bank of Ghana charges on loans it makes to banks and other financial institutions. The banks in turn use the discount rate as a benchmark for the interest it charges on the loans it makes to consumers. Therefore, most borrowers who go to the bank get to deal only with the interest rate, rather than the discount rate.

The choice of an appropriate discount rate to use in forestry investment analyses is controversial. This is because the discount rate plays a very crucial role in determining whether a given project or investment is economically feasible or not, especially the long-term investments encountered in forestry. The idea behind discounting is to reflect time preference, and the discount rate is therefore the key to comparing values accrued over time. A high discount rate assigns less weight to future benefits (and costs) compared to the present, and vice-versa. Assigning less weight to benefits in the future is a way of compensating for future consumption for the inconvenience of forgoing present consumption. A second reason for discounting is that capital has an opportunity cost. Therefore, when capital is tied up in a particular investment, the loss of use of that capital in alternative investments must be taken into account. The main controversy in natural resource analysis is not whether future costs and benefits should be discounted, but what rate to use. Using a lower discount rate has a potential to make many projects seem profitable, and so could result in more environmental damage as more projects are undertaken and vice versa. A compromise solution might be to use the social rate of time preference, which reflects the

---

[1] Note that the correct way to calculate the total interest is not a simple sum of the three rates, rather it is the multiplication of them: (1.2)(1.1)(1.05)=1.386 or 39%.

weighted average of individual preferences of present versus future consumption. A major problem in Ghana is the habitual high bank interest rates (ranging from 20 – 40%). Since discount rates show some correspondence with interest rates, this will tend to make most long-term resource investments unprofitable from a purely economic standpoint. In conducting economic analyses involving discount rates, sensitivity analyses should be performed to assess the impacts of various discount rate scenarios on the results.

7.2.2. Present and Future Values

One important characteristic of forestry investments is the long separation in time between the time the investments are made (costs incurred) and the benefits (revenues) accrue. This means that many costs and revenues occur over time and hence there has to be a way to compare these costs and benefits. As noted above, these comparisons are made using the discount rate. The comparison can be made either at the maturity of the investment (final harvesting of trees) or at the beginning of the investment (initial year of planting). In the latter case, discounting is used to bring all future costs and benefits to the present (base year), while in the former case, compounding is used to move all future costs and benefits to the maturity date of the investment.

The following equations are used to calculate future values (compounding) and present values (discounting) respectively:

Future value equation:

$$V_T = V_0(1+d)(1+d)........(1+d) = V_T = V_0(1+d)^T \quad (1)$$

Present value equation:

$$V_p = \frac{V_T}{(1+d)^T} \quad (2)$$

where V_T = value of an amount T periods into the future

V_0 = value of the initial investment

Vp = present value of future investments

T = number of periods over which the investment is evaluated

d = periodic discount rate, stated as a decimal

In forest plantation investments, V_0 would represent the costs of site preparation, seed acquisition, planting and tending operations, and T would represent the rotation age. If the initial cost of planting is Gh¢1500/ha, the rotation length is 50 years, and the discount rate is 10%, then the future value of the investment is calculated as:

$$V_T = 1500(1+0.1)(1+0.1)........(1+0.1) = V_T = 1500(1.1)^{50} = 176,086.28 \quad (3)$$

If on the other hand, we want to compare investments in the present year (year 0), then we will use the present value formula. If the future value of the investment is known (assume Gh¢200,000), then using the same discount rate and timeframe, the present value is given by:

$$V_p = \frac{200,000}{(1+0.1)^{50}} = 1,703.71 \qquad (4)$$

Costs and benefits that occur between year 0 and T can easily be incorporated into this framework, by discounting to the present or compounding to the future maturity date of the investment.

7.3. The Concept of Value and Valuation

The calculation of present values requires knowledge of the values of the forest resource either in the present or in future years. The value of a good or service is an expression of its worth to an individual or group at a given time and context. It expresses how much someone is willing to give up for that good or service, and hence often measured by the willingness to pay (WTP). The more something is worth to an individual, the more the person is willing to pay for it. In general, therefore, when someone pays GH¢5 (five Ghana cedi) for a good, it is worth more to the person than another good for which he pays GH¢2. The *process* of assigning a value to a natural resource product or service is referred to as valuation.

Because value is a measure of worth, it is no surprise that there are different types of values; including social, economic, cultural and spiritual. Not all values are, or can be, expressed in monetary terms. Even then, it is still possible to compare the worth of various things in relative terms. For example, barter trade practised before the advent of paper and coin money relied on valuation of different goods based on values of other goods. Also, how long a person is willing to travel to acquire fuelwood relative to other substitute products in close proximity would be an indication of its worth or value to the person. Therefore, the fact that the value of a good or service cannot be expressed in monetary terms does not mean it has less value than one that can. This is why it is critical that forest resource valuation should take into consideration all values produced by the ecosystem, and not be restricted to only those for which monetary values can be assigned.

Another important dimension of value is the question of who is assigning the values to whom. In other words, whose values? For example, if we are discussing the value of banana fruits in a forest, is it the value of the fruit to a monkey who badly needs them for lunch, or the value to human beings when sold in a market? In most valuation, it is the value of the product or service to humans as assigned by humans, not other living things that may also have uses for that product or service. Hence natural resource valuation is essentially anthropogenic (human-centered).

7.4. Information Requirements for Economic Analyses

Benefit /cost analysis (BCA) is the process of estimating the economic values of the benefits and costs of a particular course of action to decide if it is worthwhile to implement. It is the main tool used in the evaluation of new technology and policies in natural resources. Costs and benefits are determined based on the economic principles of consumer choice and the role of markets in expressing social preferences as reflected in prices. BCA for policies or projects enhances transparency and accountability, adds rigour and discipline to the analysis, contributes to more informed decisions by assessing the full impacts of decisions on society, and provides common measures of comparing options.

Economic analyses of forest resources rely on information from various biophysical, social and economic spheres. Biophysical data include the growth and yield patterns of the trees and other resources within the forest, such as animals, birds, other vegetation, etc. Social information includes preferences and needs of the landowner, while economic variables include prices of goods and services produced from the forest, the costs of production, opportunity costs, and interest and discount rates.

Chapter 6 discussed growth and yield models and how they can be constructed and used. The following sections describe how benefits and costs are estimated to provide the essential data to undertake economic analyses of forestry investments.

7.5. Economic Valuation of Forest Resources

Estimating the prices (or value/benefits) of forest resources involves many different techniques. These techniques can be divided into two main types: market-based and nonmarket-based valuation. Market-based valuation can further be divided into two: those based on direct market prices, and those that rely on indirect market prices.

7.5.1. Market-Based Approaches

Market-based approaches are applied to value resources that are traded in the market place. In most cases, these prices are relatively easy to obtain. In a free market, the price of a commodity is established based on the interaction of demand and supply. The assumptions of perfect competition are pretty strong and are the foundation of price theory in neoclassical economic theory of product markets. These assumptions include: profit maximisation, many buyers and sellers (so no one has market power), homogenous products, no market entry/exit barriers and perfect information. When market prices are used to value resources, it is often assumed that these assumptions are satisfied, and hence the prices reflect the economic scarcity of the resource.

A. Direct Market Prices

To estimate values of resources traded in markets, the forest resource goods and services to be valued have to be clearly identified. Once these are known, data can be collected through surveys or the use of secondary data. For example, to determine the value per kilogramme of fuelwood or charcoal, a survey in the market of interest can easily reveal this. The prices of non-timber forest products that are sold in markets can also be obtained by surveying local markets. In the forestry sector, prices of wood products exported can be derived implicitly by using volumes and values exported published by the Food and Agriculture Organisation (FAO) of the United Nations. The other two approaches for obtaining prices of marketed products are direct observation of the prices in the specific market or conducting experiments (usually called experimental markets).

B. Indirect Market Prices

When the assumptions of perfect competition outlined above are violated, market prices of resources are distorted and hence indirect prices are the appropriate alternative measures of value for resources traded in markets. The following are examples of indirect methods used to derive market prices from the prices of other goods and services under these circumstances.

***Residual* Value Method**

This method subtracts the costs of production and profits from the prices of final goods and services to arrive at an estimate of a residual value. In forest valuation, it is often the case that there is a need to estimate the value of the trees in a forest. The value of the trees as they stand in the forest (on their stumps) is called the stumpage value. The residual value method is commonly used to determine stumpage values in forestry using the formula:

Stumpage value = log price – harvesting costs - transport costs –
processing costs - operators profit - risk allowance (5)

Note that the operator's profit and risk allowance have to be subtracted from the final price because it is necessary to pay entrepreneurship its return too, as this is a cost of production. The main advantage of this method is that it uses what is known about wood products values traded in a market to logically deduce stumpage values. However, figuring out all the costs is complex as the costs vary by forest location. This is even more complex given that many forest products companies are vertically integrated. Therefore, one may need to go through several stages of processing (and estimate processing costs at each stage) to obtain residual prices of trees. Another downside of this approach is that it can be difficult to find an upstream wood product market which is competitive.

Surrogate Prices

A surrogate price is the price of a close substitute of the product being valued. In this method, the price for a close substitute of the product is used as a proxy for the unpriced good or service. In the example of valuing fuelwood or charcoal above, if it is believed that the market is not competitive, one way to indirectly deduce the price of fuelwood would be to use the value of a close substitute in that location such as kerosene and adjust for calorific values of the two fuels. For example, see Kengen (1997) for descriptions of how the value of fuelwood can be derived from that of kerosene.

Resource Replacement Cost Method (RCM)

Resource replacement cost methods (RCM) are generally used where damage has occurred to a component of the forest ecosystem. This method can be used to value improved water quality by measuring the cost of controlling effluent emissions or valuing erosion protection services of a forest or wetland by measuring the cost of removing sediment load from downstream areas (King and Mazzotta, 2000). The approach involves assessing the cost of restoring, rehabilitating or replacing a damaged or depleted forest asset to its pre-damage condition. This cost is then used as an estimate of the benefit of restoration. The simplest way to apply this method if the damage is restored is to use the actual cost of the restoration. If the damage is not restored, the cost could still be estimated if the resource has close substitutes. In this case, sample values of the substitutes can be obtained from primary or secondary data and used to estimate the cost to replace the resource. For resources that possess unique characteristics, application of this method would be quite challenging. In practice, there are several other implementation challenges. First, unless the damage attracts widespread national attention, it may never be repaired, in which case it may be difficult to estimate the benefits. Second, even when the damage is repaired, the resource may appear identical to the original state, though the ecological functions of the damaged ecosystem may not be completely restored.

Appraisal Method

Another valuation method for assessing forest resources that have been damaged is the *appraisal method*. In the case of a forest that has been burnt, or a forest land that has been damaged, a knowledgeable appraiser may be able to identify a fair market value for the damaged property using comparable undamaged land. This value should reflect, as closely as possible, the price at which the resource would actually sell in the market place at the time of the damage (Ulibarri and Wellman, 1997).

7.5.2. Nonmarket-Based Approaches

There are some forest resources that are not traded in markets, or the appropriate markets do not exist for trading in them. As a result, prices for goods and services provided by these kinds of resources cannot be determined using market prices (the interaction of market demand and supply). These resources include the contribution of forests to clean air, value of forest watersheds, or erosion control. These benefits are difficult to assess using market-based valuation techniques. The various techniques that have been developed to estimate such values rely on observed and unobserved (hypothetical) behaviour to deduce the values of these resources to individuals and society. Below are brief descriptions of the major methods.

A. Travel Cost Method (TCM)

Every year, millions of people around the world travel to many natural resource destinations (parks, forests, lakes, streams, etc.) for recreational activities such as hunting, camping and fishing. These natural systems therefore provide valuable services to people. Since the benefits of such natural resources are not traded in markets, economists have developed methods to capture the values that people derive from these services. The values of these natural ecosystems and the benefits they provide can be used to provide useful information to support multiple forest resource management decisions.

The travel cost method has become popular in the natural resource valuation literature for estimating the value of recreational activities such as changes in access costs for a recreational site, elimination of an existing recreational site, addition of a new recreational site, or changes in environmental quality at a recreational site. All the models used in estimating the value of recreation sites rely on the fact that a visit to a recreation site involves transactions costs (travel cost, entrance fee, opportunity costs of time, etc). These costs are incurred solely to gain access to, and enjoy the benefits or services provided by, the site. The basis for the estimation technique is quite simple. For a given site, different individuals will face different travel costs, while one individual will face different costs to different sites. Individuals will respond to the variation in costs (implicit prices) and this will impact whether and how often they visit the site. For instance, if we were interested in estimating the value of the Mole National Park using the travel cost method, we would note that people coming from different parts of Ghana and abroad will face different travel costs. The cost of travelling from Bunkpurugu to Mole will differ from that from Najong No. 1 or Gbingbalanchet to Mole. This observed variation in travel costs are used to estimate a travel cost function and to evaluate the willingness to pay (WTP) to visit the site. An individual demand function for a site is generally of the form:

$$D = f(T_r, M, Q) \tag{6}$$

where:

D = the number of visits to the recreation site
T_r = the full cost of a given visit
M = individual's income
Q = environmental quality of site

The value of the flow of services from the site is then aggregated over all those who visited the site. The data to implement the TCM can easily be obtained by phone, onsite visits, mail surveys, or information from site registration data. Using this method to estimate the value of changes to a recreational site is premised on the knowledge that the environmental attributes of sites influence how often people visit the sites. Therefore, changes in visitation rates may reflect changes in the quality of the site.

B. Random Utility Model (RUM)

If we are interested in estimating the value of a forest site based on its characteristics, we would not be able to use the standard TCM described in the preceding section. A discrete choice, or random utility model, explains the choice among sites as a function of the characteristics of that site, and hence is more suitable for this purpose.

The following description of the RUM follows McFadden (1981).The basis for the RUM is the concept of indirect utility functions in microeconomic theory. Indirect utility functions characterise the maximum utility that can be achieved given prices and income. Discrete choice theory follows similar reasoning except consumption can only be in specific quantities and so allows for choices of zero or "corner solutions" in consumption. When modelling recreation demand, discrete, rather than continuous, choices are more realistic since recreationists make decisions based on going to a site or not and also because sites cannot be sub-divided infinitely or continuously. Recreational demand models typically have a finite set of alternative sites or are discrete. The choice of alternative sites is dependent on the utility, U; recreationists derive from various attributes, Q, of the site:

$$U_{in} = f(M, Q_{in}, S) \tag{7}$$

where the utility is a function of income (M), the quality of the site or attributes describing site i as perceived by recreationist n (Q_{in}), and other socio-economic variables (S). The choice set is defined as C_n, and n is the number of alternative sites (or a subset of sites). If site i is chosen, we assume that the utility associated with visiting site i is higher than for any other site j, i. e.,

$$U_{in} > U(Q_{jn}) \qquad \forall i \neq j; \quad i, j \in C_n \tag{8}$$

Utility in this framework is treated as a random variable since researchers do not have perfect behavioural information (McFadden, 1981; Smith, 1989). More formally, utility is modelled to include a systematic/observable component and a random or unobservable component:

$$U_{in} = V_{in} + e_{in} \tag{9}$$

where V_{in} is the systematic component of utility and e_{in} is a random element. The random element captures any unexplained factors that are not directly modelled or observed by the researcher.

C. Hedonic Price Method (HPM)

The RUM described above models the demand for a site based on its characteristics. The hedonic price method (or hedonic travel cost method if applied to recreation) goes further by attempting to value the characteristics of the site, using information on site demand. The method can be used to estimate economic benefits or costs associated with environmental quality, including air or water pollution, and benefits of environmental amenities such as aesthetic views or proximity to recreational sites (King and Mazzotta, 2000).

The HPM is premised on the assumption that people choose goods or services because of their characteristics. It was developed for, and has been applied extensively in, the real estate sector for valuing real property. When applied to real estate, for example, the characteristics of a house that determine its value include: location, number of bedrooms, size of kitchen, etc. It is possible that by building two identical houses and changing only the number of bedrooms, the value of the additional bedroom(s) can be estimated as the difference in price between the two houses, all other things held constant.

To apply the HPM method to natural resource valuation, we determine market value differences for two similar goods or services that differ only in terms of one characteristic as a measure of the value of that characteristic (Kengen, 1997). For example, if there are two

National Parks located in the same geographical area, and these parks are identical in every aspect except that one has a good road network within the park and the other does not, then the difference in value between these two parks can be attributed to the good road network.

The main advantage of this method is that it can be adapted to consider several possible interactions between market goods and environmental quality. However, it is a relatively complex valuation method to implement and interpret, requiring a high degree of statistical expertise (King and Mazzotta, 2000). Also, large amounts of data must be gathered and manipulated, and time and expense to carry out an application depends on the availability and accessibility of data (King and Mazzotta, 2000).

D. Contingent Valuation Method

The contingent valuation method (CVM) uses structured surveys to elicit information from people about their preferences for a natural resource. This approach is suitable for the valuation of nonmarket goods and services where other approaches are deemed inappropriate. The value obtained through this method is contingent on the existence of a market for the good or service as described in the questions put to respondents (hence the name contingent valuation). Contingent valuation is therefore a hypothetical, direct (survey-based) method that does not rely on observed behaviour to elicit information, and hence is the only method suited for estimating non-use values of resources.

Contingent valuation surveys may be conducted as face-to-face interviews, telephone interviews, or mail surveys based on a randomly selected sample or stratified sample of individuals. Face-to-face interviews are the most expensive survey administration format, but they are generally considered the best, especially if visual material needs to be presented (Ulibarri and Wellman, 1997). The central goal of the survey is to generate data on respondents' WTP (or willingness to accept compensation) for some programme or plan that will impact the well-being of the respondents (Ulibarri and Wellman, 1997). The most commonly used hypothetical questions simply ask people what value they place on a specified change in environmental amenity or the maximum amount they would be willing to pay to have it occur. The responses, if truthful, are direct expressions of value and would be interpreted as measures of compensating surplus (Freeman, 1993).

The CVM has been subjected to various criticisms for biases arising from instrument choice, implementation and data analysis. Some of these biases include: strategic bias, scenario misspecification bias, aggregation bias and starting-point bias. The method requires a lot of expertise not only to implement, but to detect and avoid, or at least minimise, these biases. In situations where people are not used to purchasing a particular natural resource product or service, they find it difficult to attribute a monetary value to it (Kengen, 1997). For example, if we ask people living around the Gambaga East Forest reserve how much they are willing to pay to ensure that the soil is preserved from erosion, it would be difficult to get a reasonable value simply because it is not a concept they usually encounter in their daily lives.

Monetary value can be meaningless for many subsistence resource users. Thus, modifications are required in order to ask, for example, about relative preferences that can be easier to express than monetary valuations; such modifications require specific expertise (Kengen, 1997). According to Brown et al. (1995), CVM is more effective when the respondents are familiar with the environmental good or service and have adequate information on which to base their preferences, and hence is likely to be far less reliable when the object of the valuation exercise is a more abstract aspect, e.g., existence value. Critiques argue that the CVM is somewhat implausible for serious decision-making because it does not relate to real market conditions (Winpenny, 1992). A further problem with CVM is that because it involves a survey, it is time-consuming and costly to design and implement. Despite its shortcomings, CVM is considered the most widely accepted method for estimating total economic value, including non-use values (existence and bequest) and use values such as option values.

Assuming-Brempong (2003) applied the CVM to assess the important factors influencing the maximum WTP for agroforestry attributes and Sudan Savannah farming systems such as improved scenery, increased soil fertility, improved water retention, soil erosion prevention, gender and age of the farmer, household income, and supply of fuelwood.

E. Productivity or Factor Income Method

So far, we have looked at methods that deal with the outputs from forest resources that provide benefits to people. In this section, I turn to the question of whether forest resources used as inputs into production of other goods and services have value to people, and if so how such values can be measured. The productivity method is one way to estimate such values. While the method of factor income is not as well defined or widely referenced as the hedonic price or travel cost methods, it is recognised by some analysts and organisations such as the U.S. Department of Interior's natural resource damage assessment regulations (Ulibarri and Wellman, 1997). The method is applied in cases where the products or services of an ecosystem are used, along with other inputs, to produce a marketed good.

This method relies on the economic costs of production as an important source of information. Several types of resources for which the factor income approach is potentially well-suited have been identified by Ulibarri and Wellman (1997) to include: surface water and groundwater resources, forests and commercial fisheries. Surface and groundwater resources may be inputs to irrigated agriculture, to manufacturing, or to privately owned municipal water systems. The products in these cases (agricultural crops, saw logs, manufactured goods, and municipal water) may all have market prices. Similarly, commercial fishery resources (fish stocks) are inputs to the production of a catch of saleable fish (Ulibarri and Wellman, 1997). To apply the productivity method, data must be collected regarding how changes in the quantity or quality of the natural resource affect costs of production for the final good, supply and demand for the final good, and supply and demand for other factors of production (King and Mazzotta, 2000).

King and Mazzotta, (2000) further highlight the following disadvantages of the productivity approach: 1) when valuing an ecosystem, not all services will be related to the production of marketed goods. Thus, the inferred value of that ecosystem may understate its true value to society; 2) information is needed on the scientific relationships between actions to improve quality or quantity of the resource, and the actual outcomes of those actions. In some cases, these relationships may not be well known or understood; and 3) if the changes in the natural resource affect the market price of the final good, or the prices of any other production inputs, the method becomes much more complicated and difficult to apply.

7.5.3. Mixed Market and Nonmarket-Based Approaches

Natural resource valuation professionals have developed techniques that combine elements from market-based methods with pre-existing estimates of natural resource values based on either direct or indirect nonmarket valuation techniques. The major motivation for using these mixed approaches is the relative simplicity of using a pre-existing study based on an accepted method, as well as the cost considerations in undertaking a fresh natural resource valuation study (Ulibarri and Wellman, 1997). The two main approaches that have been applied are the benefits transfer and unit day value methods.

A. Benefits Transfer Method

Benefits or value transfer involves taking economic values for addressing a particular natural resource valuation question and applying it to another context (Pearce et al., 2006). The idea of benefit transfer emerged in the early 1980s when the U. S. Environmental Protection Agency proposed the use of desk studies as the basis for cost-benefit analysis in environmental impact assessment because of capital and time constraints (Yaping, 1999). The method uses a data set developed for addressing a particular natural resource valuation question to another context. For example, fuelwood values in savannah woodlands in Ghana may be estimated by applying measures of woodland values from a study conducted in Zimbabwe.

Studies to value natural resources may be expensive, especially in Ghana and other developing countries where research budgets are limited. Benefit transfer therefore offers a cheaper alternative to performing a full-scale study for any particular issue. The benefits estimated from expert opinion, observed behaviour or CVM can be transferred to another context. Benefits transfer should only be done under certain conditions in order to ensure that the information is useful. The analysts needs to know the purpose of the original value estimates, the user group(s) that were considered in generating the initial estimate, and other factors that could have influenced the results of the study being considered for benefits transfer (Ulibarri and Wellman, 1997).

Accuracy of the estimated benefits from the benefits transfer method is somewhat suspect, as it depends on the accuracy of the original estimates and how they were derived. This is more so if the benefits transfer technique extrapolates beyond the range of characteristics of the initial study. Moreover, it may be difficult to access and assess existing studies if they are not published or easily available to the general public.

B. Unit Day Value Method

The *unit day value method* is similar to the benefits transfer method. However, instead of relying on the benefits derived from a single estimate, the unit day value method derives an average value based on multiple value estimates from existing studies. Consequently, the unit day value of the underlying resource reflects a resource having average preference-related attributes, amenities, or qualities (Ulibarri and Wellman, 1997). The application of the unit day value method may also involve groups of experts attempting to interpret from the existing set of estimates (regardless of method used in the original study) a best estimate for each of a set of generic types of environmental resources or activities. The unit day value approach then combines and converts these estimates into a standardised unit of measure that reflects the average value of one unit of the resource on a per-day basis (Ulibarri and Wellman, 1997).

7.6. MEASURING ECONOMIC COSTS

The previous section discussed the benefits side of benefit/costs analysis. This section will focus on the measurement of costs to be included in economic analyses. Costs are easier to ignore or miss and more difficult to estimate than benefits in a BCA framework. The results of BCA can be severely biased if costs are not accurately and completely estimated and included. There are several types of costs: operations and maintenance costs, costs of regulations and social costs.

7.6.1. Operations and Maintenance Costs

Costs of production in forestry plantations include several types, such as the costs of labour to prepare the site, weed, prune, thin, harvest, etc. This also includes the costs of purchasing inputs: seedlings, working gear, transportation, seedling-production, and water and electricity costs. In addition there may be other overhead costs of operating and maintaining facilities such as road networks, which may extend over the life of the plantation project. Many of these costs are relatively easy to obtain and include in BCA, since they are usually out of pocket expenses and managers keep track of these for financial and tax purposes. It is also possible for engineers and other technical persons involved in the design and implementation of the project to provide fairly accurate estimates of costs for the project. In fact, most investors would order an ex ante economic feasibility analyses before undertaking a project, and hence all costs would be evaluated at this stage. Table 1 shows a template for calculating the costs of production in a typical plantation forest project. Other costs can be added or deleted from this template as required in order to arrive at the total cost/ha/year for a specific project.

Table 1. Template for calculation of typical costs in a plantation project

Operation	Unit costs (Gh¢/ha/yr)		Total (Gh¢/ha/yr)	Overhead (%)	Overhead plus unit costs (Gh¢/ha/yr)
	Materials*	Labour			
Site preparation (machinery or manual)					
Transportation of planting materials					
Seedlings (nursery or purchase)					
Planting					
Infrastructure maintenance (water, electricity, etc.)					
Road building and upgrading					
Fire protection					
Road maintenance					
Weed control					
Pruning					
Thinning					
Final harvest					
Total					

*Note: Material costs include vehicle running and maintenance costs.

7.6.2. Costs of Regulations

Whenever governments introduce regulations, there is a potential for the observance of the regulations to impose costs on both producers and consumers. In developing forest plantations, there are many legal and institutional requirements that impose direct and indirect costs on the plantation enterprise. If for example, a plantation owner requires a timber utilisation permit to harvest timber on his plantation, this would be a direct cost to the plantation enterprise. Direct costs are usually known with certainty, and can be added to the costs of operations described in Table 1. However, regulations that do not require direct payment of fees but which by being observed lead to changes in the cost of production of the output would be considered indirect regulatory costs. A typical example of such a regulation would be the requirement to have large-scale plantations subjected to environmental impact assessments. The cost of the assessment would be borne by the investor and would influence the cost of production and probably the price of the output. As regulations change, these costs will change, and hence they are not only difficult to measure with certainty, they are also difficult to predict.

From an economic perspective, the information needed to measure the costs of regulations include: the extent to which the marginal costs/supply function is shifted up by the regulation; and the extent of any output adjustments that the firms will make as a result of the cost changes associated with the regulations (Field, 2001). This means that to estimate the costs of regulations, we must know the costs and the demand function facing the industry (Fields, 2001). Fields (2001) recommends that one way to obtain such costs is to use cost surveys of the industry, using a sufficiently detailed questionnaire to examine the cost structures of the firms and how these might be affected by regulations. Of course, there are problems with this technique, as firms will have an incentive to inflate the costs in order to argue for reduced government regulations (Fields, 2001). Therefore, estimating the costs of regulation will continue to be challenging, although efforts should be made to identify and include these in any economic analyses of forestry plantations.

7.6.3. Social Costs

When individuals or firms measure costs and benefits to include in their analyses of their projects, it is often the case that they consider only the costs they themselves bear when making decisions, not the costs that may be borne by others. But in reality, individual decisions often have implications for society as a whole. The costs that are borne by society as a result of private decisions are called *social costs*. Social costs are incurred through two main ways: opportunity costs and external costs. When resources are used to produce forest products in plantation forestry, the same resources could have been used to produce other goods and services that are of benefit to society. Opportunity cost of using resources (inputs) in a particular way is the highest-valued alternative use to which the resource could have been put (Fields, 2001). The consideration of opportunity costs is one of the key differences between the concepts of economic cost and accounting cost, because the latter does not consider opportunity costs of using particular inputs.

If the use of the resources in plantation forestry is not the best use of the resource, then society is losing out, as their utility is not being maximised. When forest plantations are established on land, the same land has alternative uses, such as for rearing animals or producing food crops. Also, the inputs such as labour and materials could have been employed in other uses as well. If an input has no alternative use, then its opportunity cost is zero, otherwise it has a positive opportunity cost, even if it is difficult to estimate. In a well functioning and competitive market, the market price of an input is a good reflection of its opportunity cost. To use a concrete example, in undertaking economic analyses of a forest plantation, the opportunity cost of the land can be calculated by considering all other alternative uses of the land and the benefits that could have been obtained if the land were put to those uses. Assuming the same land would have yielded a net benefit (NPV) of Gh¢500/ha in food crops, Gh¢300/ha if used to rear animals, and Gh¢200/ha if left as fallow, then the best alternative use in this example would be food crops with a NPV of Gh¢500/ha. If the forestry plantation investment earns Gh¢450/ha it means that the plantation owner has lost Gh¢50/ha by not investing in the best alternative use of the land, that is, in food crops. This will constitute the opportunity cost of the land.

The second type of social cost is external costs or costs due to externalities. Externalities are side effects of the actions of economic agents. They can be negative - when the actions of one party impose costs on another party; or positive - when the actions of one party benefit another party. Let us consider a forest plantation that produces timber, but also contains wild animals of benefit to society in terms of meat production and game viewing. If the production of the timber negatively affects the number of animals (for example in a situation where some animals die as a result of the timber production activities), then those who depend on the animals for their food are negatively impacted due to the actions of the forest manager for which they have no control. This is an example of a negative externality. The killing of the animals by the timber production process imposes a cost on society that is not represented in the forest manager's production costs. In a perfectly competitive market situation, the forest manager chooses to produce where the marginal cost equal marginal revenue.

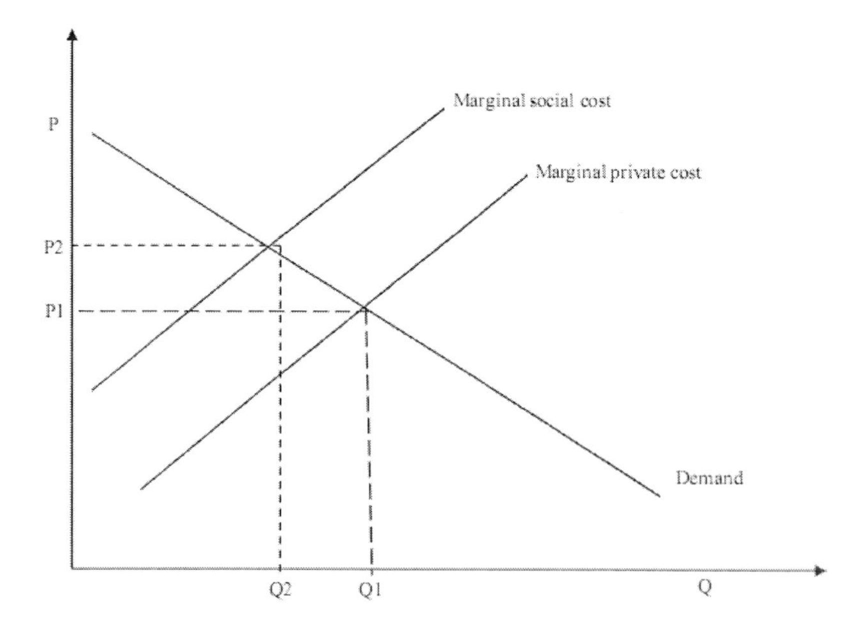

Figure 1. Production, market marginal cost and private cost for producing timber and wildlife.

Figure 1 shows the plantation timber market, with total production taking place where market marginal private cost equals market demand. This is at a quantity of Q1, and a price of P1. Now consider a marginal social cost curve that takes into account production costs plus the costs to society. The marginal social cost curve lies above the forester's marginal cost curve because killing some of the animals is a cost to society. The intersection of the marginal social cost curve with the demand curve shows that the socially optimal output of timber is Q2, at a price of P2. Society is better off because at this output fewer animals are being killed due to the reduction in the amount of timber produced.

However, in competitive markets in general, this optimal output will never be reached because firms (in this case the forest manager) do not consider the environmental concerns of the consumer. In the case of negative externalities, competitive markets will provide too much of the good at too low of a price and consequently too much externality. Competitive markets will fail to provide the efficient output. The external cost to society from Figure 1 is P2- P1, which a government agency can impose as a tax on the forest manager. With an appropriate tax it will be unprofitable for the manager to produce anything but the socially optimal output (Q2). This way, an efficient allocation of resources would occur. The external cost of P2- P1 can now be used as a cost in the BCA analysis. In this example, because the marginal social cost curve is above the marginal private cost curve, Figure 1 illustrates the case of a negative externality. If the marginal social cost curve was below the marginal private cost curve, it would be a positive externality and social optimality would require a greater output than Q1 rather than a reduction of output to Q2.

7.7. Economic Measures of Plantation Profitability

The descriptions of the various criteria that are used to conduct BCA - generally referred to as discounted cash flow (DCF) techniques- are given below.

7.7.1. Net Present Value (NPV)

The net present value (NPV) is the sum of the difference between the discounted benefits and discounted costs of an investment. The feasibility of the investment is determined based on the NPV. The computations and application of this investment criterion is very simple. If the NPV is greater than zero, it implies the benefits are larger than the costs, and so the project is economically feasible. On the other hand, if the NPV is negative, then it is an unprofitable venture. Since the benefits and costs usually occur over time, these are discounted to a chosen base year using an appropriate discount rate. The NPV is calculated using discrete time as:

$$NPV = \frac{p_t V_t - C_t}{(1+r)^1} + \frac{p_t V_t - C_t}{(1+r)^2} + ... + \frac{p_t V_t - C_t}{(1+r)^t} = \sum_{1}^{t} \left(\frac{p_t V_t - C_t}{(1+r)^t} \right) \tag{10}$$

where p_t is the price per unit of output, V_t is the volume of output resulting from the investment, and C_t is the cost of producing the output; all three variables evaluated at time t.

The discount rate is given by r, and t is the time span over which the investment is being evaluated. Usually, t is taken as intervals of one year. The time subscript is eliminated in subsequent formulae to simplify the notation, except where it is required to avoid confusion. In some circumstances, shorter discounting periods may be required, such as half a year or one month. Using the McLaurin expansion of $e^r = 1 + r + \frac{r^2}{2!} + \frac{r^3}{3!} + ...$ and taking r to be small with *continuous* discounting, e^{rt} is approximately equal to $(1+r)^t$. The NPV formula from Equation 10, written in continuous discounting form is therefore given as:

$$NPV = \sum_{1}^{t} \left(\frac{pV - C}{e^{rt}} \right) \quad (11)$$

The continuous discounting version is used in the rest of this Chapter. Note that in calculating the NPV for a particular time t, (i.e., when not summing over more than one time period) the summation operator would not apply, instead, Equation (11) would become:

$$NPV = \frac{pV - C}{e^{rt}} \quad (12)$$

In analysing forestry investments where regeneration usually occurs in the first year (taken as $t = 0$) without any income, the regeneration costs are not discounted (or would remain as C if discounted in year zero).

7.7.2. Soil Expectation Value (SEV)

The soil or land expectation value is the standard tool used in forestry for investment analyses. The NPV criterion is appropriate if we are evaluating an investment for a single rotation of crop trees. If however, we are interested in determining the sustainability of the investment (for multiple rotations), then we will need to include all costs and benefits after the first rotation. Faustmann (1849) was the first to realise the implicit value of the site as part of the incremental cost of time. The value of the bare land, also known as the Soil Expectation Value (SEV) of a forest stand is the NPV of an infinite series of rotations. Faustmann (1849) showed that if conditions remain constant for an infinite period, and if at the end of each rotation the same species is planted and managed in the same manner as previous rotations, then the soil expectation value at time t is given by:

$$SEV = -C + [pV - C]\left[\frac{1}{(1+r)^t} + \frac{1}{(1+r)^{2t}} + ... \right] = \frac{pV - C}{(1+r)^t - 1} - C \quad (13)$$

where all the variables are as defined above. With continuous discounting, this becomes:

$$SEV = \frac{pV - Ce^{rt}}{(e^{rt} - 1)} \quad (14)$$

When intermediate income and expenses such as thinning, fertilisation or insect control are incurred between initial investment and final harvest, the Faustmann (1849) model can be generalised (Chang, 1984). It can be shown that if $A(s)$ is the annual net expense incurred on year s $(0 < s < t)$, then the SEV is:

$$SEV = \frac{pV - Ce^{rt} - \int_0^t A(s)e^{r(t-s)}ds}{(e^{rt} - 1)} \quad (15)$$

If $A(s)$ is a net benefit, such as income from a thinning operation, then the arithmetic operator before the integral sign in equation 15 is positive, rather than negative.

7.7.3. Internal Rate of Return (IRR)

The internal rate of return (IRR) is the interest that is actually earned by the money put into an investment. IRR is also commonly defined as the discount rate at which the NPV = 0. Many economists, including Boulding (1955) argue that the rate of return should be maximised instead of the NPV because a rational manager should strive for a maximum rate of asset growth. In this model, the interest rate earned by the investment in the resource i, is maximised with respect to t. Here, r remains the lending and borrowing bank rate whilst i is the *actual* interest earned by the investment. The IRR principle uses the basic Faustmann or NPV models except that instead of maximising the NPV or SEV, the interest is maximised. To be able to do this, the initial value of the asset (NPV) must be known. Using the NPV formula,

$$NPV = \frac{pV - C}{e^{it}} = 0 \quad (16)$$

where i = IRR.

The solution to the maximisation problem for IRR usually leads to problems (Fortson 1972). The procedure used therefore is to select the IRR nearest zero where the discounted costs and returns are equal. This is equivalent to solving the equation:

$$i = \frac{\ln(pV - C)}{t} \quad (17)$$

If $i^* < r$, then no investment will be made as a negative NPV arises. The principal advantage of the IRR is that it does not require *a priori* specification of a discount rate. However, a potential problem with the IRR criterion is that maximising the IRR may maximise something else besides the owner's wealth. Therefore, it is possible that if investments are ranked based on the IRR alone, the largest IRR may select a project that does not maximise the NPV. Furthermore, the IRR maximisation assumes that the amount of land available for forestry is infinite and that access to all capital markets is closed. Thus, the owner will continue to invest all the returns from the original investment and never consumes from the earnings (Newman, 1988). Another problem with the IRR criterion is that the decisions relating project scale and timing may be incorrect and that changes in prices or the market interest rate do not affect the optimal relation or the level of investment. If the planting costs are equal to zero, then the IRR is undefined, and if intermediate costs and benefits are received, there could be multiple IRR's that solve the maximisation problem (Schofield, 1987). In forestry applications, if the planting costs occur in year zero and a NPV criterion is applied, then from Equation (17):

$$i = \frac{\ln\left(\frac{pV}{C}\right)}{t} \qquad (18)$$

7.7.4. Benefit/Cost Ratio (B/C)

Another version of the NPV that has commonly been used is the benefit/cost ratio. Unlike the NPV, the B/C ratio is calculated as the ratio of discounted benefits to discounted costs, rather than the difference between them. A ratio greater than one means the project is feasible, and vice versa. The B/C ratio is calculated as:

$$BC\ ratio = \frac{pV / e^{rt}}{C / e^{rt}} \qquad (19)$$

Not all projects that show a positive NPV or IRR greater than the discount rate can be carried out. There are budget constraints to be satisfied, and so limited resources require that projects or investments be prioritised. It is common for analysts to rank projects based on the highest B/C ratio or IRR. However, it is possible for the investment with the highest B/C ratio to fail to maximise the NPV (Buongiorno and Gilless, 1987).

Given the problems with the B/C ratio and IRR criteria, Buongiorno and Gilless (1987) suggest that the best way to choose among projects is to choose the one with the highest NPV. However, the B/C ratio and IRR values provide information that complements the NPV. Furthermore, the concept of rates of return is easily understood by managers and other non-technical staff.

7.7.5. Cost-Effectiveness Analysis

Cost-effectiveness analysis (CEA) is a form of economic analysis that compares the relative costs and outcomes (effects) of two or more activities. The United States Department of Defence developed cost-effectiveness as a method for adjudicating among the demands of the various branches of the armed services for increasingly costly weapons systems with different levels of performance and overlapping missions (Hitch and McKean, 1960). Since then, it has become widely used as a tool for analysing the efficiency of alternative activities and programmes outside of the military (Hitch and McKean, 1960). In short, CEA is a tool for comparing two or more alternative resource allocation options in identical units.

CEA is often used where there is general agreement on what to do, and the outstanding question is how to achieve the objective agreed upon efficiently, such as how best to comply with a legal requirement. For instance, if a community decides that they need to increase their fuelwood supply; several options could be evaluated to determine which approach would achieve the goal in the most cost-effective way. In this example, unless an estimate of the exact return on the investment or a quantitative value of the benefits is required, a full benefit/cost analysis (BCA) does not have to be conducted.

The cost-effectiveness ratio is the criterion used to calculate and compare the cost effectiveness of various options. This ratio is calculated as the cost per unit of outcome effectiveness. It is defined by Boardman et al. (2001) as the ratio of the cost of implementing alternative *i* (*Ci*) to the benefit (*Bi*) (or effectiveness) of that alternative:

$$CE_i \ ratio = \frac{C_i}{B_i} \tag{20}$$

The interpretation of this ratio is straightforward. The alternative with the lowest ratio is the most cost effective. An example of an analysis that would apply the above method is in preventing soil erosion. It may be difficult to monetise the value of soil erosion control due to a given programme or policy. In this case, the cost per kilogramme of soil saved will be an appropriate effectiveness measure of the various alternatives.

The literature also shows examples of situations where a decision has been reached to implement a policy (e.g., implement a government regulation) and there are more ways than one to achieve that goal. In such a case, an effectiveness-cost ratio (Boardman et al. 2001) can be calculated, which is essentially the reciprocal of the above formula:

$$EC_i \ ratio = \frac{B_i}{C_i} \tag{21}$$

This ratio measures the average benefits per unit cost. Some authors refer to this as the cost-effectiveness ratio. The interpretation of this ratio is that the policy alternative with the largest ratio is the most cost-effective one. If a community decides to plant 20 ha of trees to increase their wood supply, there could be several ways to achieve this goal, and the method that gives the lowest cost for the same benefit would be the preferred method. The appropriate measure of effectiveness to use is Equation (21).

To avoid confusion, analysts need to keep in mind the goal of the analysis and clearly define what the measure of cost effectiveness is for the particular study. It is important to note that in reality, CEA measures only technical efficiency, not allocative efficiency (Boardman et al., 2001). CEA is good at ranking various alternative policies, but cannot tell us whether any alternative makes economic sense to implement (i.e., has a positive net present value). Also, note that in resource allocation decisions where the benefits and costs accrue over long periods of time, the benefits and costs will have to be discounted using an appropriate discount rate. Given the drawbacks of CEA discussed above, it is usual in economic analysis, to broaden the analysis by conducting full benefit/cost analyses.

7.7.6. Real Option Theory

The traditional DCF techniques discussed above are based on the assumption that future cash flows follow a constant pattern that can be accurately predicted from regeneration up to the rotation age. The profitability or otherwise of forestry projects are then determined based on these cash flow assumptions. The uncertainty that is inherent in an afforestation or reforestation project and the management's reactions to changes in the assumptions are only dealt with superficially (Morck *et al.*, 1989). Furthermore, analyses of profitability of forestry investments have generally focused on the forest stand level, with little attention to the fact that most forest industry firms are vertically integrated, and so investment decisions at the stand level are not determined only by the profitability of the forest stand, but the overall profitability of the firm.

Therefore, in analysing investment decisions even at the stand level it is important that the forestry firm be considered as a corporate entity in which managers make decisions that commit the firm's resources, across business lines and over time. The disjunction between stand level analysis and firm level management decisions has often led to unfavourable conclusions regarding forestry investments. Another problem with the DCF forest investment analyses methods is that the DCF analysis is linear and static in nature and assumes that either the investment opportunity is reversible or if irreversible it is a now-or-never opportunity (Dixit and Pindyck, 1994). Consequently, the DCF techniques fail to adequately address business valuation of growth opportunities or strategic alternatives arising from investments in large-scale commercial projects. These limitations render the conclusions of DCF valuation of somewhat suspect.

The field of forestry entails significant amounts of uncertainties, which make strategic managerial decision-making paramount. Particularly, investments in plantations are characterised by relatively large sunk costs, as well as significant amounts of risks and uncertainties in production and prices, both input and output. Due to the irretrievable nature of most forestry investments, greater focus must be placed upon investment valuation. Previous efforts at incorporating risk and uncertainties in future prices of forestry products have focused on the determination of the optimal rotation age when prices and costs are stochastic. One of the first attempts to model changing prices and costs was McConnell et al. (1983), who developed an optimal control model of timber production, assuming that the production technology is a function of timber age, while costs and prices are functions of calendar time. An extension of this study to model the impact of evolving prices and costs on rotation length was conducted by Newman and Yin (1995). Also, Plantinga (1998) examined the role of option values in influencing the optimal timing of harvests under uncertainty and non-stationary prices. The valuation of a timber lease when prices follow a Geometric Brownian Motion (GBM) and inventory, a GBM with constant drift term was studied by Morck et al. (1989), using contingent claims analysis. Reed and Haight (1996) have estimated the mean and variance of the present value of plantations when prices and yield follow a

GBM. While these studies have incorporated uncertainties, the benefits of managerial flexibility that can be obtained by the application of the real options theory have received little attention.

Only a few studies have addressed forestry investment analysis from a real options perspective. These studies include Thorson (1999), Abildtrup (1999), and Thomson (1992). When real options are present, the traditional DCF methodology may fail to provide an adequate decision-making framework because it does not properly value management's ability to revise the initial operating strategy if future events turn out to be different from originally predicted, or to account for future investments or disinvestments (Trigeorgis, 1993).

Theoretical Forestry Investment Real Options Model[2]

The analysis presented in this section makes use of standard tools of option pricing and investment in discrete time, using the multiplicative binomial approach (Cox et al., 1979). The binomial process was chosen because of the advantages discussed below; although there are several other types of models that can be used (for example see Trigeorgis, 1999). The binomial approach is the simplest of the option pricing formulas (Elton and Gruber, 1995). With the binomial approach it is possible to price options other than European options, like American options, which in contrast to European options can be exercised at any time up to maturity. This is because the binomial model makes it possible to check at every point in an option's life (at every step of the binomial tree) the possibility of early exercise. Also the binomial model does not depend on the probability of certain outcomes. This means that the model is independent of investors that have different subjective probabilities about an upward and/or downward movement in the underlying asset. The binomial model requires relatively less mathematical background and skill to develop and use, compared to the Black-Scholes model[3].

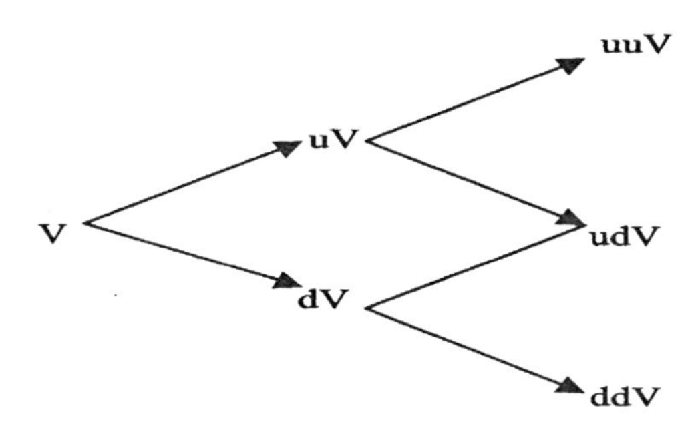

Figure 2. A Two-step binomial tree.

The binomial model breaks down the time to expiration of the option into a potentially very large number of time intervals or steps. A tree of the gross present value of the underlying asset is initially produced working forward from the present to expiration. At each step it is assumed that the value of the asset will move up or down by an amount calculated

[2] This model was first developed and published in Duku-Kaakyire and Nanang (2004).

[3] The Black-Scholes model (Black and Scholes, 1973) is a continuous time option-pricing model that is suitable for valuing European options.

using volatility and time to expiration. The tree represents all the possible paths that the asset value could take during the life of the option. At the end of the tree, that is, at the expiration of the option, all the terminal values for each of the final possible asset values are known, as they simply equal their intrinsic values. The option values at each step of the tree are then calculated working back from expiration to the present. The value of the option at each step are used to derive the option value at the next step of the tree using risk neutral valuation based on the probabilities of the project value moving up or down. Figure 2 presents an illustration of a two-step binomial tree for the paths of an asset value, V[4]. The binomial model assumes that the future values of the underlying asset follow a multiplicative binomial distribution over discrete periods. In addition, the model assumes that the volatility of the underlying asset (σ) is constant and known and uses risk neutral probabilities (p and $1-p$) rather than subjective probabilities for valuation.

In Figure 2, the gross project value of the underlying asset at the beginning of a given period, V, may increase (by a multiplicative factor u) with probability p to uV or decrease with complementary probability (1-p) to dV at the end of the period. From these, the risk neutral probability is given as:

$$p = \frac{\exp(r\Delta t) - d}{u - d} \qquad (22)$$

$$u = \exp\{\sigma(\Delta t)^{1/2}\} \qquad (23)$$

$$d = 1/u \qquad (24)$$

where Δt is the time interval or step size and r is the risk-free interest rate. It is important to note that as $\Delta t \to 0$ the parameters of the multiplicative binomial process converge to the Geometric Brownian Motion (GBM) given in Equation 25. I use this continuous approximation to determine the parameters in Equations 23-24 after which we apply the discrete binomial process to estimate the option values. The continuous GBM assumes that the value of the project, V, is determined competitively and follows the exogenously given continuous stochastic process given by the stochastic differential equation:

$$dV = \mu V dt + \sigma V d\omega \qquad (25)$$

where μ is the deterministic drift (instantaneous expected return of the project), $d\omega$ is the increment of a standard Wiener process and σ is the volatility, defined as the instantaneous standard deviation of V. The discrete version of this model is called a random walk with a drift.

If timber prices and management costs are constant over time, then we can compute the soil expectation value (SEV) of a regenerated stand at any point in time t using a continuous time version of the Faustmann (1849) formula as:

[4] Within the real options section of this Chapter, V will be used as asset value, while Q will refer to volume of timber. This should not be confused with the use of V for volume in other sections.

$$SEV[P,Q_t] = \left[\frac{PQ_t - Ce^{rt}}{(e^{rt} - 1)} \right] \tag{26}$$

where Q_t is the volume of timber harvested at time t, P is the price per /m^3 of timber, and C is the cost of regeneration. This is the classical Faustmann formula for calculating the present value of an infinite series of rotations, also known as the soil expectation value (SEV). The gross project value V, for an infinite series of rotations is calculated from Equation 27 as:

$$V_t[P,Q_t] = \left[\frac{PQ_t}{(e^{rt} - 1)} \right] \tag{27}$$

If on the other hand, timber prices change over time, and follow a binomial process, then the expected price at time t will be used instead of the constant price, and the expected SEV is given as:

$$E(SEV) = \left[\frac{pSEV[uP_t, Q_t] + (1-p)SEV[dP_t, Q_t] - Ce^{rt}}{(e^{rt} - 1)} \right] \tag{28}$$

Where $E(SEV)$ is the expected soil expectation value and $SEV[uP_t, Q_t]$ and $SEV[dP_t, Q_t]$ are the values of the stand in period t if the up and down price states have occurred, respectively. These two values of the stand are calculated using the SEV formula (Equation 26) with the up and down values respectively.

7.8. Application of Discounted Cash Flow Methods

7.8.1. Neem Plantations

Neem plantations are grown by individuals and communities in the savannah zones of Ghana to produce small timber to meet local building needs and for fuelwood. This example analyses the economic profitability of such investments, using realistic costs and variables. It compares the various criteria for assessing plantation profitability and sensitivity analyses by changing the initial assumptions in order to examine their impact on the results. A discount rate of 5% is assumed and volumes of wood output per year were determined from the yield equation for a Site Class I neem plantation from Chapter 6. The value of wood is GH¢10/m^3 and the cost of planting is assumed to be GH¢500/ha and occurs in year zero (this means it is not discounted). It is further assumed that there are no additional maintenance costs.

The results in Table 1 show that all four criteria consistently conclude that the plantations only begin to make a profit in year 5. The value of the NPV is less than that of the SEV as expected because the SEV includes the value of future rotations as well. In analysing profitability using the IRR, an investor will compare the actual interest rates earned by the investment in Table 2 with the interest she or he would have earned if the money were deposited in the bank (i.e., interest rates). Hence, unlike the NPV and SEV, IRR is a relative measure of profitability. Given the high interest rates experienced in Ghana, most economic valuation of forestry investments using the IRR criterion will indicate non-economic viability.

Table 2. Benefit/cost analysis of neem plantations in Northern Ghana at 5% discount rate

Plantation Age (yrs)	Volume (m^3/ha)	NPV (GH¢/ha)	SEV (GH¢/ha)	B/C Ratio	IRR (%)
1	1.03	-490.20	-10051.10	0.02	-388
2	13.67	-376.33	-3954.64	0.25	-65
3	32.35	-221.55	-1590.52	0.56	-15
4	49.77	-92.48	-510.19	0.82	0.0
5	64.46	1.99	9.01	1.00	5.0
6	76.58	67.32	259.73	1.13	7.0
7	86.61	110.34	373.63	1.22	8.0
8	94.99	136.72	414.72	1.27	8.0
9	102.06	150.76	416.04	1.30	8.0

Sensitivity analyses were applied to assess the impacts of changes in the variables on the NPV, SEV and B/C ratio (Table 3). A 30% discount rate was used to assess the impacts of high discount rates on the profitability of the plantations. The results showed that for all criteria, the plantation investment was not profitable at this discount rate, as the benefits were less than the costs, and the B/C ratio was less than one in all years. An increase in the price of wood from GH¢10 to GH¢20 per m^3 showed that the plantation was now profitable in year 3 and with larger NPVs and SEVs than when the price was GH¢10/m^3. A final change to the initial assumptions was to increase the planting costs from GH¢500/ha to GH¢800/ha. This increase resulted in negative NPVs and SEVs, and B/C ratios less than one for all ages.

Table 3. Sensitivity of NPV, SEV and B/C ratio, to changes in the discount rate, price of wood and regeneration costs

Plantation age (yrs)	*r=30%*			*p=* GH¢*20/m*^{3}			*C=* GH¢*800/ha*		
	NPV	SEV	B/C	NPV	SEV	B/C	NPV	SEV	B/C
1	-492.37	-1899.69	0.02	-480.40	-9850.12	0.04	-790.20	-16202.35	0.01
2	-424.99	-941.94	0.15	-252.67	-2655.12	0.49	-676.33	-7107.14	0.15
3	-368.47	-620.91	0.26	56.91	408.55	1.11	-521.55	-3744.27	0.35
4	-350.08	-500.97	0.30	315.04	1737.95	1.63	-392.48	-2165.19	0.51
5	-356.18	-458.48	0.29	503.98	2278.42	2.01	-298.01	-1347.24	0.63
6	-373.41	-447.36	0.25	634.63	2448.61	2.27	-232.68	-897.76	0.71
7	-393.94	-448.91	0.21	720.68	2440.39	2.44	-189.66	-642.24	0.76
8	-413.83	-455.12	0.17	773.45	2346.05	2.55	-163.28	-495.26	0.80
9	-431.41	-462.49	0.14	801.52	2211.87	2.60	-149.24	-411.84	0.81

The main conclusions from these sensitivity analyses are that increasing the discount rate and the planting costs decrease the profitability of the investment, while increasing the price of output increases the economic attractiveness. It is not surprising that discount rates and costs have identical impacts on profitability because a high discount rate increases the cost of borrowing and the opportunity cost of capital tied up in the plantation venture.

7.8.2. Teak Plantations

Teak plantations are grown by individuals, communities and industry for the provision of fuelwood, small timber to meet local building needs, timber for export, and for electric and telephone transmission poles. In this example, the economic profitability of teak is analysed using realistic costs and price variables for growing teak in Ghana. Volumes of wood output per year was determined from Site Class I teak plantations in the Guinea savannah and semi-deciduous vegetation zones of Ghana from Nunifu (1997). A discount rate of 15% is assumed in the analysis, with a price of teak wood of GH¢300/m^3 and the present value of the costs of planting and maintenance is assumed to be GH¢5000/ha.

Table 4. Benefit/cost analysis of teak plantations in the guinea savannah zone at 15% discount rate

Plantation Age (yrs)	Volume (m^3/ha)	SEV (GH¢/ha)	NPV (GH¢/ha)	B/C Ratio	IRR (%)
5	22.70	-3378.69	-1782.71	0.64	6.18
10	97.55	1969.49	1530.03	1.31	17.67
15	186.09	988.39	884.22	1.18	16.09
20	273.49	-963.11	-915.16	0.82	13.99
25	355.67	-2550.63	-2490.64	0.50	12.24
30	431.79	-3600.96	-3560.96	0.29	10.85
35	502.04	-4231.86	-4209.66	0.16	9.73
40	566.91	-4589.81	-4578.43	0.08	8.82
45	626.93	-4785.38	-4779.78	0.04	8.06
50	682.65	-4889.44	-4886.73	0.02	7.43
55	734.53	-4943.72	-4942.43	0.01	6.88
60	782.99	-4971.62	-4971.01	0.01	6.42
70	871.06	-4992.94	-4992.80	0.00	5.65
80	949.21	-4998.28	-4998.25	0.00	5.05

Table 5 presents the results of the economic analyses for Site Class I teak in the High forest zone, with site index 26 m at 20 years.

In Tables 4 and 5, the SEV, NPV and BCR criteria are consistent in their conclusions about the profitability of the plantations. The two plantations in the savannah and HFZ only make marginal profits in year 10 and 15 in the savannah zone and in year 10 only in the HFZ. The profitability results mainly from the high price paid for teak wood around the world that translates into higher local prices as well. The IRR criterion shows a return on investments of up to 17% in both ecological zones. From an economic investment perspective, if the interest rate at the bank or the interest that would be earned on alternative investments is actually less than what is earned on the plantations, the latter might still be a better investment than the alternatives.

Table 5. Benefit/cost analysis of teak plantations in the semi-deciduous zone at 15% discount rate

Plantation Age (yrs)	Volume (m^3/ha)	SEV (GH¢/ha)	NPV (GH¢/ha)	B/C Ratio	IRR (%)
5	28.52	-1,817.65	-959.05	0.81	10.74
10	92.27	1,514.58	1,176.63	1.24	17.11
15	155.24	-102.09	-91.33	0.98	14.88
20	211.68	-1,934.58	-1,838.26	0.63	12.71
25	261.58	-3,230.46	-3,154.48	0.37	11.01
30	305.81	-4,025.56	-3,980.84	0.20	9.70
35	345.29	-4,479.94	-4,456.43	0.11	8.66
40	380.79	-4,728.56	-4,716.84	0.06	7.82
45	412.94	-4,860.64	-4,854.95	0.03	7.13
50	442.26	-4,929.34	-4,926.62	0.01	6.56
55	469.14	-4,964.53	-4,963.23	0.01	6.07
60	493.91	-4,982.33	-4,981.71	0.00	5.65
70	538.19	-4,995.69	-4,995.55	0.00	4.96
80	576.76	-4,998.97	-4,998.94	0.00	4.43

These results show that the plantation in the savannah zone were more profitable than that in the semi-deciduous zone. Although in general, the soils in the semi-deciduous zone are more productive than those in the savannah zone, the lower profitability of the teak plantations in the HFZ examined in this example is due to their lower stocking levels compared to the savannah plantations (see Nunifu, 2010). Of course, if the price of teak is less than what is used in this example, or if there is an increase in the cost of production or an increase in the discount rates, these would all result in reductions in the profitability, as shown in the example on neem above. As an extreme example, if the discount rate is 15%, the present value of the costs of planting and tending is GH¢10,000/ha, and the price of teak wood is GH¢400/m^3, then these two plantations would not make an economic profit.

7.9. Applying Real Options Theory in Plantation Investments

The purpose of this example is to show how firm level management decisions that allow for managerial flexibility affect the value of the forest stand compared with the traditional Faustmann criterion. Consequently, I have not included the value of a wood processing plant itself, but recognise that the stand level decisions are taken in the overall interest of the processing mill. In this example, the interest is in examining the values of various options that the firm can exercise. These options include the option to delay the plantation project by up to 10 years, the option to double the processing capacity (and hence the area planted) by 20 years, and the option to abandon the plantation by 50 years if market conditions turn out to be worse than originally predicted, or the processing plant is taken over due to a corporate merger. To achieve these objectives, I describe simple models and show how the values of these options can be estimated.

It is assumed that the present cost of regenerating a hectare of land is C, which could include the present values of stand enhancement investments such as thinning and fertilisation that occur in the future. The real cost of capital is given as r, and is assumed known and constant throughout the planning horizon. The prices of timber on the other hand, are assumed to follow a GBM similar to that given in Equation [25]. We can compute the gross present value of a regenerated stand at any point in time t using Equation [27]. The volatility of timber prices, σ was estimated as 0.10 using historical prices of teak wood on the international market and the drift term, μ. The drift term and volatility are the mean and standard deviation respectively of the time series $\ln(P_{t+1} / P_t)$. The series contained drift terms of 0.04 (i. e., $\mu \approx 0.04$). This example uses teak grown in the semi-deciduous vegetation zone in Ghana, and a rotation age of 60 years.

The current price/m^3 of timber used in the analysis was Gh¢500. A step size of 1 ($\Delta t = 1$) was used in this analysis. Smaller step sizes of Δt = 0.5 years and Δt = 0.25 years did not result in significant differences from the Δt =1 case, therefore the results are reported for only Δt =1. To keep the analysis simple and in perspective, the example is restricted to a per hectare basis, which is a common form of stand level analysis in forestry. The various options were determined based on the backwards recursion solution technique described previously using Microsoft Excel. Table 6 summarises the information used in the binomial tree analysis. The options evaluated are briefly described below.

Table 6. Summary of input variables used in the binomial tree analysis

Variables	Values
Investment (regeneration) Cost (C) (Gh¢/ha)	1500.00
Present Value of Cash Flows (V) (Gh¢)	1452.16
Annual Volatility (%)	10.00
Risk Free interest Rate (%)	8.00
Δt	1.00
Starting price of timber (Gh¢/m^3)	500.00
Additional Investment Cost to Expand (C') (Gh¢)	321.82
Abandonment (salvage) value (Gh¢)	3198.18

A. Option to Delay Plantation Development by 10 Years

This option gives management the flexibility to defer forest regeneration and therefore allows management to benefit from favourable random movements in the project value, whilst at the same time they cannot be hurt by unfavourable market circumstances, because they do not have the symmetric obligation to invest. Therefore, management will wait and make the investment within 10 years if the option value within that period turns out to be favourable. In other words, the option to delay reforestation is seen as a call option on the gross project value V_t, with an exercise price equal to the required outlay in 10 years time, C_{10}. This translates into the right to choose the maximum of the project value minus the required investment, or zero, since management will simply not exercise the option if project value turns not to cover the necessary costs. This can be expressed as:

$$V_D = \max[V_t - C_{10}, 0] \tag{29}$$

The subscripts "t" on the right hand sides of Equations (29) to (31) refer to the option values at different points in time (at different steps of the binomial tree), whilst the left hand sides refer to final single values of the options (hence no reference to time).

B. Option to Expand Wood Processing Plant

Suppose that in this example, management has the option to expand their production (or sawmill) by doubling the size and scale. This means that instead of one hectare, the firm now regenerates an additional hectare by 20 years after the mill starts operations (or 20 years after afforesting the first hectare of land). Then by Year 20, management has the flexibility either to maintain the same scale of operation (one hectare) and receive project value V_t, at no extra cost, or double the scale and receive twice the project value by paying the additional cost, whichever is higher. That is:

$$V_E = \max[V_t, 2V_t - C'] \tag{30}$$

where C' is the present value of the additional cost of reforestation and V_E is the option value to the project when exercised.

C. Option to Abandon

We incorporate the option for management to abandon the sawmill (and consequently the reforestation project). This could happen if the sawmill is abandoned in exchange for the salvage value of the mill and forest if market conditions turn out to be unfavourable or there are corporate mergers or takeover. The option to abandon mill operations for any number of reasons before the rotation age (50 years) will result in a sale of the forest stand. The option to abandon the project early in exchange for the salvage value translates into the flexibility to choose the maximum of the project's present value in its present use, V_t, or its value in its best alternative use, A. We should notice that if the trees have no alternative use, then $A = 0$. However, if the processing plant is being sold to another firm, then the trees are also sold and therefore, A can have a positive value. Since it is difficult to tell what A will be in reality, we set A to zero and so the expression for the option to abandon the mill is given as:

$$V_A = \max[V_t - C, A] \tag{31}$$

Computing the Option Values

The details of calculating the various option values using the binomial method require a large number of steps that result in extremely large binomial trees. It is therefore not possible to include all details of calculations here. The backwards computations use the following equation:

$$V_{ot} = Max\left[\frac{puV_{o(t+1)} + (1-p)dV_{o(t+1)}}{(1+r)}, \quad V_t - C\right] \tag{32}$$

where V_t, refers to the stand values, V_{ot} is the value of option at time t, $uV_{o(t+1)}$ is the *up* value of the option in the previous year, and $dV_{o(t+1)}$ is the *down* value of the option in the previous year.

Table 7 shows the descriptions of the different types of models estimated, the NPVs and the option values for the real option models. The calculated SEV using the Faustmann formula for the reforestation project is negative, which suggests that the investment is not profitable to undertake. Valuing the project with its inherent options presented more favourable results. Using the binomial tree approach discussed earlier, the option to defer afforestation/reforestation by 10 years alone is valued as an American call option on the project, with an exercise price equal to the necessary investment outlays. As shown in Table 5, this option increases the project's value to \$406.52/ha, in contrast to the no flexibility base case SEV of -Gh¢60.29/ha. Alternatively, the value of this option is \$466.81/ha (project value = SEV +option value). This indicates that waiting for the 10 years prior to exercising the option might improve timber prices, but might not necessarily resolve all the risks and uncertainties associated with the plantation investment. However, this is an option that could be valuable for this investment if exercised.

Table 7. Summary descriptions, project values, and the option values of the models examined

Model #	Model description	Project value (Gh¢/ha)	Option Value* (Gh¢/ha)
1	Faustmann, with no price volatility	-60.29	0
2	Option to delay regeneration by 10 years	406.52	466.81
3	Option to double mill size by 20 years	1247.26	1307.55
4	Option to shut down mill by 50 years	1685.74	1746.03
5	Multiple options: Options 2, 3, and 4 combined	1685.74	1746.03

*Note: the option values are the differences between the project values of the other models and Model #1.

Similarly, the option to expand the operation scale is worth Gh¢1307.55/ha. This option is valued analogous to American call option to expand by paying an extra outlay as the exercise price. The present value of the additional investment required for the expansion was calculated to be approximately Gh¢321.82/ha. Including this option increases the project's expanded value to Gh¢1247.26. Should the market show improvement above management's initial expectations, this is an option that management might want to seriously consider exercising.

The option to permanently abandon operations for salvage value is also valued as an American put option with an exercise price equal to the salvage value. The salvage value was realised from the sale of the standing timber at age 50 years and was estimated as Gh¢3198.18. The option to abandon was Gh¢1746.03, which is worth a lot to this project due to the high salvage value. It is an option that should be exercised for this particular project if market conditions deteriorate severely, or if for other reasons the wood processing plant has to go out of business.

As pointed out previously, the value of the option in the presence of others may differ from its value in isolation. The presence of subsequent options increases the effective underlying asset for prior options (Trigeorgis, 1993). Moreover, exercise of a prior option may alter the underlying asset and the value of subsequent options on it. Thus, combining all the options and valuing them together could lead to a result that is different from the sum of their isolated values. It could be higher when there are positive interactions among the options or lower when interactions are negative. With this knowledge, all the three option were combined and valued together as multiple options. When all the three options are combined, their value is estimated to be Gh¢1746.03, which is equal to the option to abandon. Although the value is less than the sum of the isolated values, the combined flexibility that they afford management may be economically significant. The multiple options happened to have the same value as the abandon option because the years between the options are large and the valuation is done such that the maximum value is picked among the lot.

The results show that the inclusion of managerial flexibility can affect project value drastically and lead to different conclusions regarding the profitability of forestry investments. Stand values calculated when prices follow a diffusion process are higher than the Faustmann soil expectation value as the former method values the flexibility of forest management and investment decisions. The option theory approach holds great promise for use in forestry because static financial analyses often lead to conservative conclusions regarding the viability of investing money in forestry projects. Most forestry projects, when compared with other land use options are often at best of marginal profitability. In order to remove this bias against forestry projects the values of price uncertainties and managerial flexibility need to be incorporated into the analytical framework.

Although this analysis was restricted to three options, there are many other managerial options that can be analysed in forestry investments. For example, it should be possible to examine the option to switch the species planted after the first rotation (or after several rotations) if market conditions turn out to be unfavourable. In the traditional Faustmann framework, it is often assumed that the same species is planted on the same piece of land *ad infinitum*. Furthermore, silvicultural investments in fertilisation and thinning may also be evaluated as options for forest industry firms.

Despite the usefulness of the option theory approach, its applicability in certain circumstances requires that some difficulties be overcome. For example, how to effectively incorporate some risks, especially risks associated with security of forest tenures remains a challenge. Whilst in theory we can include risk by estimating a risk premium and adding it to the chosen risk-free interest rate, estimating the risk premium for qualitative variables such as tenure security and other legal operating environment variables is complicated at best. It should also be emphasised that these results are sensitive to the risk-free interest rate and volatility of the timber prices. Future research could focus on quantifying the interactions among the different options and determine where the option interactions are small and therefore simple option additivity could be a good approximation, and where high interactions will seriously invalidate options additivity.

7.10. CONCLUSIONS

Plantation forestry is fundamentally an economic venture and the ability to make a profit from forestry plantations may be the single most important determining factor as to whether individuals and companies will undertake plantation ventures. Ex-ante economic analyses are indispensible in evaluating the economic viability of the plantation enterprise. In cases where it is not possible to make profit, government intervention in the form of subsidies or grants may make the venture viable, but as we have seen in Chapter 3, these kinds of external support may not be sustainable, and hence in the long run, may be counter productive. Economic sustainability depends on species, site, economic conditions, markets, and management practices. Long-run sustainability of plantation forestry requires that these factors be seriously considered in the establishment and management of plantation forests to ensure that they are by themselves independently sustainable.

Chapter 8

THE FOREST HARVEST DECISION: OPTIMAL FOREST ROTATION

In this Chapter, the age-old question of the optimal rotation age in forestry is discussed, beginning with the economic foundations for determining the optimal rotation of a stand of trees. Following this, the various models in the literature for deciding the optimal rotation are described. Data from teak and neem plantations are used to present two case studies of optimal rotation ages under various combinations of prices, costs, discount rates, and risks of forest fire.

8.1. BEHAVIOUR OF ECONOMIC AGENTS

The optimal forest rotation question has been debated by foresters and economists for many decades. The optimal rotation problem can be defined as finding the optimal output and input mixes of forest management to maximise the goods and services produced by the forest. There are two main areas of economics that are relevant to solving this optimisation problem: capital theory and the theory of the firm.

Standard economic theory assumes that individuals choose actions that optimise their expected utility, which is the amount of satisfaction derived from consuming goods and services. Capital theory analyses the links among the theories of production, growth, value and distribution to explain why capital produces a return that keeps capital intact yet yields interest or a profit which is permanent (Jorgenson, 1963). According to the neoclassical theory of capital, a production plan for the firm is chosen so as to maximise utility over time. Under certain well-known conditions this leads to maximisation of the net worth of the enterprise as the criterion for optimal capital accumulation (Jorgenson, 1963).

Capital theory, as exemplified by the theory of the firm, is important in understanding how resources are allocated to produce goods and services in any forest enterprise. There is clearly a link between the decisions of economic agents and the forest rotation problem. The neoclassical economic theory of the firm is the basic starting point to understanding the kinds of economic decisions that are made by firms and individuals in forest plantation management. Even if the plantation is not managed for industrial purposes by a firm, the concept of profit (or utility/benefit) maximisation is still central to the type of decisions that are made by those who invest in forest plantations.

8.1.1. The Neoclassical Theory of the Firm

In neoclassical economic theory, the firm is an entity, which maximises profit (minimises cost) subject to constraints imposed by its technological capabilities. In a static framework, the firm may be modelled with either profit maximisation (where profit is defined as revenue minus cost) or cost minimisation. The assumption of cost minimisation is consistent with profit maximisation because if a firm operates to maximise profits, it will, after selecting an output level, select inputs to minimise costs. It is, however, a weaker assumption because cost minimisation is concerned only with the choice of inputs, whilst profit maximisation deals with the choice of both inputs and outputs. This basic model is based on several assumptions including perfect competition, the firm being a price taker in input and output markets, perfect information, a large number of buyers and sellers, etc. Given these assumptions, the firm solves the following basic problem:

$$Max \quad \pi = R - C \tag{1}$$

where π is the profit, R is revenue received from the sale of outputs, and C represents cost the firm incurred in the production process. According to this theory, firms maximise profits (Equation 1) by operating at the point where marginal revenue equals marginal cost. In a perfectly competitive market, marginal revenue equals the price of the commodity. In a monopolistic situation, however, the firm maximises profits by setting price greater than marginal revenue. Without accounting for opportunity costs of the factors of production used in Equation 1, we will be measuring the accounting profits, rather than economic profits.

In this static framework, factors of production are classified into fixed and variable factors, that is, the shorter the run, the fewer the number of factors that can be varied. One criticism of this static model is that they focus on conditions that prevail when full static equilibrium is achieved (long-run) to the exclusion of the adjustment process (Fernandez-Cornejo et al., 1992). To overcome this problem, restricted profit and cost functions were introduced, where one or more inputs are fixed. However, in a forestry context where costs and benefits occur at different points in time, and inter-temporal decisions are made, the static model is inadequate in modelling such a process and will therefore need a dynamic framework.

In the dynamic approach, a firm considers a multi-period horizon and inter-temporal allocation of resources to be an integral part of the analysis (Fernandez-Cornejo et al., 1992). According to Treadway (1971, 1974), dynamic models of the firm should incorporate both dynamic optimisation and the idea of adjustment costs. Brechling (1975) showed that multi-period profit maximisation (or cost minimisation) models that do not include adjustment costs result in static decision rules, even if the firm has multi-period objectives, whilst a model with adjustment costs lead to multi-period decision rules. Adjustment costs may be due to institutional requirements (e. g., government regulation of the forest industry), market imperfections, or internal causes. Brechling (1975) further indicates that internal adjustment causes may be associated with integrating new capital equipment into on-going operations, or re-organising production lines. In general, therefore, the long-run solution derived from dynamic models within the cost of adjustment framework is different from the long-run solution using a static optimisation (Treadway, 1971; 1974). Using this dynamic theory, we can state the problem of the firm as trying to:

$$Max \quad \beta^{t-1}\left[\sum_{t=1}^{T}(R_t)-\sum_{t=1}^{T}c_t\right] \tag{2}$$

where as before, R_t is revenue and c_t represents cost in period t. These are all discounted using a discount factor, β.

8.1.2. Extensions to the Basic Model of the Firm

The basic assumption of the forest industry firm is the same as that of any other firm; that it seeks that harvest rate through time subject to the constraints that it faces that maximises the present value of profits. Extensions of the basic models (static and dynamic) of the firm to a forest industry firm require that the institutional environment in which the firms operate be taken into consideration. When this is done, the model will be modified appropriately to allow for a correct analysis of the firm. For the purposes of this discussion, a forestry firm will be considered as a vertically integrated[1] firm that is involved in the production of forest crops, which are used as inputs in the manufacture of wood products.

There are certain characteristics that differentiate a forest industry firm from other manufacturing firms. Nautiyal (1988) discusses some of the following important decisions that face forestry firms. First, most of the production decisions of the forestry firm will bear results after a relatively long period of time. Comparisons of different costs and benefits over time mean that the notion of profit maximisation in the classical theory of the firm has to be replaced with maximisation of net present value (or net present worth). Secondly, the decision- making process is dynamic, that is, decisions taken today can have effects several years later. For example, stand management decisions and investments in silviculture (e. g., thinning) affect future growth and yield of forest crops. However, most of these effects cannot easily be determined at the time of making the decision, which leads to a considerable amount of uncertainty for the firm. The major characteristics of the forest industry firm that distinguish it from other firms and how these factors modify the classical theory of the firm are now discussed.

Joint Production

Joint production refers to a situation where a single production process results in more than one output. The process of growing forest trees, for example, may also lead to the production of wildlife, recreation opportunities, water conservation, etc. Even if one use seems dominant (usually timber production), there are often other uses, even if these are not explicitly recognised. Even for timber production alone, the idea of joint production occurs

[1] Vertical integration can be differentiated from horizontal integration. In the later, firms produce outputs at the same time from the same production process e. g., lumber and sawdust or timber and recreation.

where different species mixes are produced Nautiyal (1988). These different species and the end products they can be used to manufacture can be considered as different products. Because of the inevitability of joint production in forestry, the forestry firm must attempt to produce an optimal combination of outputs that maximises profits or minimises costs, and so it is possible to modify the theory of the firm to handle multiple outputs.

The determination of the optimal product mix requires that all products be quantifiable and measureable. The discussion on the quantification and measurement of non-timber benefits was provided in Chapter 7. We can represent the trade-off between different outputs using a production possibility frontier (PPF), given in Figure 1. The PPF shows the various combinations of the outputs that maximise the returns to the investment. For example, it shows the optimal level of timber and non-timber outputs to be X and Y respectively. The required modification to the theory of the firm will therefore be to maximise net present value over all outputs. That is,

$$Max \quad \beta^{t-1}\left[\sum_{p=1}^{P}\sum_{t=1}^{T}(R_{pt}) - \sum_{p=1}^{P}\sum_{t=1}^{T}c_{pt}\right] \tag{3}$$

where R_{pt} is revenue received from the sale of product p in period t, and c_{pt} is cost of producing product p in period t. The revenues and costs are summed over all time periods ($t=1....T$) and outputs (p=1,...,P). The problem facing forestry firms is not only that of joint production, but also, joint costs. Joint costs refer to the difficulty of attributing inputs to the production process unambiguously to one or the other output (Nautiyal, 1988). In a forestry context, it could be the use of inputs to produce both timber and recreational opportunities, or in the production of lumber and chips (which is a by-product of the lumber production process).

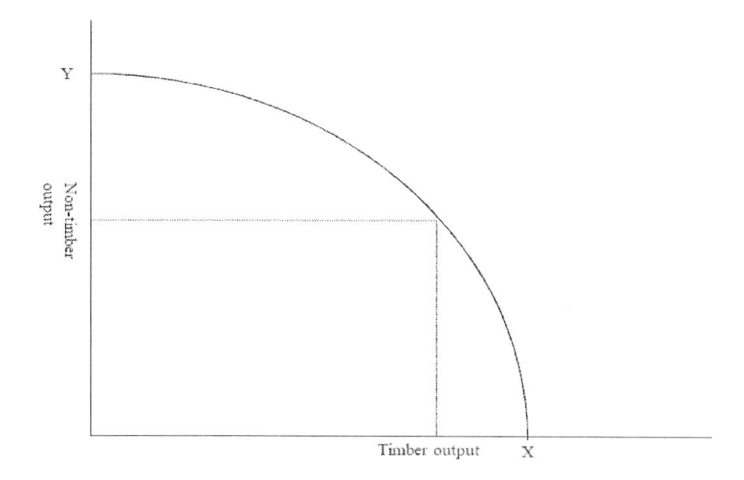

Figure 1. A production possibilities frontier curve for timber and non-timber outputs.

An interesting question, however, is whether the vertically integrated forestry firm should maximise net present value from each of the two production processes separately, or to maximise returns to both processes together. Nautiyal (1988) argues that the forestry firm will make more profit by maximising the returns from each process than by combining the entire enterprise. Therefore, vertical integration is not desirable if there are good markets for logs, for example. But given that mills have to process their logs, or sell logs in local markets due to the ban on log exports, vertical integration for these firms is inevitable in many cases. Luckert and Haley (1991), also note that in most cases, investments in log production alone is not profitable, however, by combining with the manufacturing process, the overall firm can make some profits.

Externalities

One of the cornerstones of economic theory is that under certain conditions markets will produce Pareto efficient allocation of resources. A Pareto optimal allocation is one in which at least one person is made better off and no one is made worse off. When conditions prevail such that Pareto optimal solutions are not obtained, market failure occurs as a matter of definition. One of the causes of market failures in forestry are externalities. Externalities occur when an economic agent affects the production possibilities or the utilities of others in a way that is not reflected in the market. Externalities are common in forestry due to joint production of multiple outputs. In the case of growing trees, a forestry firm may at the same time improve habitats for wildlife (external economies) or destroy the habitat that results in reduced wildlife (external diseconomies). Also, in harvesting one tree species from a land base that contains other tree species, some of the species not being harvested may be damaged leading to negative externalities. Another example of an externality relating to an integrated forestry firm is that of effluent emissions from processing (especially pulp) mills. To reduce this externality means firms have to spend extra money on pollution control. Theoretically, it is possible to eliminate most externalities in forestry by changing property rights arrangements. For example, negative externalities against wildlife can be internalised by expanding the rights of tenure holders to include wildlife, rather than the rights to just timber alone.

Nonmarket Goods and Services

As discussed above, joint production of outputs for the forest industry firm is the rule rather than the exception in many cases, and so we will have to deal with non-timber values. Associating monetary values to non-timber benefits has been a challenge to economists for some time. In terms of maximising net present value for a forestry firm, the only complication presented by the production of non-timber benefits is how to value these outputs. Once a monetary value is assigned to these benefits, the firm then simply maximises the NPV of the combined timber and non-timber benefits. That is, the firm then solves the following problem:

$$Max \quad \beta^{t-1}\left[\sum_{t=1}^{T}(RTB_t + RNTB_t) - \sum_{t=1}^{T} c_t\right] \tag{4}$$

where:

RTB_t = *revenues* received by firm in period t from *harvesting and processing* timber;

$RNTB_t$ = *revenues* received by firm in period t from non-timber resources;

c_t = joint costs of harvesting and processing timber and providing non-timber benefits in period t.

8.2. The Concept of Forest Rotation

The decision by a landowner as to when to harvest a forest plantation affects the flow of raw material to the firm, revenues from the sale of the raw material as well as future regeneration decisions. Rotation in forestry refers to the number of years between the time a stand regenerates and the time of final felling (or final harvesting). Rotation age is then the age of the plantation when final felling occurs. The rotation length is an important tool for controlling tree size in forestry: the longer the rotation, the larger a tree can be grown. But rotation length also markedly influences yield, profitability, and regeneration methods (Evans and Turnbull, 2004). The correct rotation age is the age when the utility from the forest is maximised (capital theory).

The question of the correct optimum rotation of a forest stand is as old as forestry itself and has not been without controversy. Historically, foresters have used the age at which the mean annual increment (MAI) peaks as the basis for deciding the optimal rotation age of a stand. Though the practice is still in use today, economists have argued that this rotation is too long to be efficient and that it ignores the opportunity costs of labour and capital (Anderson, 1992). Many different economic models have been proposed for the determination of the optimum rotation age varying in their basis and degree of sophistication (e.g., Faustmann, 1849; Fisher, 1930; Anderson, 1976). The large number of approaches can be taken as an indication that no one criterion has received more than partial recognition. One reason for the difficulty of the rotation question stems from the fact that investments and benefits in forestry are spread over the growth cycle of the forest. The uncertainty involved in predicting accurately inputs and outputs in monetary and physical terms (prices, costs, yields, etc.), interest rates, and other risks well into the future (sometimes about 100 years or more) are factors which compound the difficulty of the optimal rotation question. Furthermore, forests are capable of providing both timber and non-timber products. These non-timber values in most circumstances are difficult to estimate and include in the optimal decision framework.

8.3. Types of Rotations

Evans and Turnbull (2004) identified four categories of rotations used in forestry. These are: physical rotations, technical rotations, financial rotations, and rotation of maximum volume production. In this section, brief descriptions of these types of rotations are given (after Evans and Turnbull, 2004).

Physical Rotations are determined by the site or other environmental factors which may prevent a stand reaching a certain size or maturity. For example natural disturbances such as cyclones or bushfires, severe droughts and shallow soils and the build up of lethal pests or pathogens can physically limit the size a stand can reach (Evans and Turnbull, 2004).

Technical Rotations are those planned to yield the most output of a specified size and type of forest product to satisfy a particular end-use, which may include lower and upper limits (Evans and Turnbull, 2004). For example on individual and community plantations grown for fuelwood and small-scale construction timber, short rotations are preferred because the products can easily be harvested and handled with locally available implements. Moreover, these products are harvested to meet the local demand for small wood. In other tropical areas where fast grown exotic species are grown for pulpwood, the rotations are determined based on the technical qualities of the wood produced.

Rotations based on economic criteria are determined by the rotation age that yields the highest financial return under a particular set of circumstances. Economic analyses may be done in several ways using different kinds of criteria such as the soil expectation value (or the net present value), the internal rate of return or the benefit/cost ratio. However, these criteria produce different results regarding the profitability of the forest enterprise, depending on the assumptions.

The maximum sustained yield rotation (also called the biologically optimum rotation or rotation of maximum MAI) is the rotation that yields the greatest average annual production of timber. In terms of wood yield, it realises the full growth potential of the site (Evans and Turnbull, 2004). This rotation is reached when the annual growth increment of the stand falls to the level of the overall mean annual increment. In addition to the above, there are also a few other *minor* factors determining rotation age such as silvicultural reasons, opening of new markets, failure of anticipated markets, inaccessibility of stands, uneven age-class distribution, etc (Evans and Turnbull, 2004).

8.4. Optimum Rotation Models

A review of the literature reveals a wide variety of models used in determining the optimal forest rotation. For example, a major review by Newman (1988) showed at least six different approaches. However, only a handful of these are used frequently, and so only these will be discussed in this section. Unless otherwise stated, the following are some of the basic assumptions used in the analysis of the various rotation models presented in this review. These assumptions are made in part to ease the mathematical manipulations but more importantly because the world of 19th century Germany, where the models were first presented, fit many of these static assumptions (after Newman, 1988).

- *Even-aged management:* Starting from bare land, a single stand of trees is grown and all trees are cut at the same time (rotation age). The land has unchanging growth potential, and the technology available for growing trees does not change. The land is capable of being regenerated at the same fixed costs instantaneously after harvest. A regulated forest was also shown by Faustmann (1849) to give the same results as a single stand. Chang (1983) and Hall (1983), among others have shown that the optimal conditions will change under uneven-aged stand management.
- *Perfect certainty of future growth and yields:* There must be perfect certainty regarding the stand's growth function, future market prices, interest rates, and costs. All are assumed known and constant, or at least predicted with certainty.

- *Access to capital markets*: It is assumed that there is unlimited access to capital markets and money can be borrowed or lent at the same interest rate, r.
- *Constant net stumpage price*: Net stumpage price is not a function of tree quality but rather is constant per unit of volume produced. This assumption is needed because multiple optima can arise when management objectives favour different products or when discontinuous jumps in prices create non-concave sections in the total revenue function (Newman, 1988).

It is possible to vary these assumptions through sensitivity analyses to assess the impacts of changes in interest rates, costs, prices, and changing technology on the optimal forest rotation.

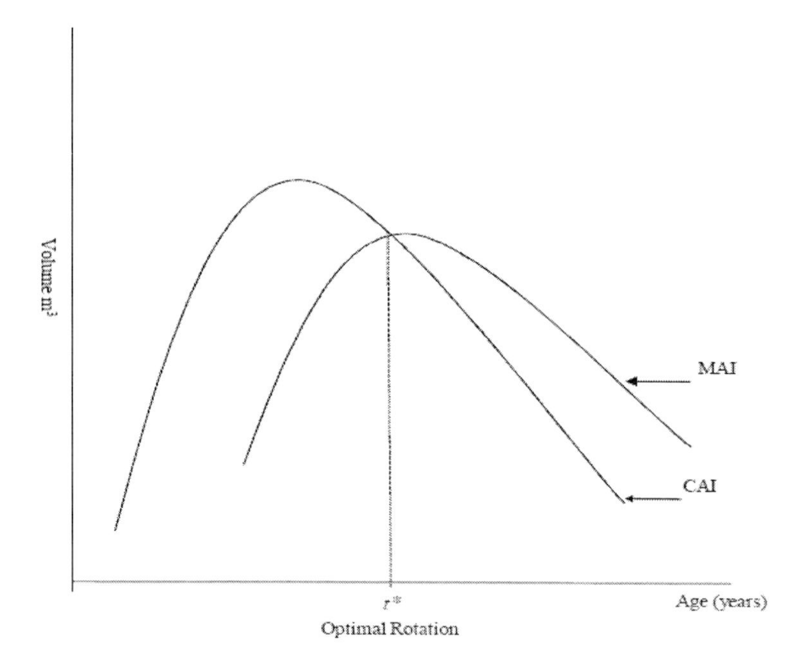

Figure 2. Illustration of the optimal rotation decision in the MSY model.

8.4.1. Maximum Sustained Yield [MSY]) or Maximum Mean Annual Increment

In forestry, a popular rule for determining the optimum rotation of a stand is the age at which the current annual increment (CAI) intercepts the mean annual increment (MAI). The annual allowable cut is often set based on the maximum sustained yield (MSY) concept. This is the point which maximises the MAI (shown by point t^* in Figure 2). The MAI is given by

$$MAI = \frac{V(t)}{t}$$

and the CAI by

$$CAI = V'(t) = \frac{dV(t)}{dt}$$

The optimum rotation according to the MSY is to cut the trees when the current annual increment equals the mean annual increment, CAI = MAI. That is, $\frac{V(t)}{t} = V'(t)$, where V(t) is the volume production at time *t* (plantation age in years) in both cases. The optimal rotation age under this framework is based only on biological considerations, and is not affected by changes in market conditions and hence is relatively easy to apply.

8.4.2. The Net Present Value (NPV) or Single Rotation (Fisher) Model

The maximum sustained yield criterion does not include any economic considerations. When regeneration and other silvicultural costs and interests on capital tied up in the forest are considered with a finite number of rotations, then the decision rule for optimal rotation changes. This calls for the maximisation of the net present value (NPV) of one rotation.

$$\max NPV = \frac{pV(t) - C}{e^{rt}} \qquad (5)$$

Using calculus, at the optimal point, the first derivative of the function equals zero, hence, this requires that at the optimum rotation age,

$$\frac{dNPV}{dt} = 0 \qquad (6)$$

$$\Rightarrow \quad \frac{dNPV}{dt} = Y'(t)e^{-rt} - re^{-rt}Y(t) = 0 \qquad (7)$$

$$\frac{Y'(t)}{Y(t)} = r \text{ or } Y'(t) = Y(t)r$$

(8)

where:

Y(t) = V(t)P(t), the stumpage value of a stand t years old, P(t) is today's stumpage price for trees of various ages; V(t) is the timber volume of the stand at age t, Y'(t) = dY(t)/dt

r = discount rate

t = rotation age

This model states that the forest should be harvested when the rate of value increment equals the interest rate.

8.4.3. Soil Expectation Value (SEV)

The NPV criterion is appropriate if we are evaluating the optimal rotation age for a single rotation of crop trees, while the SEV or Faustmann model is used when evaluating the optimal rotation age for an infinite series of rotations. The following derivation and interpretations of the optimisation and marginal analysis follow Chang (1984). To determine the optimal rotation age using the Faustmann model, we maximise the SEV:

$$\max SEV = \frac{pV(t) - C}{e^{rt} - 1} - C \text{, which requires that } \frac{dSEV}{dt} = 0: \tag{9}$$

$$\frac{dSEV}{dt} = \frac{Y'(t)(e^{rt} - 1) - re^{rt}[Y(t) - C]}{(e^{rt} - 1)^2} = 0 \tag{10}$$

$$\Rightarrow Y'(t)(e^{rt} - 1) - re^{rt}[Y(t) - C] = 0 \tag{11}$$

where $Y(t) = p_t V_t$ (total revenue). Equation (11) provides the necessary condition and a sufficient condition is that: $\frac{d^2 SEV}{dt} < 0$ (since this is a maximum point). Equation (11) can be re-written as:

$$Y'(t) = \frac{re^{rt}[Y(t) - C]}{(e^{rt} - 1)} \tag{12}$$

From (11), $Y'(t)$ which is equal to $\frac{dY(t)}{dt}$, represents the marginal revenue product (MRP) of waiting out the rotation. The right-hand side represents the marginal input cost (MIC). The marginal analysis interpretation of the optimum rotation criterion is that if the MRP of letting the stand grow one more year is greater than the MIC of doing so, one should allow the stand to grow another year. On the other hand, if the MRP of letting the stand grow one more year is less than the MIC of doing so, one should harvest the trees. The above marginal rules can be written as:

If $Y'(t) > \frac{re^{rt}[Y(t) - C]}{(e^{rt} - 1)}$, let the stand grow another year;

However, if $Y'(t) < \frac{re^{rt}[Y(t) - C]}{(e^{rt} - 1)}$, then harvest the trees.

The MIC includes the cost of holding the growing stock and the cost of holding the land for future rotations.

$MIC = \dfrac{re^{rt}[Y(t) - C]}{(e^{rt} - 1)}$ can be separated into these two costs as follows:

$$MIC = rY(t) + r\left(\frac{[Y(t) - C]}{[e^{rt} - 1]} - C\right) \tag{13}$$

where $rY(t)$ is the cost of holding the growing stock - the amount of interest payment the landowner would get if he sold the stand at year t and invested the money at an interest rate r for one year, and the last term on the right-hand side is the cost of holding the land - the soil rent the landowner should charge himself for waiting one more year (Note that this is equal to $rSEV$).

Chang (1984) notes that past attempts at explaining the optimum rotation age determination in terms of marginal analysis were unsuccessful because $Y'(t)$ has traditionally been *mislabelled* as marginal revenue (MR). Since MR means change in total revenue per unit change in total output, while MRP is defined as the change in total revenue per unit change in input, and time is an input to, rather than an output of, timber production, Chang (1984) contends that $Y'(t)$ should be called MRP rather than MR. Secondly, the use of the discrete discounting version of the SEV formula does not enable one to separate the MIC into the cost of holding the trees and the cost of holding the land (Chang, 1984).

8.4.4. Relationship between the Models

Relationship of the MSY Model to the Faustmann Model

When *p(t),* the stumpage price is a constant price (p) for all age classes, the cost of regeneration C is zero, and the interest rate is zero, maximising MAI is the same as maximising SEV.

$$\max SEV = \frac{pV(t) - Ce^{rt}}{(t-1)} = \max \frac{V(t)}{t} \tag{14}$$

In general, the optimal rotation determined from the MSY approach results in a rotation age longer than the Faustmann rotation. However, Binkley (1987) showed that if $r \leq 1/t^*_{MSY}$, where r is the discount rate and t^*_{MSY} is the optimal rotation determined from the MSY model, then, the MSY model gives a shorter rotation than the economic rotation models.

Relationship of the Single Rotation Model to the Faustmann Model

We can re-write the optimum condition of the Faustmann principle as:

$$Y'(t) = rY(t) + rSEV \tag{15}$$

We recall that the optimum condition for the NPV model is given by as:

$$Y'(t) = rY(t) \tag{16}$$

This implies that when SEV = 0, the NPV and Faustmann models are the same. The NPV model disregards all the income that could be generated by future rotations and hence is always smaller than the SEV in absolute terms. Also, the NPV model does not include the opportunity cost of holding the land, leading to a longer rotation than the Faustmann model. Finally, this model shows that the regeneration cost has no effect on the determination of the optimal rotation age. The two formulae for computing the SEV and the NPV are reproduced below.

$$SEV = \frac{pV(t) - Ce^{rt}}{(e^{rt} - 1)} \text{ and } NPV = \frac{pV(t) - C}{e^{rt}} \tag{17}$$

Further comparisons of the two formulae show that when t is small, the difference between (e^{rt} -1) and e^{rt} is large in proportion and the difference between the rotation lengths determined by the Faustmann and NPV models become very pronounced. But for large values of t, the difference may be negligible.

8.5. Effect of Changing Prices and Costs on the Optimal Rotation

Comparative statics in the Faustmann model shows that if timber prices rise, then the harvest rotation lengths shorten and regeneration inputs used in production increase. Conversely, if input costs rise, then rotations lengthen and regeneration inputs decline (Jackson, 1980; Hyde, 1980; Chang, 1984). However, if prices and costs change simultaneously over time, general insights cannot be gained from the Faustmann model because of its inability to incorporate comparative dynamics (Yin and Newman, 1995).

If a is the rate at which prices and costs change, where a is a constant not greater than 1, then it can be shown that the rotation lengths do not change over time, as there will be no shift in the production function. The only effect of this equiproportionate change is a reduction in the discount rate from r to $r_1 = r(1 - a)$. This gives the Faustmann solution with a reduced discount rate. Yin and Newman (1995) deduce the following from the above analysis, based on the assumption that the changes in prices and costs are less than the discount rate itself: (1) if prices and costs change at the same rate, then input use and rotation length do not change; (2) if prices and costs increase proportionately over time, the effective discount rate is reduced by the same proportion; (3) different proportionate changes in prices and costs correspond to different effective discount rates and therefore different rotation lengths.

8.6. INCORPORATING NON-TIMBER VALUES

The above analysis of the optimum rotation age determination assumed that the forests are only capable of producing timber. However, there is increasing interest in incorporating the non-timber benefits which forests provide during their growth cycles into optimum timber rotation models. Harvest timing models have therefore been developed that incorporate both timber and non-timber values into the rotation decision.

Calish et al. (1978) incorporated non-timber values by assuming that the standing timber volume Y(t), provides a flow of benefits each period given by Z[Y(t)]. When this is done, there are now two advantages to holding a stand of timber during the current period: the increase in stumpage value caused by the growth of forest volume and the flow of non-timber benefits. The cost of holding the stand is, therefore, the carrying cost of the timber and occupied land. The optimal harvesting point occurs when

$$Y'(t) + Z[Y(t)] = rY(t) + rSEV \tag{18}$$

In addition, the value of the bare land (SEV) on the right-hand side of Equation (18) must now include both the value of future timber harvests from bare land and the value of the future non-timber benefits that will be provided by the future sequence of forests (Anderson, 1992).

If $Z^* = Z[Y(t)]$ and the value of bare land due to non-timber benefits is A" = Z*/r, then (18) can be written as

$$Y'(t) + Z^* = r[Y(t) + SEV + Z^* / r] \tag{19}$$

The terms involving Z* cancel out so that the optimal rotation is unaffected by adding non-timber benefits to the timber optimisation framework. The only time that addition of non-timber values will lengthen the rotation is when the flow of non-timber benefits depends positively on the stage of the forest growth cycle such that forests containing larger timber volumes provide a larger stream of benefits to users (Newman, 1988). In this case, the benefits exceed the carrying cost, provided the timber growth rate is positive (Anderson, 1992).

Hartman (1976) derives an optimal harvesting rule in the situation where the amenity streams never diminish with age. If the landowner is able to capture non-timber and timber benefits, the optimal rotation is found by maximising:

$$B(t) = \frac{\int_0^t \gamma(t)e^{-rt}dt + pV(T)e^{rT} - C}{(1 - e^{-rt})} \tag{20}$$

Since γ(t) is the flow of amenity values accruing to the standing forest, the integral of γ(t) over the time between rotations sums up and discounts the value of future benefits due to delaying harvesting. Using first derivatives, the first order conditions for the optimal rotation result in the following decision rule at the optimal rotation age:

$$\frac{pV'(T)+\gamma(T)}{pV(T)-C}=r\left[\frac{1}{1-e^{-rT}}+\frac{\int_0^t \gamma(t)e^{-rt}dt}{[pV(T)-C](1-e^{-rT})}\right] \quad (21)$$

The optimal interpretation of this condition is that the trees should be harvested when the rate of value increment of both timber and non-timber benefits equal the interest rate. Hartman (1976) has shown that if the amenity values are small, then the Faustmann rotation approaches the Hartman model. If however, the non-timber benefits are large the optimum rotation would be longer or shorter than the Faustman model depending on whether the non-timber values increase or decrease with forest age. In a situation where the non-timber benefits are very large, the optimal decision may be to not harvest the trees (optimal rotation age of infinity).

Plantations established in Ghana provide a wide range of timber and non-timber benefits that can be incorporated into the optimal decision framework. With regards to timber benefits, the obvious ones are wood for timber, construction, and fuelwood and charcoal processing. Non-timber benefits include fodder for animals, fruits, gums, honey, control of soil erosion, windbreaks, provision of shade, improvement in soil fertility, carbon sequestration, etc. If these values can be monetised, they can be included in Equation 21 to determine the optimal rotation age that optimises total benefits.

8.7. Effect of the Risk of Fire on the Optimal Rotation

Wildfires are an annual occurrence in most parts of Ghana, especially in the savannah vegetation zones. In analysing the optimal forest rotation problem, the risk of forest fires is an important aspect since it affects the decisions by landowners. We can rewrite the optimal condition for the Faustmann model as:

$$\frac{pV'(t)}{pV(t)-C}=\frac{r}{e^{rt}-1} \quad (22)$$

Conrad and Clark (1987) derive the optimal rotation age decision rule when there is a constant risk of fire equal to λ per unit time. It is assumed that the probability of fire follows a Poisson distribution with a parameter λ. Using these assumptions, the optimal rotation condition is given by (Conrad and Clark, 1987) as:

$$\frac{pV'(t)}{pV(t)-C}=\frac{r+\lambda}{e^{t(\lambda+r)}-1} \quad (23)$$

This optimal rotation rule effectively adds the probability of fire to the discount rate. If the probability is zero, we get the Faustmann rotation rule. If λ is very large, the rotation length will be very short. What this means is that landowners will harvest the trees early to avoid them being destroyed by fire, and in the extreme case where $\lambda = \infty$, the rotations will be so short that the optimal decision will be to not establish plantations at all.

8.8. Case Study 1: Optimal Rotation Age Determination for Neem Plantations

8.8.1. Base Case Analysis

Using growth and yield data of neem in the Guinea Savannah zone of Ghana discussed in Chapter 6 and given in Nanang (1996), the optimal rotation ages are estimated using the three criteria described above. Table 1 gives a summary of the growth and yield data for Site Class I neem plantations in the Tamale Forest District. Although non-timber benefits are recognised as being jointly produced from the forest, these are not taken into account in these simple examples. In Table 1, the MAI and CAI intersect at approximately plantation age 5, and hence this is the optimal rotation age based on this criterion. Beyond age 5, the MAI is now greater than the CAI.

Table 1. Determination of the optimum rotation age for neem plantations in Northern Ghana using the maximum sustained yield criterion

Plantation Age (yrs)	Volume (V_t) (m^3/ha)	CAI (m^3/ha/yr)	MAI (m^3/ha/yr)
1	1.03	1.03	1.03
2	13.67	12.68	6.86
3	32.35	19.30	11.00
4	49.77	16.91	12.48
5	64.46	14.72	**12.93**
6	76.58	12.03	12.78
7	86.61	10.21	12.41
8	94.99	8.48	11.92
9	102.06	7.58	11.44

Table 2 presents the results of the optimal age determination when the single rotation (or NPV) criterion is applied. The following assumptions are used in the analysis for the NPV and SEV criteria: r =5%, p= GH¢20/m^3 and planting costs of GH¢1000/ha. It would be recalled that the optimal rotation in the NPV model occurs where $Y'(t) = Y(t)r$; i.e., if the annual increment in value of the trees is larger than the interest that the money gained by harvesting the plantation the previous year would have accrued, then do not harvest the trees in that year. In Table 2, the optimal rotation age is at year 10. Holding the plantation to year 11 will make the cost MIC [rY(t)] greater than the annual value increment MRP [Y'(t)].

Table 2. Determination of the optimum rotation age for neem plantations in Northern Ghana using the NPV criterion

Plantation Age (yrs)	Volume (V_t) (m^3/ha)	Y(t) (GH¢/ha)	MRP [Y'(t)] (GH¢/ha)	MIC[rY(t)] (GH¢/ha)	NPV
1	1.03	20.60	20.60	1.03	-980.41
2	13.67	273.34	252.74	13.67	-752.67
3	32.35	647.03	373.69	32.35	-443.09
4	49.77	995.49	348.45	49.77	-184.96
5	64.46	1289.14	293.65	64.46	3.99
6	76.58	1531.60	242.45	76.58	134.63
7	86.61	1732.22	200.63	86.61	220.68
8	94.99	1899.76	167.54	94.99	273.45
9	102.06	2041.19	141.43	102.06	301.52
10	108.09	2161.88	120.69	108.09	**311.25**
11	113.30	2265.91	104.03	113.30	307.32

When the SEV criterion was applied, the optimum rotation occurred at 9 years. From Table 3, at plantation age 8 years, the MRP is GH¢167.60 and MIC is GH¢136.46. Holding the trees for another year reduces the MRP to GH¢141.40 and increases the MIC to GH¢143.66. As discussed previously in the theoretical section, the optimum rotation occurs where the SEV is at a maximum. The last column of Table 3 shows the SEV at each age, with a maximum value occurring at age 9. In this case holding the trees for another year makes the MIC > MRP. Therefore the trees should be harvested at age 9 years.

Table 3. Determination of the optimum rotation age for neem plantations in Northern Ghana using the SEV rotation criterion

Plantation Age (yrs)	Volume (V_t) (m^3/ha)	Y(t) (GH¢/ha)	MRP [Y'(t)] (GH¢/ha)	rY(t) (GH¢/ha)	SEV (GH¢/ha)	rSEV (GH¢/ha)	MIC[rY(t)] (GH¢/ha)
1	1.03	20.60	20.60	1.03	-20102.4	-1005.12	-1004.09
2	13.67	273.34	252.74	13.67	-7909.28	-395.46	-381.80
3	32.35	647.03	373.69	32.35	-3181.03	-159.05	-126.70
4	49.77	995.49	348.45	49.77	-1020.38	-51.02	-1.24
5	64.46	1289.14	293.65	64.46	18.02	0.90	65.36
6	76.58	1531.60	242.45	76.58	519.46	25.97	102.55
7	86.61	1732.22	200.63	86.61	747.27	37.36	123.97
8	94.99	1899.76	167.54	94.99	829.43	41.47	136.46
9	102.06	2041.19	141.43	102.06	**832.07**	41.60	143.66
10	108.09	2161.88	120.69	108.09	791.03	39.55	147.65
11	113.30	2265.91	104.03	113.30	726.43	36.32	149.62

The difference in rotation age between the NPV and Faustmann models is mainly in the fact that the Faustmann model includes the opportunity costs of the land holding the trees. If this opportunity cost is high, the rotation will be shorter than that predicted by the NPV model.

The optimal rotations determined by the three approaches are 5, 10 and 9 years for the MSY, NPV and SEV models respectively. From the theoretical analysis, it was observed that the MSY usually results in a longer rotation than the remaining two models. However, it was also noted from studies by Binkley (1987) that when the reciprocal of the rotation age determined by the MSY is greater than or equal to the discount rate, then the MSY produces a shorter rotation length than the SEV or NPV models. In this example, it can be verified that this condition holds. From a practical point of view, unless fuelwood and very small dimension timber is the ultimate objective of management, most landowners will not harvest their plantations at age five. As shown, this rotation age does not optimise the economic return on the investment. The economic return is maximised at age 9 years under the SEV model.

8.8.2. Sensitivity Analyses

From Table 3, the optimal rotation age for neem plantations using r =5%, p= GH¢20/m^3 and planting costs of GH¢1000/ha was 9 years. However, this optimal rotation age would change as the assumptions in the variables used in the analyses change. Table 4 reports the results of sensitivity analyses to examine the impacts of changes in the assumed discount rates, prices, costs and the risk of fire on the optimal rotation age.

Table 4. Sensitivity of the optimal rotation age and SEV to changes in the discount rate, price of wood, regeneration costs and risk of fire

Risk of fire (λ%)	r=5%, p= 20, c=1000		r=5%, p=40, c=1000		r=5%, p=20, c= 2000		r=10%, p=20, c= 1000	
	Rotation	SEV	Rotation	SEV	Rotation	SEV	Rotation	SEV
0	9	832.07	7	4897.21	16	-1490.51	8	-265.83
10	7	-605.83	6	413.53	13	-1927.71	7	-760.34
20	7	-846.00	5	-366.24	12	-1981.33	6	-894.73

Note: p is measured in GH¢/m^3, c is in GH¢/ha, and rotation age is measured in years.

The results are consistent with the theoretical analyses. The risk of fire reduces the optimal rotation ages for all scenarios examined. Except when the price of timber is 40/m^3, all other scenarios produce negative SEVs when the risk of fire is greater than 0%. This is consistent with the reasoning that the probability of the risk of fire is equivalent to an increase in the discount rate. The main conclusions of these analyses are: Increasing the discount rate and the risk of fire each reduce the optimal rotation age, while increasing the cost of regeneration increases the rotation length for the plantation. Increases in the price paid per m^3 of wood not only decreases the rotation age, but also increases the profitability of the plantation, holding all other variables constant. Under the assumptions, the only scenarios that produce a net benefit for the plantation are when the discount rate is 5%, p = 20, and c= GH¢1000/ha, with no risk of fire. When the risk of fire is 10%, the plantation is still profitable only if the price of timber is GH¢40/m^3, with the discount rate at 5% and c= GH¢1000/ha.

8.9. Case Study 2: Optimal Rotation Age Determination for Teak Plantations

8.9.1. Basic Analysis

Table 5 gives a summary of the growth and yield data for Site Class I teak plantations in the guinea savannah zone of Ghana from Nunifu (1997). The MAI and CAI intersect at approximately plantation age 30, and hence this is the optimal rotation age based on this criterion. Beyond age 30, the MAI is now greater than the CAI. If the decision to harvest the plantation is based solely on the age that yields maximum volume growth, then it will be at 30 years.

Table 5. Determination of the optimum rotation age for teak plantations in the guinea savannah zone using the MSY criterion

Age	Volume (V_t) (m^3/ha)	CAI (m^3/ha/yr)	MAI (m^3/ha/yr)
5	22.76	11.33	4.55
10	97.8	17.21	9.78
15	186.56	17.87	12.44
20	274.17	17.06	13.71
25	356.54	15.87	14.26
30	432.85	**14.66**	**14.43**
35	503.27	13.52	14.38
40	568.28	12.5	14.21
45	628.45	11.58	13.97
50	684.3	10.77	13.69

Table 6. Determination of the optimum rotation age for teak plantations in Northern Ghana using the NPV criterion

Plantation Age (yrs)	Volume(V_t) (m^3/ha)	Y(t) (GH¢/ha)	Y'(t) (GH¢/ha)	rY(t) (GH¢/ha)
5	22.70	4540.67	2017.38	227.03
10	97.55	19510.39	3382.30	975.52
15	186.09	37218.59	3570.90	1860.93
16	203.87	40774.71	3556.12	2038.74
17	221.53	44306.91	3532.19	2215.35
18	239.04	47808.14	3501.23	2390.41
20	273.49	54697.52	3424.49	2734.88
22	307.07	61414.02	3335.44	3070.70
23	323.51	64702.29	**3288.28**	**3235.11**
24	339.71	67942.40	**3240.11**	**3397.12**
25	355.67	71133.76	3191.36	3556.69

In order to apply the NPV and SEV criteria, the following assumptions are used: r =10%, p= GH¢100/m^3 and planting costs total GH¢1500/ha. With these assumptions, the NPV criterion shows that the optimal rotation age for these plantations occurs at age 23 years.

From Table 6, the fourth column shows the marginal increase in value of the plantation in each 5-year period. The last column shows the opportunity costs of holding on to the trees for each additional 5-year period. This column basically calculates the interest earned on the proceeds if the plantation were harvested in the previous period and the money deposited in a bank at an interest rate of 10% per annum. In deciding when to harvest the plantation, the landowner compares the benefits and costs of holding the trees for an additional period. As soon as the costs outweigh the benefits, the plantation should be harvested.

Table 7. Determination of the optimum rotation age for teak plantations in Northern Ghana using the SEV criterion

Plantation Age (yrs)	Volume(V_t) (m^3/ha)	Y(t) (GH¢/ha)	Y'(t) [MRP] (GH¢/ha)	rY(t) (GH¢/ha)	rSEV (GH¢/ha)	MIC (GH¢/ha)
5	22.70	4540.67	2017.38	227.03	460.28	687.32
10	97.55	19510.39	3382.30	975.52	1313.15	2288.66
15	186.09	37218.59	3570.90	1860.93	1523.86	3384.79
16	203.87	40774.71	**3556.12**	2038.74	**1527.34**	**3566.08**
17	221.53	44306.91	**3532.19**	2215.35	1522.69	**3738.04**
18	239.04	47808.14	3501.23	2390.41	1511.33	3901.73
20	273.49	54697.52	3424.49	2734.88	1472.99	4207.86
22	307.07	61414.02	3335.44	3070.70	1419.74	4490.44
23	323.51	64702.29	3288.28	3235.11	1389.24	4624.36
24	339.71	67942.40	3240.11	3397.12	1356.88	4754.00
25	355.67	71133.76	3191.36	3556.69	1323.076	4879.764

Tables 6 and 7 show the optimal rotation ages for teak in the guinea savannah zones using the NPV and SEV models. Again, the SEV model results in a shorter rotation (16 years) than the NPV model (23 years) because of the difference in the opportunity costs of holding the land. The MSY model resulted in the longest rotation of 30 years. This is because the MSY framework does not account for interests on capital investments or the opportunity costs of holding the land and trees.

8.9.2. Sensitivity Analyses

The optimal rotation age for teak plantations in the savannah vegetation zone was used to examine the impacts of changing prices, costs, discount rates and the risk of fire on the optimal rotation and profitability of the plantation project. Table 8 reports the results of sensitivity analyses.

The difference between the effect of the risk of fire in the neem plantations and teak is because of the higher value of teak wood. The higher value of teak is able to compensate for the increased costs (risks) of fire. What this implies is that in general, a higher value for wood allows the investor additional latitude to take more risk. In these analyses, increasing the discount rate and the risk of fire all reduce the optimal rotation age while increasing the cost of regeneration decreases the rotation length for the plantation. Increases in the price paid per m^3 of wood, however, decreases the rotation age and increases the profitability of the plantation, holding all other variables constant. Due to the higher value of teak wood, the plantation is profitable under a wider range of prices, discount rates and costs, and risks of fire.

Table 8. Sensitivity of the optimal rotation age and SEV to changes in the discount rate, price of wood, regeneration costs and risk of fire

Risk of fire (λ%)	*r*=5%, *p*= 200, c=2000		*r*=10%, *p*=200, c=2000		*r*=5%, *p*=400, *c*= 2000		*r*=5%, *p*=200, *c*= 4000	
	Rotation	SEV	Rotation	SEV	Rotation	SEV	Rotation	SEV
0	16	29638.86	12	8565.10	16	62,909.64	17	26,087.71
10	10	3029.28	9	797.84	10	8632.99	11	512.19
20	8	-297.42	8	-914.73	8	1718.20	9	-257.10

Note: *p* is measured in GH¢/m^3, *c* is in GH¢/ha, and rotation age is measured in years.

The risk of fire reduces the optimal rotation ages for all scenarios examined. Even with a 20% risk of fire, the plantation still returns a profit as long as the price of timber is GH¢400/m^3.

8.10. CONCLUSIONS

The determination of the optimal rotation age is all about optimising the benefits that society derives from the resource. If the trees are harvested too early before the optimal rotation age is reached, or too late after the optimal rotation age has passed, resources are wasted. Although there are several methods to determine the optimal rotation age, the economic model proposed by Faustmann (1849) and its subsequent modifications provide the correct approach. In Ghana, the most important factor that impacts on the optimal rotation age determination is the interest and discount rates. Using the data for teak and neem, this Chapter demonstrated that it is possible to grow plantations of these two species profitably even with timber benefits only. If other non-timber benefits are considered, the economic returns would improve even further.

Chapter 9

CARBON MARKETS AND PLANTATION FORESTRY

9.1. INTRODUCTION

This chapter examines an important international environmental policy issue that has implications for plantation development and management. The Kyoto Protocol and the Clean Development Mechanism (CDM) provided thereunder have provided a market for carbon sequestered in afforestation and reforestation projects as if carbon were a tangible commodity. The importance of the CDM to plantation forestry in Ghana is that it provides incentives to develop plantations to earn carbon credits in addition to other socio-cultural and environmental benefits. A good understanding of how the Kyoto Protocol and future international climate change policy could impact on plantation forestry is important to plantation forestry investors in Ghana.

These carbon credits and the potential income that can be generated could change the economics of the plantation enterprise in some circumstances from uneconomic to profitable. Carbon sequestration considerations could also impact on the type of management that would be applied to plantations. As we saw in Chapter 8, the inclusion of non-timber benefits could change the optimal rotation age of plantations and in some cases it will be optimal not to harvest the trees at all. This Chapter begins by looking at the larger context of the role forests play in the global climate change policy development and implementation. It then looks at the Kyoto Protocol and the flexible mechanisms in general. The economic basis for introducing the flexible mechanisms is examined along with the criticisms against these market-based approaches to climate change policy. The rules governing the CDM, the administrative structures in Ghana to manage CDM projects, and application of the CDM to forestry projects are discussed. An example application of how carbon credits could be awarded to a hypothetical teak plantation in Ghana concludes the Chapter.

9.2. THE ROLE OF FORESTS IN CLIMATE CHANGE MITIGATION

There is increasing scientific evidence that carbon dioxide (CO_2) in the atmosphere contributes to the greenhouse effect. The greenhouse effect is a name given to the process whereby certain gases trap heat within the surface-troposphere system. These greenhouse gases (GHG) absorb and emit infrared radiation causing the heating at the surface of the planet. This heating of the planet is also believed to be leading to changes in global weather and climate patterns, a process known as climate change. The main source of CO_2 from the earth is through the burning of fossil fuels and other organic materials, and through other natural processes.

Through photosynthesis, trees and other vegetation absorb CO_2 from the atmosphere into their tissues and soil and store this as carbon. This ability to absorb and store carbon makes forests carbon sinks[1]. This carbon remains stored in the plant material until the material is either burnt or decays. Therefore, through growth, trees remove CO_2 from the atmosphere and the rate of removal is proportional to the rate of growth (or the rate of photosynthesis). The process of removing carbon from the atmosphere and storing it in plant tissues is known as carbon sequestration[2]. About 30% (about 4 billion ha) of the earth's surface is covered by forests (FAO, 2005) excluding other vegetation types. The worlds forests store an estimated 289 gigatonnes (Gt) of carbon in trees and vegetation, which is more than all the carbon in the atmosphere (FAO, 2010). It is clear that forests can play an important role in reducing the amount of CO_2 in the atmosphere and hence contribute to mitigating global climate change. Forests can also be sources of CO_2 when they release the stored carbon through forest fires, decaying of forest biomass/products, pests and disease infestation and through deforestation activities.

9.3. The Kyoto Protocol

9.3.1. The Kyoto Protocol and the Flexible Mechanisms

The Kyoto Protocol (KP) is an international agreement that was negotiated on 11 December 1997 in Kyoto, Japan and commits industrialised countries (defined in the KP as Annex I countries) to reduce their (GHG) emissions by an average of 5.2% compared to 1990 levels, during the first commitment period from 2008 to 2012. The KP, which came into force on 16 February, 2005, is linked to the United Nations Framework Convention on Climate Change (UNFCCC). Under the KP, each Annex I country has a binding target and countries are expected to meet these targets using domestic measures. However, three "flexible mechanisms" are included in the KP to help countries reach their targets in a cost-effective way. These mechanisms are: clean development mechanism (CDM); Joint implementation (JI); and emissions trading (ET). All three mechanisms were designed to help participating countries in meeting their domestic emission reduction targets at lower costs and to help developing countries and countries in transition in their sustainable development by encouraging technology transfer. The CDM allows emission-reduction (or emission removal) projects in developing countries to earn certified emission reduction (CER) credits, each equivalent to one tonne of CO_2. These CERs can be traded and sold, and used by industrialised countries to a meet a part of their emission reduction targets under the KP. These projects are also expected to result in sustainable development and emission reductions, as defined by the host country. Joint Implementation (JI) provides a means for

[1] In contrast, a carbon source emits GHGs into the atmosphere.

[2] Note that carbon sequestration is a general term referring to any process that increases a carbon pool other than in the atmosphere.

countries or companies to invest in GHG reduction measures and sequestration projects in other industrialised countries or countries in transition (such as Russia), and gain certified credits. Emissions trading (ET), as set out in Article 17 of the KP, allows countries that have emission units to spare (those emissions permitted them but not used) to sell this excess capacity to countries that are over their targets. Therefore, ET provides a means for emitters to purchase emissions reduction credits through a special market that will be set up for this purpose. The only mechanism of interest in this Chapter is the CDM, and it will be the main focus of the remaining discussion.

9.3.2. The Economic Basis of the Flexible Mechanisms

The main objective of the introduction of flexible mechanisms under the KP was to reduce the overall cost of meeting binding targets by industrialised countries who are signatories to the KP. There are two main ways for countries to achieve their targets under the KP: a) take national measures through command-and-control approaches such as legislation, or direct and indirect taxation; or b) use economic instruments such as trading in GHG emissions and carbon sequestration projects. A legislative approach requires everyone to reduce their emissions irrespective of the cost of doing so. For example, if there are two firms with the same legislated limit on how much GHG they can emit, both will be expected to reach the same limit, even if it costs Firm A more money than it costs Firm B to meet those obligations.

The second approach to controlling emissions creates markets for trading in carbon emissions. There are two main schemes: the cap-and-trade system and carbon offset trading system. Under the former scheme, a cap on emissions is determined by the country and those industries that emit less than the cap can sell the excess "credits" to firms that cannot meet their target with existing technology. The national cap on emissions is set for a given time period, and this cap is divided into targets for the various industries within the country. Each country develops national regulations that determine which firms are covered by a cap in their emissions and those that are not. Firms are then given permits, and they cannot emit GHGs above what their permits allow them. Firms with caps must then decide the most cost effective way to meet their targets, which could include cutting production, changing production processes through technology, or buying allowances from other firms (hence the term flexible mechanisms). What this means is that firms that have technology and can reduce their emissions at a lower cost would do so; those with higher costs of reducing emissions would buy credits from others who have excess to spare. Theoretically, this will reduce the overall costs of meeting the national target. It is expected that as time passes, the availability of permits will be reduced, thus increasing the scarcity of the permits and hence the price and cost to pollute. The largest emissions trading scheme in the world today is the European Union Emissions Trading Scheme (EU-ETS) which was launched in 2005. As at 2008, the EU-ETS was worth about $63 billion (Gilbertson and Reyes, 2009). There are other trading schemes in New Zealand, USA and Australia.

Under the carbon offset trading scheme, firms and individuals finance 'carbon offset projects' outside the industries with caps on their emissions to obtain emission credits which they can apply to offset their own emissions. Recall that from above, the economy was divided into two main sectors for purposes of emissions trading: those industries with caps on their emissions; and those without caps. There are no emission caps on forests hence forestry activities can generate offset credits. The CDM and the JI are examples of the offset trading schemes, with the CDM being the largest offset trading scheme in the world.

Critiques of emissions and offset trading schemes point out that these schemes do not lead to emission reductions. In the case of the offset scheme, they argue that these actually increase global emissions rather than decrease them (e.g., Gilbertson and Reyes, 2009). There are two main objections: leakage due to the project and emissions resulting from the buyer of the credits. Gilbertson and Reyes (2009) argue that even in cases where it can be verified by the seller that the emissions have been reduced due to the CDM project, there is increased emissions by the buyer which negates the benefits from the CDM project.

9.4. Plantation Forestry and Carbon Markets

The term *carbon market* refers to the buying and selling of emissions permits or credits that have either been distributed by a regulatory body or generated by GHG emission reduction projects. The largest market for carbon credits resulting from afforestation and reforestation (A/R) projects is within the CDM framework of the Kyoto Protocol. In this section, I examine the issues related to the CDM, important definitions related to A/R projects within the CDM and the rules governing the award and trading of credits. There may be markets for carbon credits within domestic offset trading schemes of Annex I countries. For example, Canada is developing a domestic offset trading system to support its local large industrial final emitters to meet their GHG obligations. Also, the Government of New Zealand has already enacted an emissions trading scheme (ETS) under which owners of Kyoto Protocol compliant forests will receive units for increases in carbon stocks of their plantations (Manley and Maclaren, 2010). There are several requirements, procedures and rules to be followed in order to be able to earn carbon credits through A/R projects under the CDM. These rules are not fully covered in this book, but interested readers can find information through the Environmental Protection Agency (EPA) in Ghana, or on several websites of the UNFCCC or the Intergovernmental Panel on Climate Change (IPCC).

9.4.1. General Provisions of the CDM

The Clean Development Mechanism (CDM), established under the Kyoto Protocol, is the primary international offset program in existence today, and while not perfect, it has helped to establish a global market for greenhouse gas (GHG) emission reductions. It generates offsets through investments in GHG reduction, avoidance, and sequestration projects in developing countries.

According to Article 12 of the KP, a CDM allows the countries with legally binding greenhouse gas (GHG) emissions reductions commitments to receive credits towards their obligations by investing in projects in developing countries. Article 12 stipulates that two criteria must be met for projects undertaken by Annex I countries to be eligible for crediting under the CDM:

- Projects must produce both "certified emissions reductions" that "are additional to any that would occur in the absence of the certified project activity" and contributions to the host country's sustainable development; and
- In addition, developing country parties will have access to resources and technology to assist in the development of their economies in a sustainable manner.

The KP identifies six GHGs and gas classes whose reductions qualify under the CDM: carbon dioxide, methane, nitrous oxide, hydroflourocarbon, perflourocarbon, and sulphur hexafluoride. Sectors that may qualify from CDM projects include energy, industrial processes, solvent and other product use, waste and land use, land use change and forestry.

The Marrakech Accords (UNFCCC, 2002) recognises afforestation and reforestation as the only eligible land uses under the CDM. The KP recognises that forests, forest soils and forest products all play important roles in mitigating climate change. The incentives provided under the CDM is that tree growers can generate carbon credits that can be sold to entities subject to the KP to offset their emissions, in the form of Certified Emission Reductions (CERs). Since CDM projects can only be undertaken in developing countries, this offers interesting opportunities for the management of plantation forests for sequestering carbon in Ghana.

9.4.2. Important Definitions and Concepts Related to the CDM

Articles 3.3 and 3.4 of the KP and the Marrakech Accords (UNFCCC, 2002) specified the following definitions for projects related to CDM initiatives:

Afforestation is the direct human-induced conversion of land that has not been forested for a period of at least 50 years to forested land through planting, seeding and/or the human-induced promotion of natural seed sources.

Reforestation is the direct human-induced conversion of non-forested land to forested land through planting, seeding or human-induced promotion of natural seed sources, on land that was forested but that has been converted to non-forested land. For the first commitment period (2008–2012), reforestation activities will be limited to reforestation occurring on those lands that did not contain forests on 31st December, 1989.

A *Forest* is a minimum area of land of 0.05–1.0 ha with tree crown cover of more than 10–30% with trees, with the potential to reach a minimum height of 2–5 m at maturity in situ. A forest may consist either of closed forest formations, where trees of various storeys and undergrowth cover a high proportion of the ground or open forest. Young natural stands and all plantations which have yet to reach a crown density of 10–30 per cent or tree height of 2–5 metres are included under forest, as are areas normally forming part of the forest area which are temporarily not stocked as a result of human intervention such as harvesting or natural causes but which are expected to revert to forest. For purposes of implementing the CDM, Ghana has chosen a crown cover of 15%, a minimum area of 0.1 ha and a minimum height of 2.0 m to define a forest.

A *baseline* is the scenario that reasonably represents the anthropogenic emissions by sources and removals by sinks that would occur in the absence of a project activity. The development of project baselines is important to ensuring the project emission reduction/removal is real. The impact of a project typically is assessed relative to a baseline, with certified credits issued for above-baseline carbon sequestration or emission reductions. Methodologies for determining forest project baselines include modelling, control plots, and historic benchmarks. Whichever approach (es) is (are) chosen, it is expected that the onus will be on a project developer to support their baseline choice, and the Executive Board of the Clean Development Mechanism will review proposed project baselines to ensure standards of accuracy and consistency. Standardised baseline methodologies could also be developed and approved for use by projects. For specific projects within the CDM, baseline definitions must take into account the following elements, where applicable: natural and anthropogenic GHG emissions and removals, type of practice(s) and degree of implementation (or level), trends in practices and applicable technologies, and regulations and indirect climate change measures.

Additionality: For a CDM project, emissions reductions must be beyond - *or in addition to* – the reductions that would have occurred in the absence of the project. Additionality is assessed by comparing the carbon stocks and flows of the project activities with those that would have occurred without the project (i.e., its baseline). For example, the project may be proposing to afforest farmland with native tree species, increasing its stocks of carbon. By comparing the carbon stored in the 'project' plantations (high carbon) with the carbon that would have been stored in the 'baseline' abandoned farmland (low carbon) it is possible to calculate the net carbon benefit. Additionality is important to any cap-and-trade system that allows for projects from outside the cap as in the CDM. If reductions from projects are used to offset emissions elsewhere (i.e., reductions from GHG sources and sinks not covered by a cap, and are used to offset emissions under a capped system), then those reductions should not have happened anyway, or else there will be a net increase in atmospheric emissions by that amount.

Permanence is the length of time for which carbon will remain stored after having been fixed in vegetation. In other words, the ability to store carbon stocks generated by a project under the CDM until the crediting period expires. Any future emissions that might arise from these stocks need to be accounted for. Permanence is one of the main concerns related to the use of sinks as a greenhouse gas (GHG) mitigation option. In reality, the concern is about lack of permanence, or 'reversibility' of the benefits of storage, as a result of the possible loss of carbon stocks created or conserved by a project, whether on purpose or as a result of undesirable events (e.g., natural disasters). Natural disturbances or human activities can cause a reduction in net sequestration within a project boundary, and non-permanence occurs when this net sequestration rate (or sink size) falls below the amount for which offset credits have been issued. The two elements in a system to address non-permanence of offset projects are: the preparation of a plan to manage the risk of non-permanence events, reducing the likelihood of their occurrence, and a mechanism by which the validity of an issued offset credit is maintained when a non-permanence event occurs.

Project Boundary: The project boundary encompasses all anthropogenic emissions by sources and removals by sinks of greenhouse gases under the control of the project proponents that are significant and reasonably attributable to the project activity. As it pertains to forest plantations, this favours a project boundary that encompasses not only the activity for which credits are being issued, but also other forestry activities under the direct control of the project proponent (that are significant and reasonably attributable to the project). Logically, a project proponent should not be issued credits for maintaining a plantation forest, for instance, if the resultant carbon benefits are effectively offset by a consequent increase in harvesting on other lands managed by the project proponent (the areas where this increased harvesting is occurring has to be in the leakage assessment area). Identifying afforestation and reforestation project boundaries and leakage is a two-tiered process. First, determining whether the project proponent *controls* a GHG source isolates potential leakage from potential inside-boundary emissions. Secondly, the source must be assessed in terms of its *significance, attribution,* and *cost-effectiveness,* and on this basis either dismissed or accounted for as leakage or inside-boundary emissions, as applicable.

Leakage is an increase in emissions or reduction in removals outside a project's boundary (the boundary defined for the purposes of estimating the project's net GHG impact) resulting from the project's activities. Leakage is associated with changes in reductions/removals that are significant and reasonably attributable to the project, but are not under the control of the proponent. The amount of potential leakage would depend mainly on the nature of the project and the type of activities. Afforestation/reforestation (AR) projects would likely have less potential for leakage especially if implemented on land that has few or no competing uses. For forest projects that change practices, these activities would generally not result in leakage because the underlying economic activity continues so that no activity shifting would be expected. However, leakage could occur if the changes in practices result in a reduction in the output of forest products. The potential for leakage implies that it may be necessary to: 1) identify if a project has any leakage potential; 2) try to mitigate the potential for leakage through project location and design; and 3) account for any leakage when estimating total project greenhouse gas (GHG) benefits. There is currently no standard method to identify or quantify leakage effects due to GHG mitigation projects in forestry.

9.4.3. Rules Governing CDM Forestry Projects

Below are the rules and conditions that will apply to CDM projects, especially to afforestation and reforestation projects, which provides the overall framework for approving projects and accounting for the carbon credits generated (UNFCCC, 2002).

- All CDM projects must be approved by the CDM Executive Board.
- Only areas that were not forest on 31st December 1989 are likely to meet the CDM definitions of afforestation or reforestation.
- Projects must result in real, measurable and long-term emission reductions, as certified by a third-party agency.
- Emission reductions or sequestration must be additional to any that would occur without the project. They must result in a net storage of carbon and therefore a net removal of carbon dioxide from the atmosphere.
- Projects must be in line with sustainable development objectives, as defined by the government that is hosting them.
- Projects must contribute to biodiversity conservation and sustainable use of natural resources.
- Only projects starting from the year 2000 onwards will be eligible.
- Two percent of the carbon credits awarded to a CDM project will be allocated to a fund to help cover the costs of adaptation in countries severely affected by climate change. This adaptation fund may provide support for land use activities that are not

presently eligible under the CDM, for example conservation of existing forest resources.

- Some of the proceeds from carbon credit sales from all CDM projects will be used to cover administrative expenses of the CDM (a proportion still to be decided).
- Projects need to select a crediting period for activities, either a maximum of seven years that can be renewed at most two times, or a maximum of ten years with no renewal option.
- The funding for CDM projects must not come from a diversion of official development assistance (ODA) funds.
- Each CDM project's management plan must address and account for potential leakage. Leakage is the unplanned, indirect emissions of CO_2, resulting from the project activities. For example, if the project involves the establishment of plantations on agricultural land, then leakage could occur if people who were farming on this land migrated to clear forest elsewhere.

9.4.4. Types of Carbon Credits Awarded for Plantations under the CDM

It is important that interested investors in afforestation and reforestation carbon sequestration projects under the CDM should also acquaint themselves with the types of carbon credits awarded under the CDM. Carbon credits are awarded for the CO_2 that is sequestered over and above the baseline, taking into consideration any leakage that might have occurred as a result of the project. This means that earning credits has to be based on the determination of a credible baseline for the particular project. The CDM Executive Board has published a guide book on the various methodologies available for determining baselines in A/R projects (e.g., see UNEP, 2005).

The main difference between forestry sequestration projects and emission avoidance projects (those that prevent carbon emissions) is that trees reduce CO_2 in the atmosphere by capturing and storing it in the biomass, whereas, many other projects prevent or reduce the release of CO_2 into the atmosphere. This means that CO_2 reduction from forestry projects is not permanent, as the CO_2 can be re-emitted into the atmosphere for example, through forest fires, pest attack related to dying forests, flooding of forest, etc. or human actions (for example, logging, deforestation or burning the wood) (UNEP, 2005). For this reason, CO_2 reduction from forestry sequestration is considered temporary.

As a result of this potential of non-permanence of the sequestered CO_2, carbon credits issued for forestry activities A/R under the CDM during the first commitment period of the implementation of the Kyoto Protocol from 2008-2012 are temporary Certified Emissions Reductions (tCERs) or long-term CERs (lCERs), which are different in characteristics from the CERs issued for projects that prevent the emission of GHGs. As per modalities and procedures agreed for A/R projects, the A/R project proponents can choose either tCERs or lCERs (UNEP, 2005). Temporary credits (tCER) are issued and expire at the end of the commitment period following the one during which it was issued. Once expired, a tCER can be reissued several times during the project as long as the forest exists. On the contrary, a long term credit (lCER) expires at the end of the crediting period of the overall afforestation or reforestation project (Olschewski and Benıtez, 2005), and cannot be reissued (Neef and Henders, 2007).

These different types of accounting procedures have impacts not only on the amount of CER awarded to reforestation projects but also on the risk and the value of the certificates (Olschewski and Benıtez, 2010). Theoretically, potential buyers of carbon credits would to be indifferent between buying a permanent credit today and buying a non- permanent credit (lCER or tCER) today and replacing it by a permanent one when the initial credit expires (Olschewski and Benıtez, 2010). Equation 1 reflects this difference in preference, where *t* is the lifetime of CER units, P_t is the price per tCO_2 for a credit valid for t-years, P_∞ is the price for a permanent credit, and r is the discount rate of potential buyers of the carbon credits (Olschewski et al., 2005):

$$P_\infty = P_t + \frac{P_\infty}{(1+r)^t} \tag{1}$$

From a carbon buyers' perspective, it would make sense to buy a non-permanent CER if its price is lower than that of a permanent CER minus the discounted price of a permanent CER in year *t*, when the expiring temporary CER has to be replaced (Olschewski and Benıtez, 2010). Therefore, the maximum price that an enterprise would be willing to pay for temporary carbon credits, P_t, results from Equation (1) as (after Olschewski and Benıtez, 2010):

$$P_t = P_\infty \left[1 - \frac{1}{(1+r)^t} \right] \tag{2}$$

Given that lCER and tCER credits only differ with respect to their expiring time, the price derived from Equation 2 is valid for both types of credits. CDM official accounting rules allow for temporary credits with an expiring time of 5 years and A/R projects with maximum duration of 30 years (Olschewski and Benıtez, 2005; Olschewski and Benıtez, 2010).

9.5. The Clean Development Mechanism in Ghana

Ghana ratified the United Nations Framework Convention on Climate Change (UNFCCC) in 1995 and the Kyoto Protocol in May, 2003. Because it is a developing country, it has no binding targets under the KP, but is eligible to participate in projects under the CDM. Ghana has taken steps to take advantage of the CDM by setting up a Designated National Authority (DNA) on CDM within the Environmental Protection Agency (EPA).

9.5.1. Structure and Functions of the Designated National Authority (DNA)

A legislative process has been in the works since 2005 to make the DNA operational, but it has not been passed by parliament yet. At present, the DNA in Ghana is working on the basis of ministerial declarations (CDM Ghana, 2008). CDM project evaluation is being carried out by ad hoc expert groups who make recommendations to the CDM Governing Council, which is entitled to take the final approval decision (CDM Ghana, 2008). The DNA is to be funded from fees charged for projects and also from the Ministry of Environment and donations from public institutions.

The DNA has a governing council that consists of the following:

1. Executive Director (Environmental Protection Agency) Chairman
2. Chief Director (Ministry of Environment and Science) Member
3. Chief Director (Ministry of Energy) Member
4. Chief Director (Ministry of Lands and Forestry) Member
5. Director (External Resource Mobilization Division, Ministry of Finance and Economic Planning) Member
6. Chief Director (Ministry of Trade and Industry) Member
7. National Climate Change Coordinator (Environmental Protection Agency) Member-Secretary

According to CDM Ghana (2008), the functions of the DNA include:

- Receives projects for evaluation and approval as per the guidelines and general criteria laid down in the relevant rules and modalities pertaining to CDM in addition to the guidelines issued by the Clean Development Mechanism Executive Board and Conference of Parties serving as Meeting of Parties to the United Nations Framework Convention on Climate Change. The evaluation process of CDM projects includes an assessment of the probability of eventual successful implementation of CDM projects and evaluation of extent to which projects meet the sustainable development objectives, as it would seek to prioritise projects in accordance with national priorities.
- Recommends certain additional requirements to ensure that the project proposals meet the national sustainable development priorities and comply with the legal framework so as to ensure that the projects are compatible with the local priorities and stakeholders have been duly consulted.
- Ensures that in the event of project proposals competing for same source of investment, projects with higher sustainable development benefits and which are likely to succeed are accorded higher priority.
- Carries out the financial review of project proposals to ensure that the project proposals do not involve diversion of official development assistance in accordance with modalities and procedures of Clean Development Mechanism and also ensure that the market environment of the CDM project is not conducive to under-valuation of Certified Emission Reduction (CERs) particularly for externally aided projects.
- The DNA carries out activities to ensure that the project developers have reliable information relating to all aspects of Clean Development Mechanism which include creating databases on organisations designated for carrying out activities like validation of CDM project proposals and monitoring and verification of project

activities, and to collect, compile and publish technical and statistical data relating to CDM initiatives in Ghana.

- The Member-Secretary of the Governing Council of the DNA is responsible for day-to-day activities of the Authority including constituting committees, sub groups or ad hoc committees to coordinate and conduct detailed examination of the CDM project proposals.

9.5.2. The CDM Project Approval Process

According to the draft legislation on CDM, the CDM project approval process would involve the following steps (after CDM Ghana, 2008):

Application to the DNA: An application for Initial screening must be made to the DNA. This application includes a letter signed by the project proponent and/or the project sponsors and a project Identification Note in the format provided by the DNA.

Initial Screening: This is a voluntary step that allows project developers the opportunity to receive an initial evaluation of their project from the DNA to identify any potential conflicts with the project approval criteria and other government policies. The DNA will conduct an initial evaluation of the sustainable development impacts of the project against the sustainable development criteria. Results of the initial screening will be provided within 15 working days of submission of the application form and Project Identification Note. The comments will be submitted to the project developer in the form of an informal notice of letter-of-no-objection.

Submit Validated Project Design Document (PDD) to DNA: The project proponent's request for final project approval must have: a) a Validated Project Design Document in most recent format published by the Executive Board of the CDM; b) a completed application form; and 3) An Environmental Impact Assessment (EIA) of the project if this is required.

Public Comment: The DNA will post the submitted Project Design Document (PDD) on its website for public consultation, for a period of 21 working days. Project Design Documents will also be made available to any interested parties upon request. The recommendations prepared by the DNA will be sent to the members of the Governing Council of the DNA for consideration at its next meeting. The DNA may also decide to submit the project to the Council via email circulation for comments. Any comments on the PDD received from the period of public consultation will also be circulated to the Council for consideration. The Council of the DNA will evaluate the project and submit its comments to the DNA.

Letter of Approval: The Governing Council of the DNA will make a final decision based on the feedback received from the public. The decision to approve or disapprove a project will take no more than 60 days.

Appeal: Project participants will have the right to appeal the final decision taken by the DNA. In a first step, they may appeal the decision with the Minister of Environment and Science. The Minister will verify the decision taken by the DNA and determine whether it has been produced in accordance with the approval procedures (formal and substantial determination). The Minister will notify the project participants of his/her decision within 60 days. The project participant has the right to appeal the determination of the Minister before the Administrative Courts of Ghana.

9.5.3. Potential Challenges to CDM Implementation

The main challenge to CDM projects in Ghana, especially those related to land use (forestry) is the small size of the projects. As mentioned previously, many of the plantations established in Ghana are small- to medium-sized plantations to serve individual or community needs. It would be difficult for many of these plantations to enter into the CDM offset trading scheme. First, there would likely be a conflict in priorities of local needs for wood versus the environmental need to sequester and store carbon in trees for long periods of time. Secondly, the market for carbon is not frictionless, i.e., there are transaction costs to be borne by project proponents, which these small-scale plantation owners may be unable to bear. Furthermore, there is a need for expertise to manage the plantations according to the procedures of the CDM, which is lacking in Ghana in general and at the community levels in particular. Hiring such expertise may be beyond the financial ability of communities. The obstacles to raising commercial capital to develop projects within the CDM would be identical to those already discussed for forest plantations in general. This is even more so given the potential risks of natural disturbances such as fire, pests and insect destruction and the fact that carbon assets are still being conceived by most financial institutions as "abstract assets". It is likely that only industrial plantations or community plantations with NGO or donor-funded support will be able to participate in the CDM.

There are also risks related to the regulatory environment. At present, there is no legislation that backs the DNA and large-scale investors may be concerned about the legality and long-term implications of Ministerial declarations to operate a national CDM programme. There are legal issues as well. Land and tree tenure systems do not provide incentives for long-term forestry projects, especially when there is a financial benefit for the standing trees. The ownership of carbon contained in trees cannot be divorced from the ownership of the trees. Hence, the land and tree tenure obstacles discussed in Chapter 3 are relevant here as well, which does not make Ghana attractive for international investors who seek carbon credits and a return on their investments. In addition, the ability to earn credits is based on the proponent's ability to demonstrate additionality. There is a general lack of data on plantation growth and dynamics for most of the tree species in Ghana to establish credible baselines that would demonstrate additonality and prove that leakage issues can be adequately addressed.

9.6. Carbon Sequestration and Accounting in Plantations

9.6.1. Estimating the Amount of Carbon Dioxide Sequestrated by Forest Plantations

I begin this section by describing the process of estimating the amount of CO_2 sequestered by a forest plantation. Protocols for quantifying C storage in living trees rely on the use of known allometric relationships and biomass expansion factors, based on measures of stem size, to estimate biomass and C contained in the stem wood, bark, branches, and coarse roots of living trees (Jenkins et al., 2003, Lambert et al., 2005). Under current CDM rules, accounting for changes in C in the forest floor and mineral soil is not mandatory, but could be included in future climate change negotiations. The following steps are used to estimate total above- and below-ground CO_2 in a plantation.

1. Estimate the Total (Green) Biomass Contained in One Ha of the Plantation

First, determine the total aboveground biomass for the plantation (in tonnes/ha) either through direct measurement or the use of biomass equations for the species. Most biomass equations give a relationship of the dbh and /or height to aboveground biomass. In general, the root systems of many species weigh about 20% of the aboveground biomass. This is a conservative estimate based on studies by (Cairns et al., 1997, Li et al., 2003) for conifers. The ratio of belowground to aboveground biomass was found in teak to be about 16% in Panama (Kraenzel, 2000). This ratio is smaller than the more general ratio Cairns et al. (1997) produced from a review of tropical forest biomass studies. They found that the average ratio for primary and secondary tropical forests was 24%, with a standard deviation of 14%. If the actual root system to above ground biomass ratio is available for the species, this should be used. Hence to get the total aboveground and belowground estimate of the plantation biomass, multiply the aboveground biomass by the appropriate factor (e.g., a factor of 1.2 for conifers or 1.16 for teak plantations).

If biomass equations are not available, stem wood volume can be estimated using yield equations or yield tables in m^3/ha. This stem wood is then converted to stem wood biomass by multiplying the stem volume by the mean wood specific gravity for the tree species. Stem wood biomass is then converted to total aboveground biomass (i.e., including biomass for branches, leaves, twigs, etc.) if the appropriate relationships exists between stem biomass and total tree biomass. If such equations do not exist, then this technique of going through the yield table to estimate total biomass cannot be applied.

2. Determine the Dry Weight of the Biomass

The next step is to convert the green biomass determined in Step 1 into the dry biomass. For each species, there may be equations that relate green weight to dry weight. This factor should be used if known, or assumed, if unknown.

3. Determine the Weight of Carbon

Total living tree biomass is converted to total carbon by multiplying the dry weight by 0.50, which is the average proportion of carbon in dry plant biomass. That is, about 50% of dry mass of wood is made up of carbon.

4. Determine the Weight of Carbon Dioxide (CO_2) Sequestered

Total CO_2 sequestration (i.e., CO_2e contained in living biomass C) is now estimated by multiplying total C by 3.667 (i.e., 1 kg biomass C equals uptake of 3.667 kg CO_2). The reason is that we know that CO_2 is composed of one molecule of carbon and 2 molecules of oxygen. The atomic weight of carbon is 12.00, and that for oxygen is 16.00. This gives the weight of CO_2 as 44.00 and a ratio of CO_2 to C as 3.667. To estimate the amount (weight) of CO_2 sequestered per ha/year, we divide the amount of CO_2 sequestered per ha by the age of the trees.

In cases where it is required and where the information exists, carbon sequestered in the soil carbon pool (includes various forms of soil organic carbon (humus) and inorganic soil carbon and charcoal, but excludes soil biomass, such as roots and living organisms) and litter would be included in the accounting framework. Current practice in economic analyses of impacts of carbon sequestration on optimal timber management have often assumed that carbon sequestered in mineral soil, branches, roots and litter are simply recycled into the next rotation of crop trees in order to simplify the data requirements and analyses (e.g., see Van Kooten et al., 1995).

9.6.2. Analysing Subsidies and Penalties for Carbon Management in Plantations

Benefits resulting from carbon sequestered by plantations fall into the general category of non-timber benefits. Incorporating these benefits follow similar analytical methods as discussed in Chapter 8. In order to evaluate the impact of carbon accounting and credits on plantation management decisions, I incorporate carbon sequestration benefits into economic analyses of plantation management and optimal rotation decisions in a fashion identical to Section 8.6, but introduce more specificity to the analyses. Previously, we saw that the correct optimal rotation age for a forest stand is determined by the Faustmann rotation model. It would also be recalled that when non-timber benefits are included in the analysis, the optimal rotation may change from the Faustmann rotation, depending on the behaviour of the non-timber benefits as stand age increases. Carbon uptake in forests in particular, increases proportionally with stand growth and hence carbon uptake is a function of the biomass and the amount of carbon per m^3 of biomass (van Kooten et al., 1995). Using this as a basis, van Kooten et al. (1995) derived an optimal rotation framework for analysing subsidies and taxes on carbon sequestration. Due to the ability of this framework to tease out the various incentives (and disincentives) for forest management, and its impact on the optimal rotation age, it has been adopted in this section.

Assume a forest stand of volume at time t given as V(t), with α being the amount of carbon / m^3 of volume. The carbon sequestered at any time t, is given by $\alpha V'(t)$, where V'(t) is the increase in volume per unit time (i.e., first derivative of the volume function). If the price of carbon is P_c /m^3 of carbon taken from the atmosphere and r is the discount rate, then the present value of the carbon uptake benefits is given as:

$$PV_c = \int_0^t P_c \alpha V'(t) e^{-rt} dt \tag{3}$$

In Equation (3), it is assumed that the carbon sequestered has a net benefit to society. P_c could be the price paid for each m^3 sequestered in the CDM, a domestic offset trading system or a pure subsidy from governments to encourage carbon sequestration. On the other hand, governments could decide to impose a penalty for emitting carbon from a forest through harvesting or forest fires. In this case, a penalty equal to $P_c \alpha (1-\beta)$ where, β is the fraction of the harvested timber that goes into long-term storage[3] can be imposed. The present value of this penalty is given as:

$$PV_p = P_c \alpha (1-\beta) V(t) e^{-rt} \tag{4}$$

When the timber is harvested at time t, the present value of the wood sold in markets with a price of P_F/m^3 is given as:

$$PV_F = P_F V(t) e^{-rt} \tag{5}$$

Combining Equations 3-5 gives the present value of the timber and carbon benefits for all future rotations using the Faustmann framework as:

$$SEV = \frac{PV_c + PV_F - PV_p}{1-e^{-rt}} = \frac{P_c\left[V(t)e^{-rt} + r\int_0^t V(t)e^{-rt}dt\right] + \left[P_F - P_c\alpha(1-\beta)\right]V(t)e^{-rt}}{1-e^{-rt}} \tag{6}$$

From Equation 6, van Kooten et al. (1995) derive the optimal rotation age that considers timber values, carbon benefits and penalties for releasing carbon into the atmosphere by taking the first derivative of Equation 6 with respect to t, and setting the result to zero as:

$$\left[[P_F + P_c\alpha\beta]\frac{V'(t)}{V(t)} + rP_c\alpha\right] = \frac{r}{1-e^{-rt}}\left[(P_F + P_c\alpha\beta) + \frac{rP_c\alpha}{V(t)}\int_0^t V(t)e^{-rt}dt\right] \tag{7}$$

If the price of carbon, P_c, is set to zero, the optimal condition in Equation (7) equals the Faustmann rotation model, while setting β to zero gives the general Hartman (1976) model discussed earlier (Equation 21 in Section 8.6).

While the above theoretical framework is useful for analysing the implications of subsidies and taxes, no country has as yet implemented any such policy, and hence so far, this approach has not been applied in practice.

9.7. Case Study: Optimising Joint Production of Carbon and Timber in Teak Plantations

This case study uses information about teak plantations in Ghana to evaluate the implications of accounting for carbon benefits in terms of the potential amount of credits that can be obtained and its implications on plantation management. The following information and assumptions were used. The growth and biomass functions for teak developed by Nunifu (1997) were used to estimate timber volume and above- and belowground biomass of the teak plantations. These equations are given as:

[3] Carbon dioxide is only emitted if the wood is burnt or decays. If the wood is used for construction or long-term storage, the carbon is not emitted. Therefore, industry only pays a penalty for the portion of carbon that is emitted.

Timber volume:

$$\ln V = 8.10 - 11.13t^{-0.5} \quad (8)$$

Aboveground fresh Biomass:

$$\ln B_{AG} = 7.69 - 10.683t^{-0.5} \quad (9)$$

Aboveground oven dry biomass:

$$B_{AG} = 1.284 - 0.969D + 0.314D^2 \quad (10)$$

where t is the plantation age in years, and D is the mean dbh of the trees.

Table 1. Estimated timber volume, biomass and carbon sequestered in teak plantations in the savannah vegetation zone of Ghana

Age (yrs)	Timber Volume (m^3)	Aboveground Biomass (t/ha)	Belowground* Biomass (t/ha)	Total Biomass (t/ha)	Carbon (tCO_2/ha)
1	0.05	0.87	0.14	1.01	1.86
2	1.26	0.64	0.10	0.74	1.36
3	5.33	2.41	0.39	2.79	5.12
4	12.62	6.26	1.00	7.26	13.31
5	22.70	11.75	1.88	13.64	25.00
6	35.03	18.45	2.95	21.40	39.23
8	64.39	34.07	5.45	39.53	72.47
9	80.64	42.54	6.81	49.35	90.49
10	97.55	51.24	8.20	59.44	108.99
12	132.56	68.93	11.03	79.96	146.61
14	168.24	86.59	13.85	100.44	184.16
15	186.09	95.30	15.25	110.55	202.70
18	239.04	120.76	19.32	140.09	256.85
20	273.49	137.05	21.93	158.98	291.49
25	355.67	175.21	28.03	203.24	372.64
30	431.79	209.80	33.57	243.37	446.22
35	502.04	241.21	38.59	279.80	513.01
40	566.91	269.82	43.17	312.99	573.86
45	626.93	296.00	47.36	343.36	629.55
50	682.65	320.08	51.21	371.29	680.77
55	734.53	342.32	54.77	397.09	728.07
60	782.99	362.95	58.07	421.03	771.95
65	828.40	382.16	61.15	443.31	812.81
70	871.06	400.11	64.02	464.13	850.98
75	911.25	416.94	66.71	483.65	886.77
80	949.21	432.76	69.24	502.00	920.42
85	985.14	447.67	71.63	519.30	952.14
90	1019.22	461.77	73.88	535.65	982.12
95	1051.62	475.12	76.02	551.14	1010.52
100	1082.47	487.79	78.05	565.84	1037.47

*Belowground biomass is the biomass of the root system only, calculated as 16% of the above ground biomass.

To determine the belowground biomass, the aboveground biomass estimated from Equation 9 is multiplied by a factor of 0.16, based on studies by Kraenzel (2000) for teak plantations in Panama. Using this information, the aboveground and below ground carbon sequestered by Site Class I teak plantation in the savannah zone was computed. This information is presented in Table 1. If we assume that the area to which this plantation was planted would have remained savannah grassland without trees, we can project that the amount of carbon that would have been sequestered by the grassland vegetation would be negligible. Hence we would be able to award all the carbon sequestered in the plantation as credits under the CDM (baseline = 0) for this afforestation project.

From Table 1, the temporary CER credits awarded to the teak plantations between year 5 and 30 years (the maximum number of years allowed under current CDM rules) are summarised in Figure 1. The rules imply that the last period for awarding credits is year 25 of the plantation life. The credits awarded in year 25, together with any credits that were renewed, will all expire at year 30. No credits are given for the growth in carbon in year 30 and beyond.

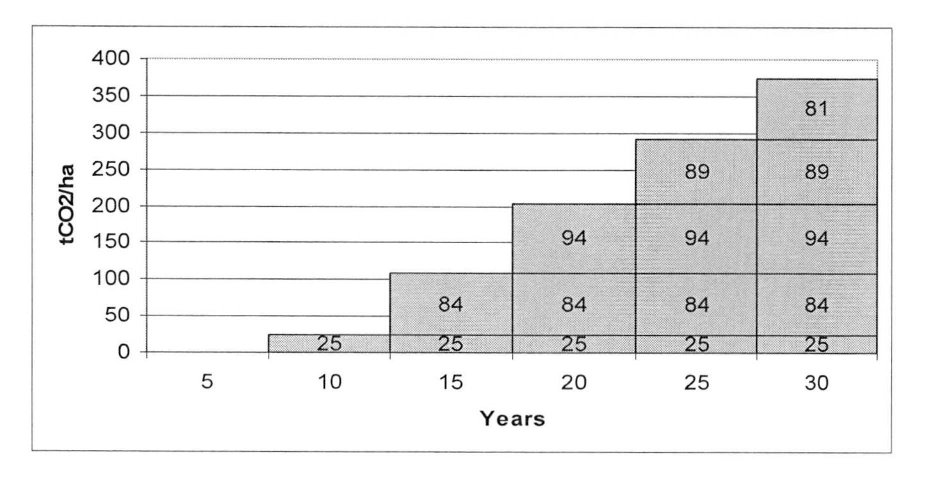

Figure 1. Accumulated carbon credits awarded to a teak plantation over 30 years.

As illustrated in Figure 1, during the first five years of the plantation life, a total of 25 tCO_2/ha is accumulated and issued as tCER to the owner at the end of year 5 of the plantation. The credits issued in year 5 expire in year 10 of the plantation life, but can be re-issued for another five years, together with the additional carbon accumulation between year 5 and 10 of 84 tCO_2/ha. This process continues as in Figure 1 until the maximum 30 years is reached.

Table 2. Present value of income from carbon credits with changes in tCER prices

Permanent price (Ghc/tCO_2)	Equivalent temporary tCER price (Gh¢/tCO_2)	Present value of income (Gh¢/ha)
10	2	1,338.95
20	4	2,677.90
25	5	3,347.38
30	7	4,686.33
35	8	5,355.80

The flow of income to the landowner can be calculated by estimating the price of temporary carbon credits from those of permanent credits using Equation 2. From the literature, estimates and forecasts of permanent carbon credits vary widely, from \$3/tCO_2 (The World Bank, 2003) to about \$35/tCO_2 floated at the European Union Emissions Trading Scheme (PointCarbon, 2008). In this analysis, various starting prices of a permanent carbon credit were used, and the corresponding tCER prices calculated using Equation 2, with a discount rate of 5% and t =5 years (Table 2).

One of the main implications of carbon credits from A/R activities is that the carbon for which credits have been awarded must be maintained till the end of the crediting period (permanence). This means that the trees cannot be harvested before the end of the crediting period. In the example in Table 2, if the landowner decides to obtain credits up to the maximum 30 years, then the trees cannot be harvested before the end of the 30 years, otherwise, he would be liable to replace the released carbon for which he has received payment.

We can analyse whether the landowner should apply for credits up to the maximum period of 30 years or for a shorter period by comparing the carbon and timber benefits at each 5-year interval. Basically, we will allow the flexibility for the landowner to decide at the end of each 5-year-crediting period, whether he should renew the credits already received or terminate the project altogether (harvest the trees). The following assumptions are used in Tables 3 and 4. A discount rate of 5%, price of wood of Gh¢100/m^3, and a present value of costs of planting and management of Gh¢4000/ha are assumed. A higher regeneration and management cost is assumed here because managing the trees to generate carbon credits entails more transaction costs (application fees, monitoring, measurements, etc.) than just managing the plantation for timber only. The costs incurred are assumed to be joint costs of producing timber and carbon.

Table 3 presents the results of the present values of carbon benefits for various years and prices of tCER. The calculations were based on the allocation of carbon credits over the crediting periods in Figure 1. Each value in the table represents the amount of money the landowner would receive for the particular year and tCER price. For example, for the tCER price of Gh¢2/tCO_2 and year 10, the present value of the amount is Gh¢336.05. The totals show the present value of the accumulated revenue over the 30 years if the landowner decides to hold the trees for that length. If we now subtract the costs of production of Gh¢4000/ha, we notice that the carbon prices up to Gh¢5/tCO_2 do not make a profit. Carbon prices have to be greater than Gh¢5/tCO_2 for this project to make economic sense. Interestingly, the project would post a loss for all prices in the range used if the full 30-year crediting period is not used. This is seen from the fact that at each 5-year crediting period, the present value of the benefits is smaller than the present value of the costs, for the full range of the tCER prices.

In Table 4, the tCER price of Gh¢5/tCO_2 is used to illustrate the impact of combined management for carbon and timber on the optimal time to terminate the project under the CDM. From Table 4, the NPV of the combined timber and carbon values is maximised in year 20. This is the optimal time for the plantation to be harvested. This suggests that although the plantation could have received additional carbon credits and timber revenue beyond this point, the returns are lower than if the owner decides to terminate the project, harvest and sell the wood and invest the money in a bank for a 5% interest on the proceeds. Even if the tCER of Gh¢8/tCO_2 is used; it results in the same optimal rotation age of 20 years.

Table 3. Present values of carbon credits for various years and prices of tCER

Year	Temporary credit carbon price Gh¢/tCO_2				
	2	4	5	7	8
5	176.04	352.08	440.10	616.14	704.16
10	336.05	672.09	840.11	1176.16	1344.18
15	363.47	726.95	908.68	1272.16	1453.89
20	339.87	679.75	849.69	1189.56	1359.50
25	299.56	599.11	748.89	1048.45	1198.23
Total	1,338.95	2,677.90	3,347.38	4,686.33	5,355.80
Net value*	-2,661.05	-1,322.10	-652.62	686.33	1,355.80

*Net value is the total benefit minus the costs of production.

Table 4. Timber and carbon values for various years and a tCER price of Ghc5/tCO_2

Year	Present value of Carbon (Gh¢/ha)	Present value of Timber (Gh¢/ha)	Net timber and carbon values (Gh¢/ha)*
5	440.10	-19,131.45	-18691.4
10	840.11	-211.39	628.72
15	908.68	5,288.54	6197.22
20	849.69	6,424.48	7274.17
25	748.89	5,872.61	6621.5

*Net values are calculated by subtracting the present value of costs (Gh¢4000) from the total benefits.

The above analysis does not consider the risk of fires to these plantations. Bush fires that burn teak plantations is an annual occurrence in the savannah zones. A landowner could either spend additional resources to ensure that the plantations are protected from these fires, or buy additional insurance to ensure he can be reimbursed if a fire were to occur during the time credits have been received for the carbon sequestered by the plantation. These analyses suggest that carbon management under the current assumptions and CDM rules can be profitable, however, it must be emphasised that just because there is a potential for carbon credits does not mean all plantations would automatically become profitable. At present, there is no approved CDM forestry project in Ghana, and hence the full range of expected costs of managing for carbon is not available. The necessary ex-ante analyses need to be undertaken in each case using the most realistic expected costs and benefits in order to make well-informed decisions regarding the economic profitability of joint management of the plantations for carbon and timber.

9.8. The Copenhagen Accord and Cancun Agreement

The 15th session of the Conference of the Parties (CoP-15) was held in December 2009 in Copenhagen, Denmark. The resulting Accord, known as the Copenhagen Accord is not legally binding and does not commit countries to agree to a binding successor to the KP, whose present round ends in 2012. The Accord endorses the continuation of the KP and agrees that the Parties to the Kyoto Protocol would strengthen their existing targets. The

Accord recognises "the crucial role of reducing emission from deforestation and forest degradation and the need to enhance removals of greenhouse gas emission by forests", and the need to establish a mechanism (including REDD-plus) to enable the mobilisation of financial resources from developed countries to help achieve this. In regards to emissions trading, the meeting decided to pursue opportunities to use markets to enhance the cost-effectiveness of, and to promote, mitigation actions.

The latest Conference of the Parties, CoP-16 was held in Cancun, Mexico in December 2010 and the major decisions re-affirmed some of the decisions of CoP-15. The decisions out of Cancun, known as the Cancun Agreement, provides for action to reduce emissions by both developed and developing countries, embeds pledges from the Copenhagen Accord in a CoP decision, sets out a framework for avoided deforestation and confirms the continuation of emissions trading and project based mechanisms under the Kyoto Protocol. It establishes a Green Climate Fund to help money disburse money needed to help developing countries combat climate change, a Cancun Adaptation Framework and a Technology Network to support the deployment of low carbon economy.

9.9. Conclusions

International climate change policies continue to recognise the important role of forests in mitigating climate change impacts. This has led to the creation of markets for carbon sequestered in forest plantations. The Kyoto Protocol has developed the Clean Development Mechanism, which provides a global market to trade in carbon credits. Ghana has already taken steps to benefit from this opportunity and has set up a designated national authority within the Ministry of Environment Science and Technology. There are other international initiatives such as REDD+ (reducing emissions from deforestation and forest degradation), which Ghana is seriously pursuing. It is surprising that the current REDD implementation framework in Ghana does not put a major emphasis on plantation forestry, given that the development of plantation forestry would contribute in a major way to the reduction of deforestation. There are still opportunities to refocus REDD+ strategy to ensure plantation forestry has a central role and thereby secure long-term funding to support their development.

Chapter 10

CONCLUSIONS: CREATING THE CONDITIONS FOR SUCCESSFUL PLANTATION FORESTRY

The forestry sector in Ghana currently relies heavily on the exploitation of the natural forest resource to meet the needs of its population. The sector remains one of the major sources of foreign income, and supplies the domestic market with a wide range of timber and non-timber forest products. Presently, plantation forestry plays only a minor role, with the majority of plantations being teak, which were established over many decades across the country. Despite several decades of attempts at plantation forestry and different versions of national plantation programmes, none of these have led to a substantial increase in the land area planted to forest plantations in Ghana. The reasons for this failure may lie in the constraints identified in Chapter 3 of this book. Plantation establishment and management are more labour and capital intensive than managing natural forests, and hence require lots of initial and continuous investments of capital, technical expertise, long-term commitment, economic management skills and forward integration to value adding secondary processing to be successful (AFORNET, 2008).

In this final chapter, I discuss the conditions that must prevail for Ghana to achieve a successful plantation programme. Some of the requirements are currently being pursued to different degrees in Ghana, whilst others are not. These recommendations are categorised into four major areas: economic, policy, technical and institutional and are required to promote, encourage and support plantation forestry development in Ghana.

10.1. Economic and Fiscal Environment

Plantation forestry is essentially an economic venture, whether they are grown as private industrial plantations, community or individual woodlots, or for environmental protection. Therefore, the economic benefits derived from the investments play a large part in determining its success. Furthermore, the economic environment and the fiscal and monetary policies in the country work to affect the outcomes of investment decisions. The economic requirements for successful plantation development centre on two main areas: financial ability to establish and manage plantations, and the ability to make a profit from the plantation enterprise. In this case, the major areas for improvement include taxation, financial assistance through incentives or loans, and market availability. The following are recommended:

- The Government should, through its fiscal and monetary policies, promote plantation development. First, economic stability makes investments in Ghana attractive and hence could encourage private sector investments in plantations. Secondly, plantation forestry can be promoted by ensuring that taxation of plantations and plantation forest products are favourable to investment. This can be achieved through providing tax and import duty exemptions for investments in forest plantations. Currently, most of the tax-related incentives are targeted at industrial plantations and foreign investors. However, the majority of plantation forests in Ghana would remain small- to medium-sized community and individual plantations. Therefore, tax exemptions should be expanded to include community, individual and NGO-supported plantations. For example, investors in plantation forests greater than a certain minimum size, say 5 ha, should be allowed by the tax laws to deduct the costs of planting and management from their annual tax obligations. Finally, tax policy on the forest industry should be targeted at achieving increased domestic processing in the country.
- There is a need for continued flow of private investment into plantation forestry. This is in recognition that the government alone cannot fund plantation forestry. This can only be achieved if the government continues to ensure a climate that promotes private investments. Private investors do not necessarily have to be foreigners. Efforts should be made to make plantation forestry attractive to local businesses, such as banks, timber companies and other financial institutions.
- The government should provide low interest rates loans or loans with zero interest, or grants to investors in plantation forestry.
- The National Forest Plantation Development Fund (NFPDF) should be streamlined to provide financial support to small-scale farmers and communities that want to establish forest plantations. Although in theory there is no limitation on who can apply, the bureaucratic bottlenecks, processes and costs involved in applying for funds from the NFPDF are beyond the abilities of most small-scale farmers. These processes should be streamlined to ensure that small-scale farmers can truly benefit from these incentives.
- The government should continue to provide seedlings and other materials to help in the establishment and management of plantations.
- The government, through the FC and the research organisations, should promote lesser-used plantation species locally and internationally to increase acceptability and marketability. Plantation products that have no chance of being marketed do little to support the economic sustainability of the enterprise. International marketability of plantation forest products would increase the value of the plantation products, and hence their economic profitability.

10.2. Policy And Legislative Changes

Policies affect how investment decisions are made by individuals and firms. Therefore, poorly designed and coordinated policies can be detrimental to any plantation development programme. It is important that the government re-examines the current forest policy and legislative framework to ensure that the policies are not only relevant, but also effective and efficient. The following recommendations are made:

- There is an urgent need for policy and legislative reform to remove inconsistencies and contradictions in the legislative framework in the forestry sector, and to harmonise these with other sectors of the economy. Forestry laws and regulations should also be consolidated to ensure efficiency. The current practice of superimposing plantation forestry laws/regulations directly on (or by amending existing) laws that were intended for natural forests creates more confusion. These amendments would remove the ambiguities in the forest legislation as it relates to plantations. For example, the kinds of fees, taxes and export levies applicable to forest plantations and plantation forest products must be clarified.
- Section 21 of the *Forest Plantation Development Fund Act*, 2000 (Act 583) requires the Minister responsible for forestry to make regulations for the effective implementation of the Act. After 10 years of passing Act 583, these regulations have not yet been made. These regulations should be made as a matter of urgency, and should include provisions that direct a prescribed percentage of the funds accruing to the Fund to support small-to-medium scale plantation forestry. Even more useful would be regulations that prescribe a streamlined application process that is less bureaucratic for these categories of applicants. In fact, farmers and communities should be able to apply for funds through their District Assembly offices, rather than in Accra.
- Section 7(b) of the *Forest Plantation Development Fund Act* (2000) that empowers the Fund Board to "invest" the Fund money should be amended to remove the wording that has led to the Fund Board investing the money received into the fund in money markets rather than in plantation forestry. Furthermore, improved monitoring of the investments made by the Fund Board will ensure that the funds are used appropriately.
- The *Timber Resources Management Act*, 1997 (Act 547, section 8(d)) should be amended to remove the requirement for TUC holders to reforest their contract areas after logging. Given that the reforestation requirement has not achieved much success, the regulations should instead impose a plantation forest levy that should go into a specific fund which should be used to hire a private firm to undertake reforestation activities on logged areas.
- Modernisation of the land and tree tenure systems in the country to provide well-defined and enforceable property rights over trees is absolutely essential. A new policy, backed by legislation that clearly defines ownership rights over timber and carbon sequestered (and other non-timber benefits) by plantation forestry under the different types of land ownership and tenurial arrangements in Ghana is urgently needed. It is obvious that people will not invest in plantations if there are ambiguities

as to who will get the benefits of the investments. Therefore, changes to the land tenure systems in Ghana are absolutely critical for the success of plantation forestry.

- The government must continue to maintain polices that promote plantation development and to develop the legislative framework to recognise market mechanisms for carbon offset trading within the CDM or a separate domestic offset trading scheme. In fact, it is time for Ghana to develop a domestic carbon offset trading scheme that will recognise and award credits for carbon sequestered by forests and other greenhouse gas mitigating activities. This will stimulate the development of low-carbon energy technologies within the industrial sector.
- Because plantations are long-term investments, it is often the case that initial interest in the projects fades over time. Governments must not relax in their determination to maintain this interest and encouragement needed to see that plantations become successful.
- The government should introduce favourable trade laws for plantation logs or products that make it easy to market plantation products in domestic and international markets.
- The government should ensure that policies related to the forestry sector minimise, rather than increase corruption. For example, policies that involve lengthy bureaucratic processes are often prone to corruption, because people prefer to pay their way to avoid following through the cumbersome processes.
- Policies that support and encourage stakeholder participation in plantation development are essential to ensure that the trees are well tended and protected. Whether communities are the owners of the plantations or live adjacent to plantations, obtaining their buy-in into the project at the beginning of the project and throughout its lifecycle are critical to the success of plantations.
- Ghana is a small country in terms of land area, and hence land use efficiency has to be central to the government's priorities. Policies that reduce urban sprawl and those that increase intensive, rather than extensive agriculture should be pursued. In short, an integrated land-use planning system approach that uses land efficiently should be adopted.

10.3. Technical Requirements

The technical needs of plantation forestry lies mainly in effective research, training and extension. Research is the foundation for developing a successful national plantation programme. Training and extension ensure that the results are effectively disseminated and utilised. These are the areas where the government can make the most direct contribution in terms of funding and support.

- The government and private sector investors should support sustained research into plantation forestry. There should be an efficient mechanism to disseminate the results to those who need it most. Some of the funds accruing to the National Forest Plantation Development Fund should be used to support research into plantation forestry.

- The research should focus on: seed production and vegetative propagation, species/site matching, species trials, testing multipurpose species for community forestry and industrial agroforestry, economic viability of plantations, local factors affecting plantation adoption, diseases and pest management, growth and yield models, fire management, social research on how to ensure true community participation, etc.
- Training forestry professionals at all levels must be improved, with a focus on plantation forestry and extension training. With the rapid decline of the natural forest resources, it is clear that forestry training that remains focused on natural forest silviculture and management would be inadequate in meeting plantation forestry requirements of the future.
- Government should provide funding to FORIG to augment its seed production and storage capabilities. The basis of any successful plantation programme relies on the availability of good seeds for the targeted species. In addition, FORIG should be funded to establish permanent nurseries that can produce good quality seedlings for plantation development programmes.
- Information dissemination and extension services for plantations are weak and need to be improved. More forestry technical officers need to be deployed to targeted communities, more forestry offices in strategic districts in the country, and the promotion of forestry as an environmentally sustainable land use option would work together to ensure continuous technical support to tree growers and increase plantation forestry acceptance.

10.4. Institutional Reform

There is a strong link between institutions and policy, as the former helps in setting, interpreting and enforcing the rules of society. Ghana's institutional framework can shape forest plantation development through the policy process; as a result, the following recommendations related to institutions are made:

- First and foremost, forest sector governance reform that promotes multi-sector and stakeholder support and participation is needed to encourage plantation development efforts.
- The government of Ghana must commit to ensuring that plantation forestry remains a national policy priority for the next several decades. In this regard, the importance of the Plantation Department of the FSD should be elevated to enable it play a much bigger role in the forestry sector.
- The planning and management functions of the Forestry Commission should be re-examined, with a view to eliminating any inefficiencies in the existing structure. The FC should remain the central agency for planning, while implementation of plantation forestry programmes can be out-sourced to another public or private organisation.
- The government of Ghana should develop a long-term plantation forest development programme. Past and present programmes (such as the NFPDP) have been ad hoc in

nature and tend to change in implementation from one government to another. The long-term programme should have the support of all political parties and should have legislated annual planting targets over at least a 20-year period. All governments should be obliged to adequately fund and implement the plan and report to Parliament each year on the achievements of the programme. Together with legislated targets, institutional changes are also required to ensure that the FSD receives adequate funding to support plantation activities. Consequently, plantation forestry funding should be allocated in the annual national budget.

- The government of Ghana should directly undertake tree planting to target environmentally sensitive and poor areas in the country. These should include ecologically fragile watersheds, major rivers and streams (such as the tributaries of the Volta River) mountainous regions and other ecologically sensitive areas.
- District Assemblies should be encouraged to use some of their Common Fund allocation to support community and individual growers as a way to improve community energy supplies and incomes. Some of their Common Fund should be set aside for tree planting to protect critical watersheds and waterways, such as streams, rivers and dams.
- The government should go into partnerships with industry or the NGO-community to develop plantations. For example, the government could pay upfront for the establishment costs and later hand over the management of the plantation to the private entity, in return for a percentage equity in the plantation enterprise.
- The government should seek international funding to support community-based and individual plantation forestry programmes.
- Industry-community partnerships should also be encouraged, whereby industrial investors can work with communities to establish and maintain forest plantations that meet the needs of the communities and those of the industrial partners.

BIBLIOGRAPHY

Abildtrup, J. 1999. Optimal thinning of forest stands: An option value analysis – preliminary results. *Scandinavian Forest Economics* 37: 2-1 – 2-13.

Adegbehin, J. O. 1982. Preliminary results of the effects of spacing on the growth and yield of *Tectona grandis,* Linn F. *Indian Forester, 108*, 423-430.

Adegbehin, J.O., J.O. Abayomi, and L. B. Nwaigbo. 1988. *Gmelina arborea* in Nigeria. *Commonwealth Forestry Review*, 67(2):159–166.

Adjei, S. and Kyereh, B. 1999. Land suitability assessment of some degraded forest reserves and headwaters for the establishment of *Ceiba pentandra* plantations in the dry semi-deciduous forest in Ghana. *Journal of the Ghana Science Association* 1(2)110-124.

Adu-Nsiah, K. 2009. Ghanaians consume $205 million of bushmeat annually. *Report of Ghana News Agency,* August 25, 2009. www.ghanaweb.com.

AFORNET, 2008. Plantation forestry in Sub-Saharan Africa. Lessons learnt on Sustainable Forest Management in Africa Policy brief No 4. African Academy of Sciences, Nairobi, Kenya.

Agyeman, V. K., Marfo, K. A., Kasanga, K. R., Danso, E., Asare, A. B., Yeboah, O. M., and Agyeman, F. 2003. Revising the taungya plantation system: new revenue-sharing proposals from Ghana. *Unasylva, 45*, 40-43.

Ahmed, S. 1984. Use of neem materials by Indo-Pakistani farmers: Some observations. *In* Saxena R.C and S. Ahmed (Eds.). Proc. Res. Planning Workshop, Botanical Pest Control Project. Int. Rice Res. Inst., Los Banos, Philippines.

Ahmed, S. A. and Koppel, B. 1985. Plant extracts for pest control: village level processing and use by limited-resource farmers. Paper presented, Amer. Assoc. Advancem. Sci., annual meeting, Los Angeles CA, May 26-31 1985.

Ahmed, S. A. and Grainge, M. 1985. Use of indigenous plant resources in rural development; potential of the neem tree. *International Journal for Development Technology* 3 (2): 123-130.

Ahmed, S. A. and M. Grainge. 1986. Potential of the neem tree (*Azadirachta indica*) for pest control and rural development. *Economic Botany* 40(2): 201-209.

Aidoo, J.B., 1996. Our Common Estate Tenancy and the Land Reform Debate in Ghana. The Royal Institution of Chartered Surveyors, London, England. 13p.

Akindele, S. O. 1991. Development of site index equations for teak plantation in South-western Nigeria. *J. Trop. For. Sci., 4*, 162-1 69.

Alam, M.K., Siddiqi, N.A., and Das, S. 1985. Fodder trees of Bangladesh. Bangladesh Forest Research Institute, Chittagong, Bangladesh. 167 p.

Alder, D. 1980. *Forest volume estimation and yield prediction*, Vol. 2: Yield prediction. FAO forestry paper 22. Rome. FAO.

Allan, G. G., Gara, R. I., and Wilkins, R. M.. 1973. Phytotoxity of some systemic insecticides to Spanish cedar. *International Pest Control* 15(l):4-7.

Allen, J.C., and Barnes, D. F. 1985. The Causes of Deforestation in Developing Countries. *Annals of the Association of American Geographers, 75*(2), 163-184.

Anane, M. 2003. Gold Rush in Ghana's Forest Reserves Resisted. Environment News Service. http://www.ens-newswire.com/ens/may2003/2003-05-13-01.asp. Accessed July, 2009.

Anderson, F.J. 1976. Control theory and the optimum timber rotation. For. Sci., 22: 242 - 246

Anderson, F. J. 1992. Natural resources in Canada. Economic theory and policy. 2nd edition. Nelson Canada. Scarborough. 292 p.

Anderson, D. 1986. Declining tree stocks in African countries. *World Development* 14 (7): 853-863.

Anderson, D. and Fishwick, R. 1984. *Fuelwood Consumption and Deforestation in African Countries*, World Bank Staff Working Paper No. 704, Washington, D.C.

Anning, A. K., and Yeboah-Gyan, K. 2006. Diversity and distribution of invasive weeds in Ashanti Region, Ghana. African Journal of Ecology. 45(3): 355-360.

Anonymous, 1952. Effect of hoeing in plantations of *Melia Azadirachta.* Report For. Adm. Nigeria. 1950/51 (22-3).

Anonymous, 1992. Summary of survey data for teak. Ghana forestry Department records, Planning Branch, Kumasi, Ghana (Unpub.).

Anonymous, 2004. Características y usos de 30 especies del bosque latifoliado de Honduras. Fundacion Cuprofor, Proecen, Proinel, EAP-Zamorano.

Anonymous, 2009. Growing teak under farm forestry for posterity http://planning.up.nic.in/innovations/inno3/fw/teak.htm.

Apetorgbor, M.M., Siaw, D., and Gyimah, A. 2003. Decline of *Ceiba pentandra* seedlings, a tropical timber species, in nurseries and plantations. *Ghana Journal of Forestry* 11(2): 51-62.

Appiah, M., Blay, D., Damnyag, L., Dwomoh, F. K. , Pappinen, A., and Luukkanen, O. 2009. Dependence on forest resources and tropical deforestation in Ghana. *Environment, Development and Sustainability*, 11 (3):1573-2975.

Asare, R. 2004. Agroforestry initiatives in Ghana: a look at research and development . A presentation made at the World Cocoa Foundation conference in Brussels April 21 –22, 2004. Danish Centre for Forest, Landscape and Planning - KVL, *Horsholm Kongevej* 11, DK-2970.

Assmann, E. 1955. Die Bedeutung des "erweiterten Eichhorn'schen Gesetzes" für die Konstruktion von Fichten-Ertragstafeln . *Forstwiss. Centralbl.* 74 , 321 – 330.

Assmann, E. 1959. Höhenbonität und wirkliche Ertragsleistung *Forstwiss. Centralbl.* 78, 1 – 20

Assmann, E. V. 1970. The principles of forest yield study. Pergamon Press, Oxford U.K. 506p.

Asuming-Brempong, S. 2003. Roles of Agriculture Project. National Report Ghana. Agricultural and Development Economics Division (ESA) Food and Agriculture

Organisation of the United Nations. International Conference held October 20-22, 2003 Rome, Italy.

Avery, T. E. and Burkhart, H. E. 1994. Forest measurements. 4th ed. McGraw-Hill Co., New York. 408 pp.

Ayamga, R. A. 1997. Evaluation of community participation in the Community Afforestation programme in Northern Ghana. A case study of the Tolon-Kumbungu District. BSc (Tech) Thesis, Department of Renewable Natural Resources, University for Development Studies, Tamale, Northern Region.

Ayine, D. 2008. Social responsibility agreements in Ghana's forestry sector. Developing legal tools for Citizen Empowerment Series, *IIED*, London.

Bailey, R. L. and Dell, T.R. 1973. Quantifying diameter distributions with the Weibull function. *Forest Science* 19(2):97-104.

Bailey, J. D., and Harjanto, N. A. 2005. Teak (*Tectona grandis* L.) tree growth, stem quality and health in coppiced plantations in Java, Indonesia. *New Forests*. 30(1): 55-65

Bailey, M. D., and Sporleder, T. S. 2000. The real options approach to evaluating a risky investment by a new generation cooperative: further processing. *A Paper Presented at the Annual Meetings of NCR-194 Research on Cooperatives*, Las Vegas, NV, Dec. 12-13.

Baldwin, C., and Ruback, R. 1986. Inflation, uncertainty, and investment. *Journal of Finance*, July, 657-669.

Baskerville, G. L. 1972. Use of logarithmic regression in the estimation of plant biomass. *Can. J. For. Res.* 2: 49-53.

B.C. Ministry of Forests. 1995. Forest practices code (1995). B.C. Ministry of Forests, Government of British Columbia, Victoria, BC, Canada.

Beard, J. S. 1942. Summary of silvicultural experience with cedar, *Cedrela mexicana* Roem. in Trinidad and Tobago. *Caribbean Forester* 3(3):91-102.

Beauchamp, J. J. and Olson, J.S. 1973. Corrections for bias in regression estimates after logtharithmic transformation. *Ecology* 54(6): 1402-1407.

Beck, D.E., and Trousdell, K.B., 1973. Site Index. Accuracy of prediction. Research paper SE-108 Asheville, North Carolina, Southeastern Forest and Range Experiment Station, U.S. Forest Service.

Becker, H., and Vanclay, F (eds.) 2003. The International Handbook of Social Impact Assessment: *Conceptual and Methodological Advances*. Edward Elgar, Cheltenham, UK.

Beers, T. W. 1962. Components of forest growth. *Journal of Forestry*, 60:245-248

Behre, C. E. 1924. Computation of total cubic contents of trees. *J. For.* 22(6): 62-63.

Bell, F. W. 1991. Critical silvics of conifer crop species and selected competitive vegetation in Northwestern Ontario. Ont. Min. Nat. Resour., *Northwestern Ont. For. Tech. Dev. Unit Rep. No.* 19. 177 p.

Benhin, J.K.A., and Barbier, E.B. 2001. The Effects of the Structural Adjustment Programme on Deforestation in Ghana. *Agricultural and Resource Economics Review, 30*(1), 66-80.

Benhin, J.K.A., and Barbier, E.B. 2004. Structural Adjustment Programme, Deforestation and Biodiversity Loss in Ghana. *Environmental and Resource Economics, 27*: 337–366.

Benneh, G., 1989. The dynamics of customary land tenure and agrarian systems in Ghana. In: Ghana, P. (Ed.), WCARRD; Ten Years of Follow-Up, The Dynamics of Land Tenure and Agrarian Systems in Africa, *Case Studies from Ghana, Kenya, Madagascar and Togo*. FAO, Rome, Italy, pp. 34– 97.

Berger, J. J. 2006. Ecological Restoration and Non-indigenous Plant Species: A Review. *Ecological Restoration* 1(2):74-82

Bergeuschbacher, R. J. 1990. Natural Forest Management in the Humid Tropics: Ecological, Social, and Economic Considerations. *Ambio* 19(5): 253-258.

Betancourt, A. 1972. The growth of *Azadirachta indica* in Cuba. *Baracoa* 2(2): 17 - 23.

Bhati, U. N., Klijn, N., Curtotti, R., Dean, M., and Stephens, M. (1991). Impediments to the development of commercial forest plantations in Australia. The Role of Trees in Sustainable Agriculture, National Conference, Albury, New South Wales, 30 September - 3 October 1991. Australian Bureau of Agricultural and Resource Economics. *GPO Box* 1563, Canberra 2601.

Bjerksund, P., and Ekern, S. 1993. Contingent claims evaluation of mean-reverting cash flows in shipping. In L. Trigeorgis (Ed.). Real options in capital investment (New Contributions) New York, NY, Praeger.

Black, F., and Scholes, M. 1973. The pricing of options and corporate liabilities. *Journal of Political Economy,* 3:637-654.

Bliss, C. I. and Reinker, K. A. 1964. A lognormal approach to diameter distributions in even-aged stands. *Forest Science* 10: 350-360.

Binkley, C. S. 1987. When is the economic rotation longer than the rotation of maximum sustained yield? *Journal of Environmental Economics and Management.* 14:152 - 158.

Boadu, F. O. 1992. Contingent valuation for household water in rural Ghana. *Journal of Agricultural Economics*, 43(3), 458-465.

Boardman, A. E., Greenberg, D. H., Vining, A. R., Weimer, D. L. 2001. Cost-benefit analysis: concepts and practice. 2nd Edition. Upper Saddle River, NJ. Prentice-Hall.

Boateng, E. A. 1966. *A geography of Ghana.* 2nd Ed. Cambridge. Cambridge Univ. Press. 212 p.

Bodie, Z., and Merton, R. C. 2000. Finance. Prentice Hall. Upper Saddle River, New Jersey.

Bolfrey-Arku, G. E. K., Onokpise, O. U., Carson, A. G., Shilling, D. G., Coultas, C. C. 2006. The Speargrass (Imperata cylindrica (L) Beauv.) menace in Ghana: Incidence, farmer perceptions and control practices in the forest and forest-Savannah transition Agro-ecological Zones of Ghana. *West African Journal of Applied Ecology.* 10(1).

Boni, S. 2006. Ghanaian Farmers' Lukewarm Reforestation: Environmental degradation, the timber option and ambiguous legislation. Colloque international "Les frontières de la question foncière – At the frontier of land issues", Montpellier, 2006.

Borota, J. 1991. *Tropical forests: Some African and Asian studies of composition and structure.* New York. Elsevier Publishers.

Bosu, P.P. and Krampah, E., 2005. Triplochiton scleroxylon K.Schum. In: Louppe, D., Oteng-Amoako, A.A. and Brink, M. (Eds.). Prota 7(1): Timbers/Bois d'œuvre 1. [CD-Rom]. PROTA, Wageningen, Netherlands.

Boulding, K. 1955. Economic analysis. 3rd Edition. New York. Harper and Brothers.

Braathe, P. 1957. Thinnings in even-aged stands: *A summary of the European literature. Fac. For. Univ.* New Brunswick, Fredericton. 92 p.

Bradley, R. T. 1963. Thinning as an instrument of forest management. *Forestry* 36:181-194.

Branch, K., Hooper, D., Thompson, J., and Creighton, J. (1984). Guide to Social Assessment Boulder, Colorado. Westview Press.

Brealey, R., Myers, S., and Marcus, A. 2001. *Fundamentals of corporate finance.* Boston, MA. McGraw-Hill Irwin.

Brechling, F. 1975. Investment and employment decisions. Manchester University Press. Manchester, UK.

Brennan, M. J., and Schwartz, E. S. 1985. Evaluating natural resource investments. *Journal of Business*, 58(2), 135-157.

Briscoe, C. B., and Ybarra-Conorodo, R. 1971. Increasing the growth of established teak. Res. Notes. ITF-13, Rio Piedras, PR. USDA For. Serv. Inst. of tropical forestry.

Brown, C., 2000. The Global Outlook for Future Wood Supply from Forest Plantations. *Global Forest Products Outlook Study Working Paper Series*. FAO, Rome. 164 pp.

Brown, T.C., Peterson, G.L., and Tonn, B.E. 1995. The value jury to aid natural resource decisions. *Land Economics*, 71(2) (May), 250-260.

Browne F.G. 1968. *Pests and Diseases of Forest Plantation Trees*. Clarendon Press, Oxford, UK.

Brownlee, K. A. 1967. Statistical theory and methodology in science and engineering. 2nd ed. John Wiley and Sons, New York. 400 pp.

Bruce, D. and Schumacher, F. X. 1950. *Forest mensuration*. McGraw-Hill Co., New York. 484 pp.

Bruce, D. and Reineke, L. H. 1931. Correction alignments charts in forest research. *U.S. Dept. Agric. Tech. Bull.* 210. 87 pp.

Budelman, A. 1988. The performance of the leaf mulches of *Leucaena leucocephala, Flemingia macrophylla* and*Gliricidia sepium* in weed control. *Agroforestry Systems* 6(1): 137-145.

Buongiorno, J., and Giless, J. K. 1987. *Forest management and economics*. New York. MacMillan Publishing Co.

Burdge, R. 1994. *A community guide to social impact assessment*. Middleton, Wisconsin. Social Ecology Press.

Bury, K. V. 1975. *Statistical models in applied science*. John Wiley and Sons Inc., 625 pp. New York, U. S. A.

Buschbacher, R. J. 1990. Natural Forest Management in the Humid Tropics: Ecological, Social, and Economic Considerations. *Ambio* 19(5): 253-258.

Business News. 2009. Malfeasance hits forest plantation fund. Business News of Tuesday, 18 August 2009. .http://www.ghanaweb.com/GhanaHomePage/NewsArchive/artikel.php?ID=166993.

CAB International. 2004. Forestry Compendium Global Module. Wallingford, UK: CAB International. Accessed online at: http://www.cabi.org/compendia/fc/index.asp.

CAB International, 2005. *Forestry Compendium*. Wallingford, UK: CAB International.

Cabaret N. and Nguessan K. 1988. *Gmelina arborea spacing trial* 1985. CTFT-CI.

Cairns, M.A., Brown, S., Helmer, E. H., and Baumgardner, G.A. 1997. Root biomass allocation in the world's upland forests. *Oecologia* 111: 1-11.

Calish, S., Fight, R., and Teeguargen, D. 1978. "How do non-timber values affect Douglas-fir rotation?" *Journal of Forestry*, 76 (4): 217 - 221.

Camirand, R. 2002. Guidelines for Forest Plantation establishment and management in Jamaica. Jamaica: Trees for Tomorrow Project Phase II. *Tecsult International* 4700, Boulevard Wilfrid-Hamel Québec, Québec Canada.

Cannell M. G. R. and Last, F. T. (Eds.). 1976. *Tree physiology and yield improvement*. Academic Press, London.

Carle, J., Holmgren, P. 2003. *Definitions Related to Planted Forests.* Forest Resources Assessment Programme. Forest Resources Development Service Forestry Department. *FAO Working paper* No. 79. FAO. Rome, Italy.

Carpenter, J.F. 1998. Internally Motivated Development Projects: A potential tool for biodiversity conservation outside protected areas. *Ambio* 27: 3, pp.211- 216.

Carr, P., 1988. The valuation of sequential exchange opportunities. *Journal of Finance,* December, 1235-1256.

Castedo-Dorado, F., Crecente-Campo, F., Álvarez-Álvarez, P., and Barrio-Anta, M. 2009. Development of a stand density management diagram for radiate pine stands including assessment of stand stability. *Forestry*, 82(1): 1-16.

Cavers, S., Navarro, C. and Lowe, A.J. 2004. Targeting genetic resource conservation in widespread species: a case study of *Cedrela odorata* L. Forest Ecology and Management, 197 (1-3): 285-294.

CDM Ghana. 2008. The CDM process. http://www.epa.gov.gh/cdm/DNA/dna1.htm.

Centeno, J. C. 2009. The management of teak plantations. http://www.treemail.nl/ teakscan.dal/files/mngteak.htm#stan. Accessed January, 2010.

Chachu, R. 1989. Allowable cut from the forest. *In* Wong, J., ed., *Ghana Forest Inventory Seminar proceedings,* 29–30 Mar 1989, Accra, Ghana.UK Overseas Development Administration, London, UK.

Chacko, K. C. 1995. Silvicultural problems in management of teak plantations. Proc. 2nd Regional Seminar on Teak 'Teak for the Future' Yangon, Myanmar May 1995 *FAO* (Bangkok) 91-98.

Chang, S. J. 1984. Determination of the optimal rotation age: a theoretical analysis. *Forest Ecology and Management,* 8,137 – 147.

Chaplin, G. E. 1980. Progress with provenance exploration and seed collection of *Cedrela spp. In Proceedings, Commonwealth Forestry Conference, Port-of-Spain, Trinidad,* September 1980. 17 p.

Chapman, D.G. 1961. Statistical problems in population dynamics. *In Proc. Fourth Berkeley Symp. Math Stat. and Prob. Univ. Calif.* Press, Berkeley.

Chapman, G. W., and Allan, T. G. 1978. Establishment techniques for forest plantations. Forest Resources Division. Forestry Department. *FAO Forestry Paper* 8. Rome.

Chapman, H. H., and Meyer, W.H. 1949. *Forest mensuration.* McGraw-Hill Co., New York. 522 pp.

Charkraborty, M. 1994. An Analysis of the Causes of Deforestation in India. In K. Brown, and D.W. Pearce (Eds.), *The Causes of Tropical Deforestation* (pp. 226-238). London: UCL Press.

Chavangi, N., 1992. Household based tree planting activities for fuelwood supply in rural Kenya; the role of the Kenya Woodfuel Development Programme. In *Development form Within: survival in rural Africa.* Taylor, D.R.F., and F. Mackenzie, 1992. Routledge, London and New York. pp. 148-169.

Chen, C. M. and Rose, D. W. 1978. Direct and indirect estimation of height distributions in even-aged stands. *Minnesota Forestry Research Notes.* No. 267. January, 1978. 3 pp.

Chundamannii, M. 1998. Teak plantations in Nilambur - an economic review. KFRI Research Report No. 144. Kerala Forest Research Institute, Peechi, Kerala, India. 71pp.

Cintron B.B. 1990. *Cedrela odorata* L. *Cedro hembra*, Spanish cedar, pp. 250-257. *In:* Burns R.M.H.and Barbara H. (Eds.), Silvics of North America 2: Hardwoods. *Agricultural*

Handbook 654. United States Department of Agriculture, Washington, DC. Vol. 2. pp 250-257.

CITES. 2007. Convention on International Trade in Endangered Species of wild Fauna and Flora. Fourteenth meeting of the Conference of the Parties *The Hague* (Netherlands), 3-15 June 2007.

Cline-Cole, R.A., Main, H.A.C., Nichol, J.E. 1990. On Fuelwood Consumption, Population Dynamics and Deforestation in Africa. *World Development* 18: 4, pp.513-527.

Clutter, J. L. 1963. Compatible growth and yield models for loblolly pine. *For. Sci.* 9:354-371.

Clutter, J. L. and Bennett, F. A. 1965. Diameter distributions in old-field slash pine plantations. *Georgia For. Res. Council Rep.* 13: 9pp.

Clutter, J. L., Fortson, J. C., Piennar, L. V., Brister, G. H. and Bailey, R. L. 1983. *Timber management - a quantitative approach.* John Wiley and Sons, Toronto. 333 pp.

Coates, D., and Haeussler, S. 1986. A preliminary guide to the response of major species for competing vegetation to silvicultural treatments. British Columbia Min. For., Info. Serv. Br. 88 p.

Cochran, W. G. 1953. Sampling Techniques. John Wiley and Sons Inc., New York. 330 pp.

Cochran, W. G. 1977. *Sampling Techniques.* New York. John Wiley and Sons.

Cohen, A. C. Jr. 1965. Maximum likelihood estimation in the Weibull distribution based on complete and censored samples. *Technometrics* 7: 579-588.

Coilie, T. S. 1938. Forest classification of forest sites with special reference to ground vegetation. *Journal of Forestry.* 36: 1062-1066.

Conrad, J. M, and Clark, C. W. 1987. *Natural resource economics: notes and problems.* Cambridge University Press. Cambridge. 231pp.

Cossalter, C., Pye-Smith, C., 2003. Fast-Wood Forestry—Myths and Realities. *Center for International Forestry Research*, Jakarta, Indonesia.

Cox, J., Ross, R., Rubinstein, M. 1979. Option pricing: a simplified approach. *Journal of Financial Economics*, 7(4): 71-90.

Cox, J., and Ross, S. 1976. The valuation of options for alternative stochastic processes. *Journal of Financial Economics,* January, 145-166.

CQFA, 2009. Centrel Queensland Forest Association. Plantation Establishment http://www.cqfa.com.au/resources/agroforestry/plantation-establishment. Accessed January, 2010.

Craib, I. J. 1939. Thinning, pruning and management studies on the main exotic conifers grown in South Africa. *Dept. of Agric. And For. Sci. Pretoria. Bull.* No. 196. 179 p.

Crow, T. R. 1971. Estimation of biomass in an even-aged stand - regression and "mean tree" techniques. pp. 35-50 *in* Young, H. E (ed.) Forest Biomass Studies. XVth IUFRO Congress, Univ. of Florida, March 15-20 1971. 205 pp.

Crush, J.S., and Namasasu, O. 1985. Rural rehabilitation in the Basotho Labour Reserve. *Applied Geography* 5.

CSPS. 2008. Canada School of Public Service. *Cost Benefit analysis and risk analysis.* Unpublished course notes.

CSPS. 2009. Canada School of Public Service. *Regulatory performance measurement and evaluation.* Unpublished course notes.

Csurhes, S. 2008. Pest plant risk assessment: the neem tree (*Azaradirachta indica*). *The State of Queensland,* Department of Primary Industries and Fisheries. PR08-3685.

Cunia, T. 1964. Weighted least squares method and the construction of volume tables. *For. Sci., 10*, 180-191.

Curtis, R. O. 1967. Height-diameter and height-diameter-age equations for second growth Douglas fir. *Forest Science* 13(4): 365-375.

Daniel, T. W., Helms, J. A., and Baker, F. S. 1979. Principles of silviculture. 2nd Ed. New York: McGraw-Hill.

Daniel, W. W. 1978. *Applied nonparametric statistics*. London. Houghton Mifflin Co.

Darkwa, E. O., Johnson, B. K., Nyalemegbe, K., Yangyuoru, M., Oti-Boateng, C., Willcocks, T. J., and Terry, P. J. 2001. Weed management on Vertisols for small-scale farmers in Ghana. *International journal of pest management* 47(4): 299 – 303.

Datta, S. K. de. 1978. Fertiliser management for efficient use in wetland rice soils. *IRRI, Soils and Rice* 671-701.

Dauda, C. 2009. Interview transcript. Interview with the Minister of Lands and Natural Resources. Upper Reach. www.upper-reach.com.

Davidson, J. 1985. Assistance to the forestry sector of Bangladesh. Species and sites What to plant and where to plant. *Field Doc. No.* 5, UNDP/FAO/BGD/79/017. 50 p.

Davis, L.S., Johnson, K. N., Bettinger, P.S., and Howard, T. E. 2001. *Forest management: to sustain ecological, economic, and social values*. 4th Edition. McGraw Hill. New York. 804pp.

Day, R. J. 1985. Crop plans in silviculture. *Canadian Pulp and Paper Association*, Woodlands Section Index 2975. 55p

Day, R. J. 1996. A manual of silviculture. School of Forestry, Lakehead University, Thunder Bay, ON, Canada. 344 p.

Day, R. J. and Nanang, D. M. 1997. Principles of thinning for improved growth, yield, and economic profitability of lodgepole and jack pine. Pages 1-13 *in* Proceedings of a Commercial Thinning Workshop, Whitecourt, Alberta, 17-18 October 1996. *FERIC, Vancouver. Spec. Rep.* SR-122.

De Jong, I.L. 1991. Social Forestry in Mali: is participation by the rural population difficult to realise? *AT-Source* 17:1, pp.17-20.

Deacon, R.T. 1994. Deforestation and the rule of law in a cross-section of countries. *Land Economics 70*, 414-430.

Dean, T.J. and Baldwin, V.C. 1996 Crown management and stand density. In *Growing Trees in a Greener World: Industrial Forestry in the 21st century; 35th LSU Forestry Symposium* . M.C. Carter (ed.) Louisiana State University Agricultural Center, Louisiana *Agricultural Experiment Station* , Baton Rouge, LA. pp. 148 – 159.

Dei, G.S. 1990. Deforestation in a Ghanaian Rural Community. *Anthropologica, 32*, 3-27.

Dei, G.S. 1992. A Forest Beyond the Trees: Tree Cutting in Rural Ghana. *Human Ecology, 20*(1), 57-88.

Dixit, A., and Pindyck, R. S., 1994. *Investment under uncertainty*. Princeton University Press. Princeton, New Jersey.

Djarbeng, V. and Ameyaw, D.S., 2002. *ADRA's agroforestry development programme in Ghana gives farmers new chances.* Project document.

Donkor, B. N. 2003. Evaluation of government interventions in Ghana's forest product trade: a post-intervention impact assessment and perceptions of marketing implications. *Unpublished Ph.D. Thesis*. The School of Renewable Natural Resources, Louisiana State University. 178 pp.

Donkor, B. N., Vlosky, R. P., and Attah, A. 2006. *Evaluation of government interventions on increasing value-added wood product export from Ghana.* Louisiana *Forest Products Development Center. Working Paper #77.* 16p.

Drechsel, P., and Zech, W. 1994. DRIS evaluation of teak (*Tectona grandis* Linn F.) Mineral nutrition and effects of nutrition and site quality on teak growth in West Africa. *For. Ecol. and Manage., 70,* 121 - 133.

Drew , T.J. and Flewelling , J.W. 1979 Stand density management: an alternative approach and its application to Douglas-fir plantations . *For. Sci.* 25, 518 – 532.

Drew, T.J., and Flewelling, J.W., 1977. Some recent Japanese theories of yield–density relationships and their application to Monterey pine plantations. For. Sci. 13, 39–53.

Duke, J. A. 1983. Handbook of energy crops. Unpublished. Available on http://www.hort.purdue.edu/newcrop/duke_energy/Gmelina_arborea.html#Yields%20and%20Economics.

Duku-Kaakyire, A. and Nanang, D. M. 2004. Application of real option theory to forest investment analysis. *Forest Policy and Economics*, 6, 539– 552.

Duvall, C.S., 2009. *Ceiba pentandra* (L.) Gaertn. [Internet] Record from Protabase. Brink, M. and Achigan-Dako, E.G. (Editors). PROTA (Plant Resources of Tropical Africa / Ressources végétales de l'Afrique tropicale), Wageningen, Netherlands. < http://database.prota.org/search.htm>. Accessed 21 February 2010.

Dye, T. R. 1998. Understanding public policy. 9th Edition. New York. John Wiley.

Dzanku, F. M. 2004. Contribution of wildlife to the economy of Ghana. ISSER, University of Ghana. Unpublished notes. Available on: http://are.berkeley.edu/~dwrh/IPALP_Web/Meetings/Dakar0506/Presentation-Dzanku-Ghana.pdf.

Echenique-Marique, R. and Plumptre, R.A. 1990. A guide to the use of Mexican and Belizean timbers. *Tropical Forestry Papers*, 20. Oxford Forestry Institute.

Elton, E.J., Gruber, M. J., 1995. *Modern portfolio theory and investment analysis.* Fifth edition. John Wileyand Sons, Inc.

Enters, T. 2001. Incentives for soil conservation. In: *Response to land degradation*, E.M. Bridges, I.D. Hannam, L.R. Oldeman, F.W.T. Penning de Vries, S.J. Scherr and Samran Sombatpanit (Eds.), 351-360. New Delhi and Calcutta: Oxford and IBH Publishing Co. Pvt. Ltd.

Enters, T., Durst, P. B., and Brown, C. 2003. What Does it Take? The role of incentives in forest plantation development in the Asia-Pacific Region. *UNFF Intersessional Experts Meeting on the Role of Planted Forests in Sustainable Forest Management,* 24-30 March 2003, New Zealand.

Enters, T., Durst, P. B., Brown, C., Carle, J., and McKenzie, P. 2003. What Does it Take? The Role of Incentives in Forest Plantation Development in the Asia-Pacific Region. *FAO. RAP PUBLICATION* 2004/28. Rome.

European Commission/VPA. 2009. *FLEGT Voluntary Partnership Agreement Between European Commission Delegation in Ghana and VPA Secretariat in Ghana.*

Evans, J. 1989. Community Forestry in Ethiopia: The Bilate Project. *Rural Development in Practice* 1: 4, 7-8, 25.

Evans, J. 1992. Plantation forestry in the tropics. 2nd ed. Oxford Univ. Press. 403 pp.

Evans, J. 2001. Sustainability of productivity in successive plantations. In: *Proceedings of the International Conference on Timber Plantation Development*, Manila, the Philippines, 7-9 November 2000. *Quezon City, the Philippines, Department of Environment and Natural Resources.*

Evans, J and Turnbull, J. W. 2004. Plantation forestry in the tropics: *The role, silviculture and use of planted forests for industrial, social, environmental and agroforestry purposes.* Third edition. Oxford University Press. 488 pp.

FAO and UNEP. 1981. *Forest resources of Tropical Africa. Part II: Country briefs.* Tropical forest resource assessment project. GERMS, UN 23(6). *Tech. Report 2.* Rome. FAO.

FAO. 1956. *Tree planting practice in tropical Africa.* Rome, Italy. FAO.

FAO, 1981. *Map of the Fuelwood Situation in the Developing Countries.* Scale 1:25 000 000.

FAO. 1983. *Growth and yield of plantation species in the tropics.* Rome. FAO. Forest Resource Div. Rome.

FAO. 1985. Intensive multiple-use forest management in the tropics: Analysis of case studies from India, Africa, Latin Arnerica and the Caribbean. Rome. FAO.

FAO. 1992. *Mixed and pure forest plantations in the tropics and subtropics.* FAO Forestry Paper no. 102. Rome. FAO.

FAO. 1993. Vegetative propagation. *Field Manual No.5. UNDP/FAO Project* RAS/91/004. Rome, Italy.

FAO. 1995. Forest resource management - Project findings and recommendations. Terminal Report. Project FO:UTF/GHA/025/GHA. Rome.

FAO. 1997. Update on sustainable forest management and certification: Example from a developing country – Ghana. *Advisory committee on paper and wood products.* Thirty-eighth session. Rome, 23 - 25 April 1997.

FAO. 1999. *Incentive systems for natural resource management. Environmental Reports Series 2.* FAO Investment Centre. Rome, Food and Agriculture Organisation of the United Nations.

FAO. 2002a. *Hardwood plantations in Ghana.* Forest Plantations Working Paper 24. *Forest Resources Development Service, Forest Resources Division.* Rome.

FAO. 2002b. Forest plantation productivity. Report based on the work of W.J. Libby and

C. Palmberg-Lerche. Forest Plantation Thematic Papers, Working Paper 3. *Forest Resources Development Service, Forest Resources Division. FAO*, Rome (*unpublished*).

FAO. 2005. *The State of the World's Forests.* 2005. Rome. Italy.

FAO. 2009. Forestry Statistics. http://www.fao.org/corp/statistics/en/.

FAO. 2011. Global Forest Resource Assessment 2010. http://www.fao.org/news/story/en/item/40893/icode/. FAO. Italy. Rome.

Farmer, R. H. 1972. *Handbook of hardwoods.* 2nd Ed. London. H. M. Stationery Office.

Faustmann, M, 1849. Calculation of the value which forest land and immature stands possess for forestry (Transl. in M. Gane and W. Linnard in "Martin Faustmann and the evolution of discounted cash flow: two articles from the original German of 1849". Comm. For. Inst, Univ. Oxon, Inst. Pap. No. 42, 1968). *Reprinted in Journal of Forest Economics* 1996: 1 (1): 7 - 44.

Fernandez-Cornejo, J., Gempesaw II, C. M., Elterich, J. G., and Stefanou, S. E. 1992. Dynamic Measures of Scope and Scale Economies: An Application to German Agriculture *American Journal of Agricultural Economics*, Vol. 74, No. 2 (May, 1992), pp. 329-34.

Field, B. C. 2001. *Natural resource economics: An introduction.* New York. McGraw-Hill.

Finney, D. J.1941. On the distribution of a variate whose logarithm is normally distributed. *R.J. Statist. Soc. Supplement.* 7: 155-158.

Fisher, I. 1930. *The theory of interest rates.* MacMillan, New York, NY 566 pp.

Fisher, N. M. 1984. The impact of climate and soil on cropping systems and the effect of cropping systems and weather on the stability of yield. pp. 55-70 *in* Steinner, K. G. (ed.) *Report on the On-Farm-Experimentation Training Workshop*, Nyankpala Ghana, July 3-13 1984. 120 pp.

Foley, G. 1987. Exaggerating the Sahelian woodfuel problem? *Ambio* 16(6):367-371.

Ford-Robertson, F. C. (Ed.). 1971. Terminology of forest science, technology, practice and products. *Multilingual For. Terminol. Ser. No. 1. Soc. Am. For.*, Washington, DC.

Forest Research Programme. 2006. Chainsaw milling and logging in Ghana: *Background study report.* Available on: http://www.illegal logging.info/uploads/FRP_Chainsaw_ Logging_Ghana.pdf.

Forest Watch Ghana. 2006. Forest governance in Ghana. An NGO perspective. *A report produced for FERN by Forest Watch Ghana, March 2006.*

Forestry/Fuelwood Research and Development Project (F/FRED). 1994. Growing multipurpose trees on small farms, module 9: *Species fact sheets* (2nd ed.). Bangkok, Thailand: Winrock International. 127p.

Fortson, J. C. 1972. Which criterion? *Effect of the choice of criterion on forest management plans. For. Sci.,* 18, 292 - 297.

Frank, R. M. 1973. The course of growth response in released white spruce-10 year results. US Dep. Agric., For. Serv., Northeastern Forest Experiment Station, Broomall, PA. *Research Paper NE*-268. 6 p.

Freeman, A. Myrick. III. 1993. The measurement of environmental and resource values: *Theory and methods*. Washington D. C. Resources for the Future.

Friday, K. S. 1987. Site index curves for teak (*Tectona grandis* Linn F.) in the limestone regions of Puerto Rico. *Commonw. For. Review*, 66, 239-252.

Fries, J. 1991a. From village forestry towards farming systems – the development of the Swedish Sahel Programme. *IDRCurrents* No.1, pp32-35.

Fries, J. 1991b. Management of natural forests in the semiarid areas of Africa. *Ambio* 20:8, pp.395-400.

FSC. 2010. Principles and criteria for forest certification. Forest Stewardship Council. http://www.fsc.org/fsc-rules.html. Accessed March, 2010.

Furnival, G. M. 1961. An index for comparing equations used in constructing volume tables. *For. Sci.* 7: 337-341.

Gadow, K. von. 1983. Fitting distributions in *Pinus patula* stands. *South African Forestry Journal* 126: 20-29.

Gayfer, J., Sarfo-Mensah, P., and Arthur, E. (2002). Gwira Banso – Joint Forest Management Project: Mid Term Review. *Technical Report Prepared for CARE Ghana.*

Ghana Export Promotion Council. 2008. Ghana: *Non-traditional exports post strong growth.* Reported on www.modernghana.com.

Ghana Forestry Commission. 1998. Manual of procedures: Forest resource management planning in the HFZ. Section A - strategic planning. http://www.fcghana.com/ publications/manuals/hfz/StratPlanMan.htm#_Toc526929766.

Ghana Forestry Commission. 2005. *National forest plantation development programme. 2005-Annual Report.*

Ghana Forestry Commission. 2006a. Report on export of wood products. *Timber Industry Development Division, Ghana Forestry Commission.* Unpublished Report.

Ghana Forestry Commission. 2006b. *National forest plantation development programme.* 2006-Annual Report.

Ghana Forestry Commission. 2007a. Report on export of wood products. *Timber Industry Development Division, Ghana Forestry Commission.* Unpublished Report.

Ghana Forestry Commission. 2007b. *National forest plantation development programme.* 2007-Annual Report.

Ghana Forestry Commission. 2008a. *National Forest Plantation Development Programme-Anuual Report*, 2008. http://www.fcghana.com/publications/index.htm.

Ghana Forestry Commission. 2008b. Report on export of wood products. *Timber Industry Development Division, Ghana Forestry Commission.* Unpublished Report, 2008. http://www.fcghana.com/publications/index.htm.

Ghana Forestry Commission. 2009a. Report on export of wood products. Jan-May, 2009. Timber Industry Development Division, Ghana Forestry Commission. http://www.fcghana.com/publications/industry_trade/export_reports.htm/year_2009/may_2009.pdf.

Ghana Forestry Commission. 2009b. New procedures for stumpage collection and disbursement. (Published in conjunction with the in Association with the Office of Administrator of Stool Lands). http://www.fcghana.com/library_info.php?doc=55 and publication: New Procedure for Stumpage Disbursement.

Ghana Forestry Commission. 2010. Report on export of wood products. Jan-Dec, 2009. *Timber Industry Development Division, Ghana Forestry Commission.* http://www.fcghana.com/publications/industry_trade/export_reports.htm/year_2009/jan_2010.pdf.

Ghana Forestry Commission. 2011. Report on export of wood products. Jan-Dec, 2010. Timber Industry Development Division, Ghana Forestry Commission. http://www.fcghana.com/publications/industry_trade/export_reports.htm/year_2010/jan_2011.pdf

Ghana News Agency. (2006). Forest Plantation Development Fund Board calls for support. http://www.modernghana.com/news/99414/1/forest-plantation-devt-fund-board-calls-for-suppor.html.Ghana.

Ghana News Agency, 2009. Thousands of jobs to be created through reforestation. News Report from the GNA. November 30, 2009. http://www.ghanaweb.com/GhanaHomePage/NewsArchive/artikel.php?ID=172684andcomment=5308073#com.

Ghana Statistical Service. 2000. Ghana Living Standards Survey IV. Accra. Available on http://www.worldbank.org/html/prdph/lsms/country/gh/docs/G4Qprice.pdf.

Ghebremichael, A., Nanang, D. M., and Yang, R. 2005. Economic analysis of growth effects of thinning and fertilisation of lodgepole pine in Alberta, Canada. *Northern Journal of Applied Forestry*, 22(4): 254-261.

Ghosh, R. C., and Singh, S. P. 1981. Trends in rotation. *Indian Forester, 107*, 336-347.

Gilbertson, T., and Reyes, O. 2009. Carbon trading: How it works and why it fails. Critical currents No. 7. November 2009. *Dag Hammarskjöld Foundation Occasional Paper Series*. Uppsala.

Gonçalves, J.L.M., Barros, N.F., Nambiar, E.K.S. and Novais, R.F. 1997. Soil and stand management for short rotation plantations. *In*: Nambiar, E.K.S. and Brown, A.G. (Eds.) Management of soil, water and nutrients in tropical plantation forests, 379-417.

Australian Centre for International Agricultural Research (ACIAR), Monograph 43, Canberra.

Gove, J.H and Fairweather, S. E. 1989. Maximum-likelihood estimation of Weibull function parameters using a general interactive optimiser and grouped data. *For. Ecol. Manage.* 28:61-69.

Gove, J.H. 2004. Structural stocking guides: a new look at an old friend. *Can. J. For. Res.* 34 , 1044 – 1056 .

GPRS, 2002. *Ghana Poverty Reduction Strategy* 2002-2004.

Grainge, M., S. Ahmed, W.C. Michell, and J.W. Hylin. 1985. Plant species reportedly possessing pest control properties: *An EWC/UH Database*. Resource Systems Institute, East-West Centre, Honolulu, HI.

Greenhill, A. G. 1881. Determination of the greatest height consistent with stability that a vertical pole or post can be made, and of the greatest height to which a tree of given properties can grow. *Proceedings of the Cambridge Philosophical Society IV*. Part II, 65-73. Cambridge, England.

Gregersen, H.M. 1984. Incentives for forestation: a comparative assessment. In K.F. Wiersum, (Ed.) *Strategies and designs for afforestation, reforestation and tree planting.* Wageningen, the Netherlands, Wageningen Agricultural University.

Grijpma, P. 1976. Resistance of *Meliaceae* against the shootborer *Hypsipyla* with particular reference to *Toona ciliata* M. J. Roem. var. *australis* (F. v. M.) DC. *In:* Tropical trees. Variation breeding and conservation. J. Burley, and B. T. Styles, (Eds.) p. 69-79. Academic Press, Oxford.

Hafley, W. L. and Schreuder, H. T. 1976. Some non-normal bivariate distributions and their potential for forest application. *International Union of Forest Research Organisations. XVI World Congress Proceedings*, Div. VI, 104-114 (Oslo, Norway; June 20 – July2, 1976).

Hafley, W. L. and Schreuder, H. T. 1977. Statistical distributions for fitting diameter and height data in even-aged stands. *Canadian Journal of Forest Research* 7:481-487.

Hall, D. O. 1983. Financial maturity for even-aged and all-aged stands. *For. Sci.*, 29 (4): 833 - 836.

Hall J.B., Pandey D., and Hirai, S. 1999. Global Overview of teak plantations. Paper presented to the Regional Seminar *Site, Technology and Productivity of Teak Plantations* Chiang Mai, Thailand 26-29 January 1999.

Hall, J. B., and Bada, S. O. 1979. The Distribution and Ecology of Obeche (*Triplochiton Scleroxylon*) *Journal of Ecology*, Vol. 67 (2): 543-564.

Hall, J. B., and Swaine, M. D. 1981. *Distribution and ecology of vascular plants in a tropical rainforest. Forest Vegetation in Ghana.* Geobotany 1. The Hague. Springer.

Haltia, O., and Keipi, K. 1997. *Financing forest investments in Latin America: the issue of incentives.* Washington, DC, USA, Inter-American Development Bank.

Harper, J. L. 1977. *Population biology of plants*. Academic Press. London. 892p.

Hartemink, A.E. 2003. Soil fertility decline in the tropics with case studies on plantations. 360 pp. *ISRIC-CABI, Wallingford.*

Hartman, R. 1976. The harvesting decision when a standing forest has value. *Economic Inquiry* 14, 52-58.

Hawthorne, W. D., and Abu-Juam, M. 1995. *Forest Protection in Ghana. IUCN, Gland,* Switzerland and Cambridge. 203 pp.

Hawthorne, W.D., 1995. Ecological profiles of Ghanaian forest trees. *Trop. For. Pap.* 29, Oxford Forestry Institute, 345 pp.

Healey, S. P., and Gara, R. I. 2003. The effect of teak (*Tectona grandis*) plantation on the establishment of native species in an abandoned pasture in Costa Rica. *For. Ecol. and Manage., 176*, 497-507.

Hearne, D. A. 1975. *Trees for Darwin and northern Australia*, Department of Agriculture, Forestry and Timber Bureau, Australian Government Publishing Service, Canberra.

Hedegart, T. 1976. *Breeding systems, variation and genetic improvement of teak (Tectona grandis Linn F.) in tropical trees*. London. Academic Press Inc.

Hepburn, G. 1989. Pesticides and drugs from the neem tree. *Ecologist* 19(1): 31-32.

Hiley, W. E. 1930. *The economics of forestry*. The Clarendon Press, Oxford, U.K., 256p.

Hitch, C. J., and McKean, R. N. 1960. *The economics of defence in the nuclear age*. Cambridge, Massachusetts. Harvard University Press.

Holdridge, L. R. 1976. Ecología. de las Meliáceas Latinoamericanas. Studies on the shootborer *Hypsipyla grandella* Zeller. vol. 3. J. L. Whitmore (Ed.), Centro Agronómico Tropical de Investigación y Enseñanza, Miscellaneous Publication 1. Turrialba, Costa Rica. p.7.

Holm L. R., Pluncknett D. L., Pancho J. V. and Herberger J. P. (1977). *Imperata cylindrica* (L.) Beauv. In *The world's worst weeds: Distribution and biology*. pp. 62–71. University Press of Hawaii, Honolulu, USA.

Honu, Y.A.K. and Dang, Q.L. 2000. Response of tree seedlings to the removal of *Chromolaena odorata* Linn. in a degraded forest in Ghana. *For. Ecol. Manage*. 137: 75-82.

Honu, Y.A.K. and Dang, Q.L. 2002. Spatial distribution and species composition of tree seed and seedlings under the canopy of the shrub, *Chromolaena odorata* Linn. in Ghana. *For. Ecol. Manage.* 164: 185-196.

Horne, J.E.M. 1966. *Teak in Nigeria. Nigerian Information Bulletin* (New Series) No.16.

Hossain, K. L.1999. *Gmelina arborea*: A popular plantation species in the tropics. A quick guide to multipurpose trees from around the world. *Fact Sheet* 99-05.

Howland, P., Bowen, M.R., Ladipo, D.P., and Oke, J.B., 1977. The study of clonal variation in *Triplochiton scleroxylon* K. Schum, as a basis for selection and improvement. *Proceedings of the Joint IUFRO Workshop*, pp. 6–13.

Hunter, J. L. Jr. 1990. Wildlife, forests, and forestry. *Principles of managing forests for biological diversity*. Englewood Cliffs. New Jersey: Prentice Hall.

Husch, B. 1963. Forest mensuration and statistics. The Ronald Press Co., New York. 474 pp.

Husch, B., Miller, C. I., and Beers, T. W. 1982. *Forest mensuration*. 3rd ed. John Wiley and Sons. New York. 402 pp.

Hyde, W. F. 1980. *Timber supply, land allocation and economic efficiency*. The Johns Hopkins University for Resources for the Future. Baltimore, MD.

Ingersoll. J. E., and Ross, S. A. 1992. Waiting to invest: investment and uncertainty. *Journal of Business*, 65(January): 1-29.

IPCC, 2007: Climate Change 2007: Synthesis Report. Contribution of Working Groups I, II and III to the Fourth Assessment Report of the Intergovernmental Panel on Climate Change [Core Writing Team, Pachauri, R.K and Reisinger, A. (Eds.)]. *IPCC*, Geneva, Switzerland, 104 pp.

ITTO. 2005a. *Status of tropical forest Management. ITTO Technical Series* Note 24: Ghana (pp. 98-104).

ITTO. 2005b. Revised ITTO Criteria and Indicators for the sustainable management of tropical forests including reporting format. *ITTO Policy Development Series* No 15.

ITTO, 2005c. *Tropical timber market report*, 1–15 February 2005. International Tropical Timber Organisation, Yokohama, Japan.

ITTO. 2008. Developing forest certification: Towards increasing the comparability and acceptance of forest certification systems. ITTO Technical Series No 29.

Jack, S.B., and Long, J.N., 1996. Linkages between silviculture and ecology: an analysis of density management diagrams. *For. Ecol. Manage.* 86, 205–220.

Jacobson, M. 1958. Insecticides from plants: A review of the literature, 1941-53. *USDA Handbook* 154. Washington, DC.

Jacobson, M. 1975. Insecticides from plants: A review of the literature, 1952-72. *USDA Handbook* 461. Washington, DC.

Jackson, D. H. 1980. *The microeconomics of the timber industry*. Westview Press, Boulder, Co.

Jadhav, B. B. and Gaynar, D. G. 1994. Effect of Tectona grandis (L.) leaf leachates on rice and cowpea. *Allelopathy Journal, 1*, 66-69.

James, T., Vege, S., Aldrich, P. and Hamrick, J.L. 1998. Mating systems of three tropical dry forest tree species. *Biotropica* 30 (4): 587-594.

Jenkins, J.C., D.C. Chojnacky, L.S. Heath and R.A. Birdsey. 2003. National-scale biomass estimators for United States tree species. *For. Sci.* 49: 12-35.

Jensen, M. 1995. Trees commonly cultivated in Southeast Asia; Illustrated field guide. *RAP Publication:* 1995/38, FAO, Bangkok, Thailand. p. 93.

Joet, A., Jouve, P., Banoin, M. 1998. Le defrichement ameliore au Sahel. Une pratique agroforestiere adoptee par les paysans. (Improved forest clearance: an agroforestry method adopted by the people in the Sahel) *Bois-et-Forets-des-Tropiques* 225, pp.31-44.

Johnson E and Miyanishi, K. 2007. *Plant disturbance ecology: the process and the response*. Academic Press/Elsevier, Burlington, MA.

Johnson, N. L. 1949a. Systems of frequency curves generated by methods of translation. *Biometrika*. 36: 149-176.

Johnson, N. L. 1949b. Bivariate distributions based on simple translation systems. *Biometrika*. 36: 297-304.

Johnson, N. L. and Kotz, S. 1972. *Distributions in statistics: Continuous multivariate distributions*. John Wiley and Sons Inc., New York, U. S. A. 333 pp.

Johnson, N. L. Kotz, S. and Balakrishnan, N. 1994. *Continuous univariate distributions*. Vol. I. 2nd Edition. John Wiley and Sons Inc., New York, U. S. A. 756 pp.

Johnstone, W. D. 1997. The effect of commercial thinning on the growth, and yield of lodgepole pine. Pages 13-23 *in* Proceedings of a Commercial Thinning Workshop, Whitecourt, Alberta, 17-18 October 1996. *FERIC, Vancouver. Spec. Rep.* SR-122.

Jøker, D., and Salazar, R. 2000. Seed Leaflet No. 22. *Ceiba pentandra* (L.) Gaertn. Danida Forest Seed Centre. CATIE.

Jones, N., 1974. Records and comments regarding the flowering of *Triplochiton scleroxylon* K. Schum. Commonwealth For. Rev. 53, 52–56.

Jones, R. 1969. Review and comparison of site evaluation methods. U. S. Forest Service Research paper RM-51.

Judd, M. P. 2004. Introduction and Management of Neem (*Azadirachta indica*) in Smallholder's Farm Fields in the Baddibu Districts of The Gambia, West Africa. *MSc in Forestry Thesis*. Michigan Technological University.

Kadambi, K. 1972. *Silviculture and management of teak*. Bul. 24. Nacogdoches, TX, Stephen F. Austin State University School of Forestry.

Kahurananga, J., Alemayehu, Y., Tadesse, S., and Bekele, T. 1993. Informal surveys to assess social forestry at Dibandiba and Aleta Wendo, Ethiopia. *Agroforestry Systems* 24:57-80.

Kant, S., and Redantz, A. 1997. An econometric model of tropical deforestation. *Journal of Forest Economics, 3*(1), 51-86.

Kassier, H. W. and Bredenkamp, B. V. 1994. Modelling diameter and height distributions through dispersion statistics in even-aged pine plantations. *South African Forestry Journal* 171: 21-27.

Kemma, A. 1993. Case studies on real options. *Financial Management*, Autumn, 259-270.

Kengen, S. 1997. Forest Valuation for decision making: *Lessons of experience and proposals for improvement.* Rome. Food and Agriculture Organisation.

Kensinger, J. 1987. Adding the value of active management into the capital budgeting equation. *Midland Corporate Finance Journal*, 5(Spring), 31-42.

Keogh, R. M. 1982. Teak (*Tectona grandis.* Linn F.) provisional site classification chart for the Caribbean, Central America, Venezuela and Colombia. *For. Ecol. and Manag., 4*, 143-153.

Kerr, G. 1999. The use of silvicultural systems to enhance the biological diversity of plantation forests in Britain. *Forestry* 72(3): 191-205.

Kester, W. C. 1984. Today's options for tomorrow's growth. *Harvard Business Review.* March-April, 153-160.

Kester, W. C., 1993. *Turning growth options into real assets*. In: Aggarwal, A. (Ed.), Capital budgeting under uncertainty Englewood Cliffs, NJ, Prentice-Hall: 187-207.

Ketkar, C. M. 1976.Utilisation of neem (*Azadirachta indica*) and its by-products. Directorate of Non-Edible Oils and Soap Industry, Khadi and Village Industries Commission, Pune, India.

Ketkar, C. M. 1984. Crop experiments to increase the efficiency of urea fertiliser nitrogen by use of neem products. pp. 507-518 *in* Schmutterer, H. and K. R. S Ascher (eds.) *Proc. 2nd. Int. Neem Conf. Rauischholzhausen,* Federal Republic of Germany, May 25-28 1983. 587 pp.

Khandiya, S. D. and V.L. Goel.1986. Patterns of variability in some fuelwood trees grown on sodic soils. *Indian Forester* 112 (2): 118-123.

Kimmins, J. P. 2003. *Forest Ecology*, 3rd edition. Prentice-Hall, Upper Saddle River, NJ.

King, D. M., and Mazzotta, M. J. 2000. Ecosystem valuation. U.S. Department of Agriculture, Natural Resources Conservation Service and National Oceanographic and Atmospheric Administration. Website: http://www.ecosystemvaluation.org/default.htm. Accessed December, 2008.

Kira, T., Ogawa, H., and Shinozaki, K. 1953. *Interspecific competition among higher plants.* 1. J. inst. Polytech. Osaka City University. D. 4: 1-16.

Knoebel, B and Burkhart, H. 1991. A bivariate distribution approach to modelling forest diameter distributions at two points in time. *Biometrics* 47: 241-298.

Korankye-Gyamera, Y. 1997. Preliminary studies into methods of propagating dawadawa (parkia biglobosa). *BSc (Tech) Thesis, Department of Renewable Natural Resources,* University for Development Studies, Tamale, Northern Region.

Kouch, T., Preston, T. R., and Hieak, H. 2006. Effect of supplementation with Kapok (*Ceiba pentandra*) tree foliage and Ivermectin injection on growth rate and parasite eggs in faeces of grazing goats in farmer households. Livestock Research for Rural Development. *Volume 18, Article #87.* Retrieved February 23, 2010, from http://www.lrrd.org/lrrd18/6/kouc18087.htm.

Kraenzel, M. B. 2000. Carbon Storage of Panamanian Harvest-Age Teak *(Tectona grandis)* Plantations. *Masters Thesis. Department of Biology*, McGill University, Montreal, Canada.

Krishnapillay, B. 2000. Silviculture and management of teak plantations. *Unasylva 51*: available at http://www.fao.org/docrep/x4565e/x4565e04.htm#P0_0. Accessed July 2009.

Kulatilaka, N. 1993. The value of flexibility: the case of a dual fuel industrial steam boiler. *Financial Management*, 22 (Autumn), 271-280.

Kulatilaka, N., and Trigeorgis, L. 1994. The general flexibility to switch: real options revisited. *International Journal of Finance*, 6(Spring): 778-798.

Kumar, B. M., Long, J. N., and Kumar, P. 1995. A density management diagram for teak plantations of Kerala in peninsular India. *Forest Ecology and Management*. 74:125-131.

Lamb, A.F.A. 1968. *Fast growing timber trees of the lowland tropics*. No. 2 *Cedrela odorata*. Commonwealth Forestry Institute, Dept. of Forestry, University of Oxford. pp. 46.

Lambert, M. C., Ung, C.H., and Raulier, F. 2005. Canadian national tree aboveground biomass equations. *Can. J. For. Res*. 35: 1996-2018.

Lamers J. P. A., Michels, K., and Vandenbeldt, R. J. 1994. Trees and windbreaks in the sahel: establishment, growth, nutritive, and calorific values. *Agroforestry Systems* 26: 171-184.

Lamprecht, H. 1989. Silviculture in the tropics: *tropical forest ecosystems and their tree species..*

Landsberg, J.J. 1997. The biophysical environment. *In*: Nambiar, E.K.S. and Brown, A.G. (eds.) Management of soil, water and nutrients in tropical plantation forests, 65-96. *Australian Centre for International Agricultural Research (ACIAR)*, Monograph 43, Canberra.

Langyintuo, A.S., Ntoukam, G., Murdock, L., Lowenberg-DeBoer, J., and Miller, D.J., (2004). Consumer preferences for cowpea in Cameroon and Ghana. *Agricultural Economics,* 30(3), 203–213.

Laurie, M. V. 1974. Tree planting practices in African savannas. *Forestry Development Paper* No. 19. FAO. Rome.

Laurie, M. V., and Ram, B. S. 1940. Yield and stand tables for teak (*Tectona grandis* L. F) plantations in India and Burma. *Indian Forest Records*. No.1. Vol. IV-A. 115pp.

Lawson, G. W. 1968. Ghana. pp 81-86 In Inga, H. And H. Olov (Eds.). *Conservation of vegetation in Africa south of the Sahara*. Proc. of a symposium of the 16th planetary meeting of the "Association pour l'etude Taxanomique de la Flore d'Afrique Tropicale" in Uppsala, Sweden, Sept 12-16. 1966.320pp.

Lawson, G. W., Jenik, J., and Armstrong-Mensah, K. O. 1968. A study of a vegetation catena in the Guinea Savannah at Mole Game Reserve (Ghana). *J. Ecol*. 56: 505-522.

Leach, G. and Mearns, R. 1988. *Beyond the Woodfuel Crisis: People, Land and Trees in Africa*. Earthscan Publications. London.

Li, Z., Kurz, W.A., Apps, M.J. and Beukema, S. J. 2003. Belowground biomass dynamics in the Carbon Budget Model of the Canadian Forest Sector: recent improvements and implications for the estimation of NPP and NEP. *Can. J. For. Res.* 33: 126-136.

Lim, T. T., and Huang, X. 2006. Evaluation of kapok (*Ceiba pentandra* (L.) Gaertn.) as a natural hollow hydrophobic-oleophilic fibrous sorbent for oil spill cleanup. *Chemosphere.* 66(5): 955-963.

Little, E.L. Jr. 1983. *Common fuelwood crops: a handbook for their identification*. McClain Printing Co., Parsons, WV.

Logu, A. E.. Brown, S. and Chapman J. 1988. Analytical review of production rates and stemwood biomass of tropical forest plantations. *For. Ecol. and Manag., 23*, 179-200.

Long, J.N. 1985. A practical approach to density management. *For. Chron.* 23 , 23 – 26 .

Long, J.N., and Shaw, J.D. 2005 A density management diagram for even-aged ponderosa pine stands . *West. J. Appl. For.* 20, 205 – 215.

Lowe. R. G. 1976. Teak (*Tectona* grandis. Linn F.) Thinning experiment in Nigeria. *Commonw. For. Rev., 55*, 189-202.

Lowry, R. F., and Gjerstad, D. H. 1991. Chemical and mechanical site preparation. In M. L. Duryea and P. M. Dougherty (Eds.), *Forest Regeneration manual*, pp. 251-261. The Netherlands: Klumer Academic Publishers.

Luehrman, T. A, 1998. Investment opportunities as real options: getting started on the numbers. *Harvard Business Review*, July-August.

Lugo, A.E., and Brown, S. 1982. Conversion of Topical Forests: A Critique. *Interciencia, 7*(2), 89-93.

Lutz, E., Pagiola, S. and Reiche, C. 1994. Cost-benefit analysis of soil conservation: the farmers' viewpoint. *The World Bank Research Observer* 9: 273-295.

Mackay, J. H. 1952. Notes on establishment of neem plantations in Bornu Province of Nigeria. *Farm and For.* 11: 9-14.

Mackenzie, J.A., 1959. Phenology of *Triplochiton scleroxylon*. Technical Note 1. Department of Forest Research of Nigeria, 5 pp.

Madsen, K. H., and Streibig, J. C. 2003. Benefits and risks of the use of herbicide-resistant crops. In: R. Labrada (Ed.), *Weed management for developing countries*. Pp 245-269. FAO. Rome.

Malende, Y. H., and Temu, A. B. 1990. Site Index curves and volume growth of teak (*Tectona grandis*) at Mtibwa, Tanzania. *For. Ecol. and Manage., 31*, 91-99.

Malimbwi, R. E. 1978. *Cedrela* species international provenance trial (CFI at Kwamsambia, Tanzania). *In* Progress and problems of genetic improvement of tropical forest trees. p. 910. Commonwealth Forestry Institute, Oxford.

Manley, B., and Maclaren, P. 2010. Potential impact of carbon trading on forest management in New Zealand. *Forest Policy and Economics. doi*: 10.1016/j.forpol.2010.01.001

Manshard, W. 1992. Problems of deforestation in tropical Africa: Fuelwood extraction, agroforestry and sustainable development. *Agricultural Change, Environment and Economy*. Mansell, New York. Pp. 203-222.

MarMoller, C. 1954. The influence of thinning on volume increment. 1. Results of investigations. Pages 5-32 *in* Thinning Problems and Practices in Denmark. SUNY Coll. For. at Syracuse, *World For. Sev. Bull.* No. 1, Tech. Pub. No.76.

Mason, G. 2004. Overview of cost-effectiveness analysis and cost-benefit analysis and their application to labour market and social development policies. *Conference on cost effectiveness in evaluation* held on June 17, 2004. Ottawa, Ontario. Human Resources and Social Development Canada.

Mason, S. P., and Merton, R. C. 1985. The role of contingent claims analysis in corporate finance. In: *Recent advances in corporate finance.* Altman, E., Subrahmanyam, M. (Eds.), Homewood, IL, Richard D. Inin : 7-54.

Matthews, J. D. 1989. *Silvicultural systems.* Oxford University Press, Oxford.

Maydell, H-J. von. 1990. *Trees and shrubs of the sahel and their characteristics and uses.* Verlag Josef Margraf Scientific Books. 600pp.

McCarter , J.B. and Long , J.N. 1986. A lodgepole pine density management diagram . *West. J. Appl. For.* 1, 6 – 11.

McConnell, K. E., Daberkow, J. N., Hardie, I. W. 1983. Planning timber production with evolving prices and costs. *Land Economics* 59(3): 292-299.

McDonald, R., and Siegel, D. 1985. Investment and the valuation of firms when there is an option to shut down. *International Economic Review*, June: 331-349.

McFadden, D. 1981. Econometric models of Probabilistic Choice. In C. F. Manski and D. McFadden (Eds.). *Structural Analysis of Discrete Data with Econometric Applications.* Cambridge, Mass. MIT Press.

Mengel, K., and Kirkby, E. A. 1978. *Principles of Plant Nutrition.* International Potash Institute.

Mensah, J. A. 2009. Tourism raked in $US1.3 in 2008 – Minister. Ghana News Agency report of April 29, 2009. Ghana Business News. Accessed April, 2009.

Miller, A. D. 1969. *Provisional Yield tables for teak in Trinidad.* Trinidad and Tobago Government Printery.

Miller, J. J., Perry, Jr. J. P., and Borlaug, N. E. 1957. Control of sunscald and subsequent Buprestid damage in Spanish cedar plantations in Yucatan. *Journal of Forestry* 55:185-188.

Ministry of Environment and Science 2002. *National Biodiversity Strategy for Ghana.* Ministry of Environment and Science, Accra.

Ministry of Forests and Range. 2009. Growth and yield modelling: about GandY prediction models. Research Branch. Government of British Columbia, Canada http://www.for.gov.bc.ca/hre/gymodels/GY-Model/about.htm.

Ministry of Forests. 1998. The management and prospects of teak in Vietnam - Ministry of Forestry. Forest Science Sub-Institute of Southern Vietnam. Teak for the future. *Proceedings of the second regional seminar on teak.* Kashio, M. and K. White (Eds.) Rap Publication -1998/05 249 pp.

Ministry of Lands and Forestry. 2004. Criteria and indicators for sustainable management of natural tropical forests. Reporting Questionnaire for Indicators at the National Level – *Report for Ghana.* Submitted to ITTO, March 2004. Ghana Forestry Commission, Ministry of Lands and Forestry, Accra, Ghana. Unpublished.

Mitra, T and Wan, Jr. H. Y. 1985. On the Faustmann solution to the forest management problem. *Journal of Economic Theory.* 40 (2): 229 - 249.

Mitzutani, J. 1999. Selected allelochemicals. *Crit. Rev. Plant Sci., 18*, 653-671.

MOFA/AFU 1986. *The National Agroforestry Policy.* Ministry of Food and Agriculture Accra, Ghana

Morck, R., Schwartz, E., Stangeland, D. 1989. The valuation of forestry resources under stochastic prices and inventories. *Journal of Financial and Quantitative Analysis* 24: 473-487.

Munslow, B., Katerere, Y., Ferf, A., and O'Keefe, P. 1988. *The Fuelwood Trap: A Study of the SADCC Region.* Earthscan Publications. London.

Murugan, K. and Kumar, N. S. 1996. Host plant biochemical diversity, feeding, growth and reproduction of teak defoliator *Hyblaea puera* (Cramer). *Indian Journal of Forestry, 19,* 253-257.

Myers, S. C. 1977. Determinants of corporate borrowing. *Journal of Financial Economics,* November, 147-176.

Nagaveni, H. C., Ananthapadmanbha, H. S. and Rai, S. N. 1987. Note on extension of viability of *Azadirachta indica.* Myforest 23(4): 245-250.

Nair, K. S. S. 2007. Tropical forest insect pests: *ecology, impact and management.* Cambridge University Press.

Nanang, D. M. 1996. The silviculture, growth and yield of neem (*Azadirachta indica* A. Juss) plantations in Northern Ghana. Unpublished M.Sc.F Thesis, Lakehead University, Ontario, Canada.

Nanang, D. M., 1998. Suitability of the normal, lognormal and Weibull distributions for fitting diameter distributions of neem (*Azadirachta indica* A. Juss.) plantations in Northern Ghana. *Forest Ecology and Management* 103 (1): 1–7.

Nanang, D. M. 2010. Analysis of export demand for Ghana's timber products: A multivariate co-integration approach. *Journal of Forest Economics* 16(1): 47-61.

Nanang, D. M., and Owusu, E. H. 2010. Estimating the economic value of recreation at the Kakum National Park, Ghana. In D. M. Nanang and T. K. Nunifu (Eds.). *Natural resources in Ghana: Management, policy and economics.* New York. Nova Science Publishers.

Nanang, D. M., and Nunifu, T. K. 1999. Selecting a functional form for anamorphic site index curve estimation. *Forest Ecology and Management* 118: 211-221.

Nanang, D. M., and Asante, W. 2000. Effect of neem (*Azadirachta indica* A. Juss) on food crop production and quality in Northern Ghana. *Final Project Report.* National Agricultural Research Programme, Accra, Ghana

Nanang, D. M., and Yiridoe, E.K., 2010. Analyses of the Causes of Deforestation in Ghana: An Econometric Approach. In: Nanang, D. M. and Nunifu, T.K. (Eds.). *Natural resources in Ghana: management, policy and economics.* Nova Science Publishers, New York.

Nanang, D. M., Day, R. J. and Amaligo, J. N. 1997. Growth and yield of neem (*Azadirachta indica* A. Juss.) plantations in Northern Ghana. *Commonwealth Forestry Review.* 76(2):103 -106.

National Academy of Sciences. 1980. Firewood crops. *Shrubs and tree species for energy production.* National Academy of Sciences. Washington D.C. 237 pp.

Nautiyal, J. C. 1988. Forest Economics: *Principles and Application,* Canadian Scholars' Press Inc.

Navrud, S., and Vondolia, G. K. 2005. Using contingent valuation to price ecotourism sites in developing countries, *Tourism,* 53 (2), 115-125.

Neef, T., and Henders, S., 2007. *Guidebook to Markets and Commercialisation of Forestry CDM projects.* Tropical Agricultural Research and Higher Education Centre (CATIE).

Nelson, T. C. 1964. Diameter distribution and growth of loblolly pine. *Forest Science* 10(1): 105-114.

Newman, D. H. 1988. The optimal forest rotation: a discussion and annotated bibliography. Gen. Technical Report SE - 48. Asheville, NC: U.S Dept. of Agric., *Forest Service.* Southeastern Forest Experiment Station. 47p.

Newman, D. H., and Yin, R. 1995. A note on the tree-cutting problem in a stochastic environment. *Journal of Forest Economics* 1:181-190.

Newton, P.F. 1997. Stand density management diagrams: review of their development and utility in stand-level management planning. *For. Ecol. Manage.* 98, 251 – 265.

Newton, P.F., and Weetman , G.F. 1994. Stand density management diagram for managed black spruce stands. *For. Chron.* 70, 65 – 74 .

Newton, P.F., Lei, Y. and Zhang, S.Y. 2005. Stand-level diameter distribution yield model for black spruce plantations . *For. Ecol. Manage.* 209, 181 – 192.

Nketiah, T., Newton, A. C., and Leakey, R. R. B. 1998. Vegetative propagation of *Triplochiton scleroxylon* K. Schum in Ghana. *Forest Ecology and Management* 105 _1998. 99–105.

NRC. 1992. National Research Council. *Neem: a tree for solving global problems*, report of an ad-hoc panel of the Board on Science and Technology for International Development National Academy Press, Washington, DC.

Ntiamoa-Baidu, Y. 1998. *Sustainable harvesting, production and use of bushmeat.* Wildlife Department, Ministry of Lands and Forestry, Accra, Ghana.

Nunifu K. T., and Murchison, H.G. 1999. Provisional yield models of Teak (*Tectona grandis* Linn F.) plantations in Northern Ghana. *For. Ecol. and Manage., 120*, 171-178.

Nunifu, K. T. 1997. The growth and yield of teak (*Tectona grandis* Linn F.) *plantations in Northern Ghana*. M.Sc.F Thesis, Faculty of Forestry, Lakehead University, 101 pp.

Nunifu, T. K. 2010. Growth and Management of Teak (*Tectona grandis* Linn F.) Plantations in Ghana. *In*: Nanang, D. M. and Nunifu, T.K. (Eds.). Natural Resources in Ghana: management, policy and economics. Nova Science Publishers, New York.

Nwoboshi L.C. 1994. Development of Gmelina arborea under the Subri Conversion Technique: First three years. *Ghana Journal of Forestry* 1:12–8.

Nyland, R. D. 1996. Silviculture: *Concepts and Applications*. McGraw-Hill, New York, 631 p.

Nyland, R. D. 2007. Silviculture: *Concepts and Applications*. Second edition. McGraw-Hill, New York.

O'Hara, K. L. 2004. Forest Stand Structure and Development: Implications for Forest Management. *USDA Forest Service Gen. Tech. Rep.* PSW-GTR-193.

O'Keefe, P. and Raskin, P. 1985. Crisis and opportunity: fuelwood in Africa. *Ambio* 14: 4-5, pp.220-224.

Oboho, E. E. G and Ali, J. Y. 1985. Preliminary investigation of the effect of seed weight in early growth characteristics of some savannah species. pp. 144-159 *in* Okojie, J.A and Okoro, O.O (eds.) *Proc. 15th Annual Conf. of the Forestry Association of Nigeria*, Yola Nigeria, Nov. 25-29 1985.

Odoom, F.K., 1999. Securing land for forest plantations in Ghana. *International Forestry Review* 1 (3), 182– 188.

Odoom, M., and Vlosky, R. P. 2007. A Strategic Overview of the Forest Sector in Ghana. Louisiana Forest Products Development Center Working Paper #81. School of

Renewable Natural Resources, Louisiana State University Agricultural Center. Baton Rouge. USA.

OECD. 2006. *Good practice guidance on applying strategic environmental assessment (SEA) in development co-operation. Final Draft.* Organisation for Economic Co-operation and Development. Paris. 116 p.

Oliver, C. D, and Larson, B. C. 1996. *Forest stand dynamics*, 2nd edition. John Wiley and Sons, New York.

Olschewski, R., and Benıtez, P. 2010 Optimising joint production of timber and carbon sequestration of afforestation projects. *Journal of Forest Economics* 16, 1–10.

Olschewski, R. and Benitez, P., 2005. Secondary forests as temporary carbon sinks? The economic impact of accounting methods on reforestation projects in the tropics. *Ecological Economics* 55 (3), 380–394.

Olschewski, R., Benıtez, P., de Koning, G.H.J., and Schlichter, T., 2005. How attractive are forest carbon sinks? Economic insights into supply and demand of certified emission reductions. *Journal of Forest Economics* 11, 77–94.

Omoyiola, B. O. 1973. Initial observation on *Cedrela odorata* provenance trial in Nigeria. *In Tropical provenance and progeny research and international cooperation.* p. 250-254. Commonwealth Forestry Institute, Oxford.

Orwa, C., Mutua, A., Kindt, R. , Jamnadass, R., and Anthony, S. 2009 Agroforestree Database:a tree reference and selection guide version 4.0 (http://www.worldagroforestry.org/sites/ treedbs/treedatabases.asp).

Ott, S. H., and Thompson, H. E. (1996). Uncertainty outlay in time to build problems. *Management and Decision Economics*, 17(1), 1-16.

Owen, D. B. and Wiesen, J. M. 1959. A method of computing bivariate normal probabilities with application to handling errors in testing and measuring. *The Bell System Technical Journal* 38: 553-572.

Owubah, K., LeMaster, D.C., Bowker, J.M., and Lee, J.G., 2001. Forest tenure systems and sustainable forest management: the case of Ghana. *Forest Ecology and Management* 149, 253–264.

Owusu, J. H. 1998. Current Convenience, disparate deforestation: Ghana's adjustment programmes and the forestry sector. *Professional Geographer,* 50(4), 418-436.

Paddock, J. L., Siegel, D. R., Smith, J. L. 1987. Valuing offshore oil properties with option pricing models. *Midland Corporate Finance Journal,* 5(spring): 22-30.

Pandey, D., and Brown, C. 2000. Teak: a global overview. *Unasylva 51*: available at http://www.fao.org/docrep/x4565e/x4565e03.htm#P0_0. Accessed July 2009.

Pearce, D., Atkinson, G., and Mourato, S. 2006. *Cost benefit analysis and the environment: recent developments.* Organisation for Economic Co-operation and Development. Paris. OECD Publishing.

Pearse, P. H. 1992. *Introduction to forestry economics.* University of British Columbia Press. Vancouver.

Pearson, K. 1931. *Tables for statisticians and biometricians.* Part II. Cambridge University Press. Cambridge, England. 262 pp.

Pelissier, F. and Souto, X. C. 1999. Allelopathy in northern temperate and boreal semi-natural woodland. *Crit. Rev. Plant Sci., 18*, 637-652.

Perera, W. R. H. 1962. The development of forest plantations in Ceylon since the seventeenth century. *Ceylon Forester, 5*, 142-147.

Perhutani, P. 1992. Teak in Indonesia. In Wood H (ed.) Teak in Asia FORSPA publication 4. *Proc. regional seminar Guangshou*, China March 1991. FAO (Bangkok).

Phillips, H. 2004. *Thinning to improve stand quality*. Silviculture / Management No. 10. Coford Connects.

Pienaar, L. V. and Turnbull, K. J. 1973. The Chapman-Richards generalisation of von Bertalanffy's growth model for basal area growth and yield in even-aged stands. *For. Sci.* 19: 2-22.

Pindyck, R. 1988. Irreversible investment, capacity choice, and the value of the firm. *American Economic Review* 78(December), 969-985.

Plantinga, A. J, 1998. Optimal harvesting strategies with stationary and non-stationary prices: An option value analysis. *Forest Science* 44:192-202.

PointCarbon, 2008.EUAHistoricPrices.Availableat: /http://www.pointcarbon.com.

Prah, E. A. 1994. *Sustainable management of tropical high forests of Ghana*. London. Commonwealth Secretariat, IDRC, Vol. 3.

Price, C. 1989. *The theory and application of forest economics*. Basil Blackwell Ltd. London. 402 p.

Price, T., and Wetzstein, M. 1999. Irreversible investment decision in perennial crops with yield and price uncertainty. *Journal of Agricultural and Resource Economics*, 24(1), 173-185.

Radosevich, S. R., and Osteryoung, K. 1987. Principles governing plant-environment interactions . In: J. D. Walstad and P. J. Kuch (Eds.), *Forest vegetation management for conifer production*. Ch. 5. pp 105-156. New York. Wiley.

Radwanski, S. and Wickens, G.E. 1981. Vegetative fallows and potential value of the neem tree (*Azadirachta indica*) in the tropics. *Econ. Bot.* 35: 398-414.

Reed, W. J., and Haight, R.1996. Predicting the present value distribution of a plantation investment. *Forest Science* 42:378-388.

Reineke, L. 1933. Perfecting a stand density index for evenaged forests. *J. Agric. Res., 46*, 627–638.

Reineke, L. H. 1926. The determination of tree volume by planimeter. *J. For.* 24: 184-189.

Rendle, B.J. 1969. *World timbers*. Volume 2, North and South America. University of Toronto Press.

Reynolds, M. R., Burk, T. E. and Haung, W. C. 1988.Goodness-of-fit tests and model selection procedures for diameter distribution models. *Forest Science* 349(2):373-399.

Rhodey-Bowman, T. 2007. *Mechanical High Pruning*. Agriculture Notes. State of Victoria, Department of Primary Industries. AG 1014.

Rice, R.E., Sugal, C.A., Ratay, S.M., and Fonseca, G.A. 2001. Sustainable forest management: A review of conventional wisdom. *Advances in Applied Biodiversity Science*, No. 3, p. 1-29. Washington, DC: CABS/Conservation International

Richards, F. S. 1959. Flexible growth function for empirical use. *J. Exp. Botany, 10*, 290-300.

Richards, M. 1995. The role of demand side incentives in fine grained protection: a case study of Ghana's tropical high forest. *Forest Ecology and Management, 78*, 225-241.

Richards, P.W. 1996. The Tropical Rain Forest, 2nd Edition. Cambridge University Press, 525 pp.

Ross, D. W. and Walstad, J. D. 1986. *Vegetative competition, site preparation and pine performance: a literature review with reference to south-central Oregon*. Oregon State Univ., Corvallis. Paper 21 p.

Rudolf, P.O. 1974. *Larix* Mill. larch. Pages 478-485. In C.S., Schopmeyer, (Ed.) Seeds of woody plants in the United States. 450. U.S. Department of Agriculture, *Forest Service.* Washington, DC. Agric. Handb. 478-485.

Salifu, K. F. 1997. Physico-Chemical Properties of Soil in The High Forest Zone of Ghana Associated with Logged Forest and with Areas Converted to Teak *(Tectona grandis* Linn. F). MscF Thesis, Lakehead University. Thunder Bay, ON. Canada. 105 pp.

Sampong, E. 2004. A Review of the Application of Environmental Impact Assessment in Ghana. *A Report prepared for the United Nations Economic Commission for Africa* in December, 2004.

Sarfo-Mensah, P. 2005. Exportation of timber in Ghana: the menace of illegal logging operations. The Fondazione Eni Enrico Mattei Note di Lavoro Series Index: http://www.feem.it/Feem/Pub/Publications/WPapers/default.htm. Nota Di Lavoro 29.2005.

Savill, P., Evans, J., Auclair, D. and Falk, J. 1997. *Plantation silviculture in Europe*. Oxford University Press, Oxford.

Saxena, A. K., Nautiyal, J.C., and Foot, D.K. 1997. Analysing Deforestation and Exploring Policies for its Amelioration: A Case Study of India. *Journal of Forest Economics, 3*(3), 253-289.

Sawyer, J. 1993. *Plantations in the tropics – environmental concerns. The World Conservation Union (IUCN),* Gland, Switzerland and Cambridge, UK in collaboration with UNEP and WWF.

Schiffel, A. 1899. Form and Inhalt der Fichete (Form and volume of spruce). Mitt.ausd. forstl. Versuchsan. *Osterreiche* 24.(cited in Husch et al., 1982).

Schlaegel, B.E. 1981. Testing, reporting and using biomass estimation models. pp. 95-112 *in* Greshan, C. A (ed.) Proc. 1981 Southern *Forest Biomass Workshop*. The Belle W. Baruch Forest Science Institute of Clemson University, Clemson USA, June 11-12 1981. 127 pp.

Schmutterer, H. 1982. Ten years of neem research in the Federal Republic of Germany. pp 21-32. *In* Schmutterer, .H.,K.R.S. Asher, and H. Rembold (eds.) Natural pesticides from the neem tree. *Proc.First Int. Neem Conf. Rottach-Egern*., Eschborn Germany

Schofield, J. A. 1987. *Cost-benefit analysis in urban and regional planning*, Boston, MA. Urwin and Allen.

Schreuder, H. T. and Hafley, W. L. 1977. A useful bivariate distribution for describing stand structure of tree heights and diameters. *Biometrics* 33: 471-478.

Schumacher, F. X. 1939. A new growth curve and its application to timber yield studies. *J. of For., 37*, 819-820.

Seth, S. K., and Yadav, J. S. P. 1959. Teak soils. *Indian Forester, 85*, 2 – 16.

Shifley, S., and Lentz. E. 1985. Quick estimation of the three-parameter Weibull to describe tree size distribution. *For. Ecol. and Manage*., 13, 195-203.

Shinozaki, K. and Kira, T. 1956. Interspecific competition among higher plants. VII. Logistic theory of the C-D effect. *J. Inst. Polytech.,* Osaka Cy Univ. 7:35-72.

Sinha, K. C., Riar, S. S., Tiwary, R. S., Dhawan, A. K., Bardhan, J., Thomas, P., Kain, A. K. and Jain, R.K. 1984. Neem oil as a vaginal contraceptive. *Indian J. Med. Res*. 79: 131-136.

Skoupy, J. 1991. People's Participation in Planting Trees. Insert Focus on reforestation by Skoupy, J. T. Dida, T. Cecchini, G. Nasser Al Homaid, M.H. Khan and M. Sadiq (1991). *In Desertification Control Bulletin* 19 pp.33-60.

Skovsgaard, J. P., and Vanclay, J. K. 2007. Forest site productivity: a review of the evolution of dendrometric concepts for even-aged stands. *Forestry* 81(1): doi:10.1093/forestry/cpm041.

Smith D E, Larson, B. C., Kelty, M. L. and Ashton, P. M .S. 1997. *Practice of silviculture: applied forest ecology*. John Wiley and Sons, New York. 537 pp.

Smith, D. M. 1986. *The practice of silviculture*. 8th Edition. New York. Wiley

Smith, E. K. 1999. Developments and setbacks in forest conservation: The new political economy of forest resource use in southern Ghana. Natural Resources Management Programme, *Ministry of Lands and Forestry Technical paper*, Accra, Ghana.

Smith, J. 1939. Germination of neem seed. *Indian Forester* 65(3): 457-459.

Smith, J., and McCardle, K. 1998. Valuing oil properties: integrating option pricing and decision analysis approaches. *Operations Research*, March-April: 198-217.

Smith, P. 1998. The use of subsidies for soil and water conservation: a case study from Western India. Network Paper No. 87. *Agricultural and Research Extension Network*. London: Overseas Development Institute.

Smith, V. K. 1989. Taking stock of progress with travel cost recreation demand models: Theory and implementation. *Marine Resource Economics*, 6, 279-310.

Smith, V.G., 1984. Asymptotic site-index curves, fact or artifact? *For. Chron.* 60, 150-156.

Sokal, R. R. and Rohlf, F. J. 1981. *Biometry*. 2nd Edition. W. H. Freeman and Co., San Francisco. U. S. A. 857 pp.

Solomon, D. S., and Zhang, L. 2002. Maximum size–density relationships for mixed softwoods in the northeastern USA. *Forest Ecology and management*. 155: 163-170

Spaargaren, O.C. and Deckers, J. 1998. The world reference base for soil resources - an introduction with special reference to soils of tropical forest ecosystems. *In*: Schulte, A and Ruhiyat D. (eds.) *Soils of tropical forest ecosystems - characteristics, ecology and management,* 21-28. Springer, Berlin.

Spurr, S. H. 1952. *Forest inventory*. The Ronald Press Co., New York. 476 pp.

Stanton, K. 2003. Parliament needs law monitoring mechanism. *Ghana News Agency report* of March, 23, 2003.

Stone, S. W., Kyle, S. C. and Conrad, J. M.1993. Application of the Faustmann principle to a short-rotation tree species: an analytical tool for economists, with reference to Kenya and leuceana. *Agroforestry Systems* 21: 79-90.

Strang, W. 1984. On the optimal harvesting decision. *Economic Inquiry*. 14: 466-492.

Streets, R. J. 1962. *Exotic forest trees in the British Commonwealth*. Claredon Press, Oxford. 765 pp.

Styles, B. T. 1972. The flower biology of the *Meliaceae* and its bearing on tree breeding. *Silvae Genetica* 21:175-183.

Sutton, R. F. 1985. Vegetative management in Canadian forestry. *Can. For. Serv., Great Lakes For. Res. Cr., Info. Rep. \no.* O-X-369. 35 p.

Swaine, M.D., and Hall, J.B., 1988. The mosaic theory of forest regeneration and the determination of forest composition in Ghana. *J. Trop. Ecol.* 4, pp. 253–269.

Taylor, C. J. 1952. *The vegetation zones of the Gold Coast*. Government Printer, Accra, Ghana. 57 pp.

Teeguarden, D. E. 1982. Multiple services. *In*: Duerr, W. A., Teeguarden, D. E., Christiansen, N. B and Guttenberg, S. *Forest resource management*: decision-making principles and cases. Corvallis, Oregon. Pp. 276-290.

Tepper, H. B. and Bamford, G. T. 1959. Hardwoods on poorly drained sites do not respond to low thinning. Forest Research Note NE-92. Upper Darby, PA: U.S. Department of Agriculture, *Forest Service*, Northeastern Forest Experiment Station. 3 p.

Thérivel, R. and Partidario, M. 1996. *The practice of strategic environmental assessment.* London. EarthScan Publications.

Thompson, D. 2009. *Frequently Asked Questions (FAQs) on the use of herbicides in Canadian forestry*. Unpublished Notes.

Thomson, T. A, 1992. *Optimal forest rotation when prices follow a diffusion process*. Land economics 68: 329-342.

Thorson, B. J, 1999. Afforestation as a real option: some policy implications. *Forest Science* 45(2): 171-178.

Tiarks, A., Nambiar, E.K.S., and Cossalter, C. 1998. Site management and productivity in tropical forest plantations. *CIFOR Occasional Paper* No. 16.

Tilander, Y. 1993. Effects of mulching with *Azadirachta indica* and *Albezia lebbeck* leaves on the yield of sorghum under semi-arid conditions in Burkina Faso. *Agroforestry Systems* 24(3):277-293.

Timbilla, J.A. and Braimah, H. 2000. Successful biological control of *Chromolaena odorata* in Ghana: the potential for regional programme in Africa. *In*. Zachariades C, Muniappan R and Strathie L.W. (Eds.), *Proceedings of the fifth international workshop on biological control and management of Chromolaena odorata,* Durban, South Africa, 23-25 October 2000, pp 66-70.

Tomforde, M. 1995. *Compensation and incentive mechanisms for the sustainable development of natural resources in the tropics*: their socio-cultural dimension and economic acceptance. Eschborn: Gesellschaft für Technische Zusammenarbeit (GTZ).

Tosi, J. A., Jr. 1960. Zonas de vida natural en el Perú. Memoria explicativa. sobre el mapa ecológico del Perú. Instituto Interamericano de las Ciencias Agriicolas de la E.E.A. Boletín Técnico 5. Zona Andina, Lima, Penú. 271 p.

Tourinho, O., 1979. *The option value of reserves of natural resources*. Working Paper No. 94, University of California at Berkeley.

Townson, I.M. 1995. Patterns of Non-Timber Forest Products Enterprise activity in the Forest Zone of Southern Ghana. *Report to the ODA Forestry Research Programme*. Oxford Forestry Institute, Oxford.

Treadway, A. B. 1971. On the multivariate flexible accelerator. *Econometrica* 39: 845–855.

Treadway, A. B. 1974. The globally optimal flexible accelerator. *Journal of Economic Theory* 7:17–39.

Treasury Board of Canada. 2009. Regulatory Impact Analysis Template. http://www.regulation.gc.ca/documents/nriast-nmrir/nriast-nmrir-eng.asp.

Trigeorgis, L, 1993. Real options and interactions with financial flexibility. *Financial Management*, 22(3): 202-224.

Trigeorgis, L. 1998. A conceptual options framework for capital budgeting. *Advances in Futures and Options Research*, 3, 145-167.

Trigeorgis, L., 1999. *Real options: managerial flexibility and strategy in resource allocation.* The MIT Press, Cambridge, Massachusetts.

Trigeorgis, L., and Mason, S. P. 1987. Valuing managerial flexibility. *Midland Corporate Finance Journal,* 5 (Spring): 14-21.

Tripathi, S., Tripathi, A., and Kori, D. C. 1999. Allelopathic evaluation of *Tectona grandis* leaf, root and soil aqueous extracts on soybean. *Indian Journal of Forestry, 22*, 366-374.

Troup, R. S. 1921. *The silviculture of Indian trees*. Volume 1. Claredon Press, Oxford. 678 pp.

Turnbull, K. J. 1963. *Population dynamics in mixed forest stands*. Ph.D. dissertation, Univ. of Washington.186 pp.

Ulibarri, C. A. and Wellman, K.F. 1997. *Natural resource valuation: A primer on concepts and Techniques.* United States Department of Energy 86 p.

UNEP. 2005. Baseline Methodologies For Clean Development Mechanism Projects: A guide book. *The UNEP project CD4CDM*. UNEP Risø Center, Denmark.

UNFCCC. 2002. *Report of the conference of the parties on its Seventh Session*, held at Marrakesh from 29 October to 10 November 2001. FCCC/CP/2001/13/Add.2.

United Nations. 1992. Report of the United Nations Conference on Environment and Development - *Agenda 21, Chapter 11: Combating Deforestation.*

Unnikrihnan, K. P. and R. Singh. 1984. Construction of volume tables - a general approach. *Indian Forester* 110(6): 561-576.

USDA. 2010. Wood Technical Fact Sheet: *Triplochiton scleroxylon. USDA Forest Service* Forest Products Laboratory.

Van Cleve, K. and Zasada, J. C. 1976. Response of 70-year-old white spruce to thinning and fertilisation in interior Alaska. *Can. J. For. Res*. 6:145-152.

Van Kooten, G. C., C. S. Binkley, and G. Delcourt. 1995. Effect of carbon taxes and subsidies on optimal forest rotation age and supply of carbon services. *American Journal of Agricultural Economics* 77 (2): 365-374.

Vega, L. 1974. Influencia de la silvicultura sobre el comporta miento de *Cedrela* en Surinam. Instituto Forestal Latinoamericano de Investigación y Capacitación, *Boletín* 46-48. Mérida, Venezuela. p. 57-86.

Verinumbe, I. 1991. Agroforestry development in northeastern Nigeria. *For. Ecol. Manage.* 45: 309-317.

Wadsworth, F. H. 1960. Datos de crecimiento de plantaciones forestales en México, Indias Occidentales y Centro y Sur América. Segundo Informe Anual de Is Sección de Forestación, Comité Regional sobre Investigación Forestal, Comisión Forestal Latinoamericana, Organiziación de las Naciones Unidas para la Agricultura y Alimentación. *Caribbean Forester* 21 (supplement). 273 p.

Wagner, M. R., Cobbinah, J. R., and Bosu, P. P. 2008. *Forest Entomology in West Tropical Africa: Forests Insects of Ghana.* Springer. Netherlands.

Walters, D.K., Gregoire, T.G., and Burkart, H.E. 1989. Consistent estimation of site index curves fitted to temporary plot data. *Biometrics* 45(1), 24-33.

Warthen, J. D. Jr. 1979. *Azadirachta indica*: A source of insect feeding inhibitors and growth regulators. *Agric. Rev. and manuals ARM-NE-4, USDA*, Washington DC.

Watterson, K. G. 1971. Growth of teak under different edaphic conditions in Lancetilla valley, Honduras. *Turr., 21*, 222-225.

Weaver, P. L. 1993. Teak (*Tectona* grandis Linn F.). Res. Notes SO-ITF-SM-64. Rio Piedras, *PR. USDA Forest Serv. Inst. of Tropical forestry*. 18pp.

Weibull, W. 1951. A statistical distribution function of wide applicability. *Journal of Applied Mechanics*. 18: 293-297.

Wengert, E. 2001. Eliminating wood problems - An industry review: 10 ways of eliminating wood problems. Wood Processing Department of Forestry, University of Wisconsin, Madison (www.woodweb.com).

West, P. W. 2003. *Tree and Forest measurement*. Springer. 167 pp.

Westoby, M. (1984). The self-thinning rule. *Adv. Ecol. Res., 14*, 167–225.

White, F. 1983. *The Vegetation of Africa.* 3 maps, 1 chart. AETFAT Vegetation Map Committee, UNESCO, AETFAT, UNSO.

Whiteman, A., 2003. Money doesn't grow on trees: a perspective on prospects for making forestry pay. *UNASYLVA* 54 (212), 3 – 10.

Whitmore, J. L. 1976. Myths regarding *Hypsipyla* and its host plants. *In:* Studies on the shootborer *Hypsipyla grandella* Zeller Lep. Pyralidae. vol. 3. p. 54-55. Centro Agronómico Tropical de Investigación y Enseñanza, Miscellaneous Publication 1. Turrialba, Costa Rica.

Whittington, D. L., Donald, T., Wright, A. M., Choe, K., Hughes, J. A., and Swarna, V. 1993. Household demand for improved sanitation services in Kumasi, Ghana: A contingent valuation study. *Water Resources Research*. 29(6):1539-1560.

Wickens, G. E., Sief-El-Din, A. G., Sita, G., and Nahal, I. 1995. Role of Acacia Species in the rural economy of dry Africa and the Near East. *FAO Conservation Guide*, No. 27, Rome, Italy, pp.56.

Williams, J. 2001. Financial and other incentives for plantation forestry. In: *Proceedings of the International Conference on Timber Plantation Development*, Manila, the Philippines, 7-9 November 2000, (pp. 87-101). Quezon City, the Philippines, Department of Environment and Natural Resources.

Williams, T. M. 1989. Site preparation on forested wetlands of the south-eastern coastal plains. In: D. D. Hook and R. Lea (Eds.). The Forested wetlands of the Southern United states, pp.67-71. Proc. Symp., July 12-14, 1988. Orlando, *FL. US Forest Ser. Gen. Tech. Rep.* SE-50.

Wilson, E. R and Leslie, A. D. 2008. The development of even-aged plantation forests: an exercise in forest stand dynamics. *Journal of Biological Education* 42(4): 170-176.

Wilson, F.G. 1946. Numerical expression of stocking in terms of height . *J. For.* 44 , 758 – 761

Wingfield, M. J., and Robison, D. J. 2004. Diseases and insect pests of *Gmelina arborea*: real threats and real opportunities. *New Forests*. 28(2-3): 227-243

Winpenny, J.T. 1992. The economic valuation of tropical forests: Its scope and limits. *In* F. R. Miller and K. L. Adam (Eds.). *Wise management of tropical forests*. (pp. 125-138). Proceedings of the Oxford Conference on Tropical Forests, Oxford, 30 March-1 April 1992, Oxford: Oxford Forestry Institute, University of Oxford.

Wong, C.Y., and Jones, N. 1986. Improving tree form through vegetative propagation of *Gmelina arborea*. *Commonwealth Forestry Review*, 65(4):321–324.

World Agroforestry Centre (ICRAF). 1993. International Centre for Research in Agroforestry: *Annual Report* 1993. Nairobi, Kenya. pp 208.

World Agroforestry Centre. 2007. http://www.worldagroforestrycentre.org/sea/Products/.

World Bank, 1988. Staff appraisal report. *Ghana Forest Resource Management Project.* Report No.7295-GH. Nov., 17, 1988.

World Bank. 1998. *Integrated Coastal Zone Management Strategy for Ghana*. Findings. Africa Region. Number 113. Washington D. C. The World Bank.

World Bank. 2003. *Basics of the BioCarbon Fund for Project Proponents*. Available at: /www.unfccc.intS.

World Bank. 2004. *Implementation completion report on a Loan/Credit/Grant to the Republic of Ghana for a Natural Resource Management Project*, Phase I. January 2003. Report No: 27231

World Bank. 2007. *World Development Indicators*. The World Bank. Washington, D. C. www.un.org/esa/sustdev/documents/agenda21/english/agenda21chapter11.htm (accessed May, 2009).

Yaping, D. 1999. The Use of Benefit Transfer in the Evaluation of Water Quality Improvement: *An Application in China. Economy and Environment Programme for Southeast Asia.* International Development Research Centre, Ottawa, Canada. http://www.idrc.ca/eepsea/ev-8426-201-1-DO_TOPIC.html

Yin, R. and Newman, D. H. 1995. Optimal timber rotations with evolving prices and costs revisited. *For. Sci.* 41(3): 477 - 490.

Yoda, K. Kira, T., Ogawa, H., and Hozumi, K. 1963. Interspecific competition among higher plants XI. Self-thinning in overcrowded pure stands under cultivated and natural stands. *J. Biol.* Osaka City University 14:107-129.

Zankis, S. H. 1979. A simulation study of some simple estimation for the three parameter Weibull distribution. *J. Stat. Comp. Simul., 9*, 260-116.

Zarnoch, S. J. and Dell, T. R. 1985. An evaluation of percentile and maximum likelihood estimators of Weibull parameters. *For. Sci.* 31: 260-268.

Zeide, B., 1987. Analysis of the 3/2 power law of self-thinning. *For. Sci.* 33, 517–537.

Zhang, D., and Owiredu, E. A. 2007. Land tenure, market, and the establishment of forest plantations in Ghana. *Forest Policy and Economics* 9: 602-610.

Zhang, D., and Pearse, P.H., 1996. Differences in silvicultural investment under various types of forest tenure in British Columbia. *Forest Science* 42 (4), 442– 449.

Zobel, B. J. van Wyk, G., and Stahl, P. 1987. *Growing exotic forests*. New York. Wiley.

Zobel, B. J., and Talbert, J. 1984. *Applied forest tree improvement*. New York. Wiley

Zöhrer, F. 1969. Ausgleich von Haufigkeitsverteilungen mit Hilfe der Beta-Funktion. *Forstarchiv* 40(3): 37-42.

About the Author

Dr. David Mateiyenu Nanang was born and raised in Northern Ghana. He received his elementary education in Bunkpurugu and secondary school education at Notre Dame Minor Seminary in Navrongo and Nandom Secondary School in the Upper West Region. After earning a first class B.Sc. in Natural Resources Management from the Kwame Nkrumah University of Science and Technology (KNUST), Kumasi, Ghana in 1992, he subsequently pursued a M.Sc. in forestry, specialising in silviculture and forest biometrics, which he completed in 1996 from Lakehead University in Canada. In 2002, he obtained a Ph.D. in forest economics (with a minor specialisation in natural resources and environmental economics) from the University of Alberta in Edmonton, Alberta, Canada.

Dr. Nanang has extensive experience in several aspects of forestry: silviculture, ecology, forest management, forest and climate change policy, forest economics and research. He has worked in Ghana with the erstwhile Forestry Department and taught forestry at the University for Development Studies in Tamale. He was a sessional lecturer of forest policy and research associate at the University of Alberta. Dr. Nanang has worked as a forest economist for the Canadian Forest Service, as a senior policy advisor on land management issues with Indian and Northern Affairs Canada and is currently science research director with the Canadian Forest Service of Natural Resources Canada. He is an Adjunct Professor of forest economics at the Faculty of Forestry at the University of Toronto, Canada. He has published extensively in internationally reputable journals on forestry and natural resource economics, including being the senior editor of the seminal book on natural resources in Ghana.

INDEX

A

B

C

D

E

F

G

H

I

J

K

L

M

N

O

P

Q

R

S

T

U

V

W

Y